Hierarchy

Hierarchy

Perspectives for Ecological Complexity

SECOND EDITION

T. F. H. ALLEN AND THOMAS B. STARR

THE UNIVERSITY OF CHICAGO PRESS CHICAGO AND LONDON

The University of Chicago Press, Chicago 60637
The University of Chicago Press, Ltd., London

Published 2017
Printed in the United States of America

26 25 24 23 22 21 20 19 18 17   1 2 3 4 5

ISBN-13: 978-0-226-48954-4 (cloth)
ISBN-13: 978-0-226-48968-1 (paper)
ISBN-13: 978-0-226-48971-1 (e-book)
DOI: 10.7208/chicago/9780226489711.001.0001

Library of Congress Cataloging-in-Publication Data

Names: Allen, T. F. H., author. | Starr, Thomas B., author.
Title: Hierarchy : perspectives for ecological complexity / T. F. H. Allen and Thomas B. Starr.
Description: Second edition. | Chicago : The University of Chicago Press, 2017. | Includes bibliographical references and index.
Identifiers: LCCN 2016059828 | ISBN 9780226489544 (cloth : alk. paper) | ISBN 9780226489681 (pbk. : alk. paper) | ISBN 9780226489711 (e-book)
Subjects: LCSH: Ecology—Methodology. | Ecology—Mathematical models. | Ecology—Philosophy.
Classification: LCC QH541.15.M3 A55 2017 | DDC 577—dc23
LC record available at https://lccn.loc.gov/2016059828

♾ This paper meets the requirements of ANSI/NISO Z39.48-1992 (Permanence of Paper).

Contents

Introduction: The Nature of the Problem

Bruce Milne, the landscape ecologist, when he was young read the first edition of this book, and admits he could not get it. He found it kept changing the discussion in puzzling ways that did not let him get traction. But the book was a success (several thousand citations) and the buzz around it eventually forced him grudgingly to pick it up again. In the interim somehow his view had matured, and the second time he breezed through it, seizing its new insights. Another scientist, young at the time, was Anthony King, the carbon cycling ecosystem scientist at Oak Ridge National Laboratory. He complained to his mentor at the Lab, R. V. O'Neill, that he could not get the first edition of this book. O'Neill said he was reading it as a scientific discourse when he should be reading it as poetry. With coming maturity of the role of narrative in the emerging science of complexity, and with reductionism in ecology becoming manifestly bogged down, even for its own practitioners, maybe ecology at large has had time to mature in its view such that it too can breeze through this second edition as did Milne on his second crack at it. A poetic reading of it will come easier now that *American Scientist* has seen fit to publish a whole number giving the material therein a narrative spin. One of their leading contributors is the civil engineer Henry Petroski,[1] who publishes on risk and disaster in building. His earlier piece on the regularity of bridge collapse is a rattling good tale. His piece for the *American Scientist* narrative number was titled, "The story of two houses." Perhaps a poetic tolerance will have emerged in our audience; we hope so.

Some scientific and technological endeavors have been singularly fruitful. Physics and electronics are spectacular in this regard. Even protein chemists begin to develop a clear view of their material. Then there are the poor cousins, the disciplines which have recruited no less dedication and intelligence but which seem still to be primitive, whose secrets still are well kept. Ecology is one such discipline. That ecology is young is not reason enough, for the early part of the twentieth century saw much better descriptions of prairies[2] than proteins. It must be something else. Essentially it is the problem of scaling, hierarchy, prediction, and levels of analysis. This section will unpack that complex of ideas to set up the reader for the rest of the book. To deal with the issue of the apparent lack of power in some disciplines we turn to hierarchy as a device.

There is some resistance to the term *hierarchy* because of postmodern reaction to hegemony in social hierarchies. The reaction is to preempt a hierarchical conception giving validation to unfair privilege as being somehow natural, and intended by God. Being significantly postmodern ourselves, we understand sensitivities surrounding that position, but retain the term so as not to muddy the waters with political postures, worthy as such concerns may be. By showing the generality of the concept of hierarchy, we are able to blunt its usage in the cause of hegemony. Sometimes we do use social hierarchies, not because they are the general condition, but rather because they are familiar, and we can use them as examples that the reader would readily understand when some arcane aspect of hierarchy needs an example. We define what we mean at this early point so we are not mistaken as approving of male and imperial privilege. In our general conception hierarchies can flip top to bottom so that the weak, seen in different but appropriate terms, have a certain control. In hierarchies the weak can be on top, often because of sheer numbers or their fulfilling crucial roles. There is power in getting work done, and that becomes apparent when labor is withheld. Tyrants have it coming, and in the end mostly get their comeuppance. When that happens the hierarchy changes order, but it remains a hierarchy still, just one that is more egalitarian. There is power in being subjected, although it is not worth the discomfort.

First, we see hierarchies everywhere and consider human social hierarchies as an extreme special case that is not justification enough to taint the whole general concept of hierarchical order. Let us build the concept from the neutral general notion of sets. We have not heard of complaints from the left about set theory. In hierarchies entities sit inside sets. Sets alone do not make for hierarchies because hierarchies consist of levels. Sets become levels only when they are partially ordered. Partial ordering is a mathematical concept that gives a most austere conception. Putting it simply, partial ordering arises when sets are ordered one relative to another. The ordering creates an asymmetry between the sets in some version of up and down. But this does not mean that higher level sets get all the control. There are two types of control; one does come from above but the other comes from below. The levels below are constrained from above; nevertheless levels below often limit the possibilities for higher levels. The whole will not do what the parts cannot. Upper levels constrain, but that only limits the situation to what is locally allowed inside a larger limit of what is possible.

So hierarchy with its levels is a special case of general systems theory, an idea that is coming into its own now that we are learning to deal with complexity. While modernist mechanism tries to break things down until some ultimate level of causality appears, hierarchists are much more cognizant that what we experience and how we see causality comes significantly from how we look at things. Subjectivity has always been there even in the most dispassionate reductionist treatment, it is just that in a complex situation we cannot sweep the value-laden observer under the rug. Big problems cannot be solved or even addressed until the observer as part of observation is taken seriously. Hierarchy theory is significantly the study of the observer addressing a complex system.[3]

The Concept of Medium Number Systems

The whole subject of this book is to spell out the subjectivity of what appears real and objective. Even the number of parts in a system is a subjective choice. It is not given by nature, it is decided when humans make judgments as to good ways to deal with some situation. There are questions that do not yield to normal science, and those are questions that invoke systems called "medium number." A microcosm of problem solving in general is the idea of a medium number system, so we address them now in the opening pages. The concept of medium number involves the subjective decision of number of parts, and shows how that decision comes at us with a vengeance. The less insightful disciplines are encouraged to ask a certain style of question because of the way their material presents itself. Some of that is how appearances invite decisions about numbers of parts.

The idea of medium number and failed prediction was developed by Gerald Weinberg, originally in 1975.[4] The notion cleaves systems into

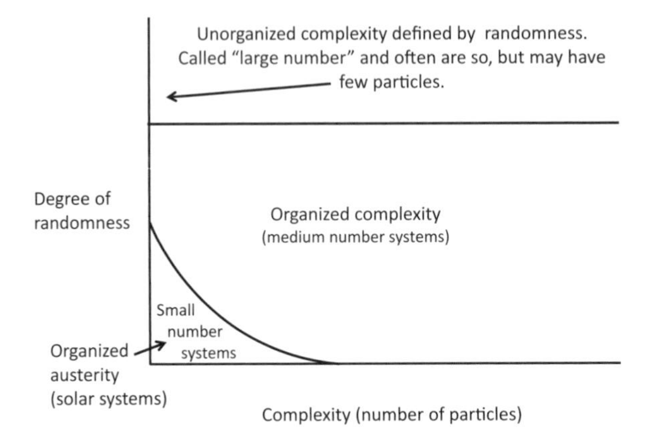

FIGURE I.1. After but modified from Weinberg's graph (Weinberg 2001) comparing large, small, and medium number systems. Large and small number systems are predictable, medium number systems are not. Low randomness means good behavior, like planets. High randomness gives reliable statistical values. Medium number systems are statistically unreliable. Weinberg emphasizes number of particles, but all three types can have a small number of particles. Even so, one can see what he means, because small number systems always have a few particles, and large number systems do in fact usually have large numbers of particles. In both, prediction is possible. Few particles in unorganized complexity is caught by unreliable small number statistics. A large number of particles in unorganized complexity does not settle down to reliable averages because constraints keep switching.

three categories (Figure I.1). Two are predictable, and occur in the more successful disciplines. The third is medium number system and they are intrinsically not predictable. Medium number system specifications do not yield even to experimentation, because the results are different every time. While the number of parts does to an extent characterize the three categories, the issue turns in fact on the stability of the parts in relation to the whole, the stability of the hierarchical conception for the system as it is specified.

Small number systems must have a small number of parts, but some examples of the other two types may also have a small number of parts, so the issue is not number of parts per se, Weinberg's nomenclature notwithstanding. For small number prediction it is critical that the system specification has so few significant particles that it is possible to write an equation for each one. Solar systems would be one such example, where one takes the same equation and adjusts the terms for planet size, distance from the sun, and so on (Figure I.2). The number of particles is not given by nature, it is chosen in the way the scientist specifies what are the parts of the system. Parts are a decision not a material reality. Once we have defined what constitutes a planet, then observation finds that nature appears to throw up just a few of them. It is dangerous to assert small number structures independent of how they are defined. There will of course be iteration in the process, as observations suggest it is worth deciding on planets as a type, because choosing them gives regularity and rewards. It was not an accident that Copernicus and Kepler chose planet-sized units. If they could have seen asteroids, they would have been well advised not to choose them as units. But their choices do not remove the fact that there is choice in all that, and choices are ours, not nature's.

The other predictable type of system is large number systems. Here there needs to be a very large number of particles (Figure I.2). In many cases trillions of particles are not enough. Physical solutions may behave regularly because of the enormous number of particles in solution. If one tries to deal with very dilute solutions, the number of particles in solution goes down. With enough dilution predictions based on the average particle begin to fail as individual particles come to exert themselves. The remarkable behavior of some monomolecular sheets arises because the layer is too thin for average behavior to overcome individual electrons jumping around in funny ways that affect the whole. With aggregate numbers, large enough averages and representative values are reliable. The gas laws would specify such systems. Even when the particles are mythical, prediction is good. There is no such material thing as a perfect gas particle, which is asserted to have no mass and no volume and exhibits no friction. But the gas laws work. Frictionless pendula are once again unreal, but are still really useful.

Weinberg works his way up to the unpredictable systems, called medium number (Figure I.1). He uses complexity in a slightly unfortunate way to mean number of particles (there is a lot more to complexity than that), but we can follow his argument using his definitions. Small number systems may be highly nonrandom, but we can calculate around that issue by using sufficiently detailed specification of individuals, one model for each. In terms of the hierarchies that are the subject of this book, small number hierarchies have a narrow span; the number of particles at the bottom form a narrow set. In large number systems, Weinberg refers to disorganized complexity (Weinberg 2001). "Disorganization" there means there is not much differentiation needed between particles. The particles have to be sufficiently similar for us to be able to get away with

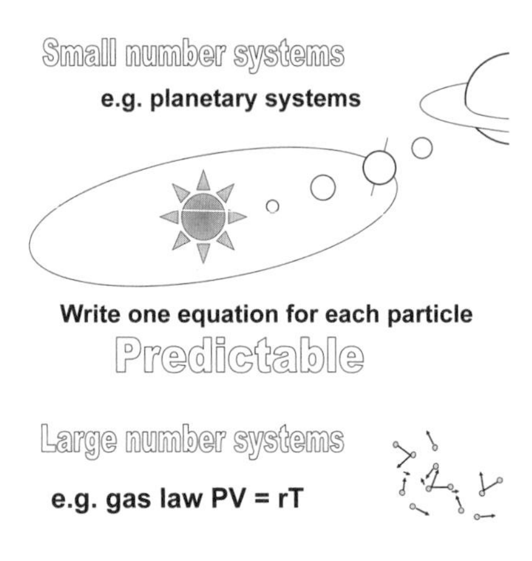
Small number systems
e.g. planetary systems
Write one equation for each particle
Predictable
Large number systems
e.g. gas law PV = rT
Avogadro's number (6.022 x 10²³) of particles or more give representative statistical numbers that are predictive.
Predictable

FIGURE I.2. a) Small number systems have so few parts that it is possible to model each one with its own equation. b) Large number systems have a huge number of particles, and so offer representative aggregates that behave well.

averages that treat the collection as randomly behaving individuals. Hierarchies for these systems are wide in scope. Gas particles meet the criterion of, "When you have seen one you have seen them all." If there are enough essentially undifferentiated particles, the predictions are powerful and specific. True, there are some situations where we cannot say that one electron is not the same as others, but that amounts to a change in the level of discourse. All specifications and predictions are confined to a level of discourse. One problem with naïve realism comes from its saying "In reality . . ." Of course we do not deal with reality directly, and only have discourses, and one at a time, and each specified. When measuring voltage, the distinctiveness of particular electrons belongs to a different discourse at a different level, which we are not addressing. Reality invokes infinities and so is undefined. Therefore ultimate reality cannot be used in practice as a specification. It might still have use in ontology, but that is a different matter.

Medium number systems exist in a region Weinberg calls organized complexity (Figure I.1; Weinberg 2001). What he calls organization is nonrandom, distinctive behavior of the particles. Thus aggregation is not available as a tool for generalization; there averages do not work. Distinctive behavior that matters is commonplace, and is manifested in behavior that is unexpected at the level of the whole. The "complexity" part of organized complexity for Weinberg means too large a number of particles for us to model the quirks of individuals. With a small number of planets, the fact that Mercury is a distinctively small planet can be put into the two-body equation for that planet and the sun. Size can be applied in exactly the same way to the rest of the planets in the full set of equations for the solar system. In organized complexity there is important individual behavior of the particles. The "complexity," in Weinberg's terms, means that treating particles individually is simply an overwhelming task (Weinberg 2001). Medium number systems have critical differences between particles so that an average cannot be substituted.

In the vicinity of a medium number system specification, there is often another specification that is not medium number, and is predictable. For instance, the particular form of a snowflake is not predictable since they are all different in detail, depending on the details of the irregular particle that seeded the growth of the crystal. However, all snowflakes are predictably hexagonal, because of the fact that snow is made of water. The distinctiveness of the particles in medium number system specifications means that the details of any one of them can come to take over the whole system, the whole snowflake and its patterns.

A dictator is just one human particle in a society, but one able to direct macro-societal behavior. If all people could be seen as the same (like gas particles), dictators would not have any special effect, and for some specifications that is true. For example, the food consumed by a society is not particularly influenced by what the dictator eats or in what quantities. The hierarchy for medium number systems is fairly wide (many items at lower levels), but the distinctive character that makes the hierarchy medium number is that the level structure of the whole, the relative position of levels, is unstable. In a medium number specification, a low level particle can change position in a hierarchy so as to move several level upwards, as it becomes an influence on the behavior of the whole. Dictators are people who do that. Hitler was at first an insignificant low level officer, discharged from the German army at the end of WWI. As the National Socialists rioted, the German police shot into the crowd, killing the protester next to him. He himself could easily have died in oblivion. His death would have been something that would have made a big difference, but a difference that an observer at the time would not have been able to detect, let alone understand.

The specification that notices Hitler in retrospect is like the one that takes into account the piece of grit that seeds the snowflake to make it particular. The form of that snowflake is not generalizable. On the other hand, there is often a large number system specification in that general neighborhood that are predictable; with snowflakes it is that they are predictably hexagonal. In the vicinity of Hitler and at the end of WWI there is prediction that could be made and was captured in 1919 by Wil Dyson in a cartoon in the *Daily Herald*. It showed the four leaders of the United States, France, Italy, and Britain returning from the Treaty of Versailles. Clemenceau, the French prime minister, labeled "The Tiger," is saying, "Curious, I seem to hear a child weeping." The child behind a shattered pillar is labeled "1940 Class": a prediction of WWII to the year (Figure I.3). The general principle is that if winners in a conflict (WWI) brutally suppress an intelligent and industrious people, the losers will be back in about a generation.

The war Dyson predicted so accurately would have been a different war with Hitler shot in 1923; perhaps a fascist Germany, but equally likely one that was socialist, or democratic capitalist. With no "Führer," sides in WWII would have been different; Count Alfonso in the middle of the cartoon was the Italian player at Versailles, because Italy was an Allied Nation in WWI. The outbreak of WWI in the first place was another example of an unpredictable small event cascading upscale. That time the person/particle was shot, but in a fiasco of lack of security and happenstance. Archduke Ferdinand, likely to be the next emperor of that Austro-Hungarian Empire, was assassinated. The perpetrators were dismayed when they found some months later what their act had precipitated in the carnage of WWI. Britain and Germany were generally friendly with cousins on the respective thrones. Without the assassination, Germany would probably have become the preeminent power in Europe, and the world would be very different, and not Anglophone as it is now. French might still be the lingua franca. German would have been in more general use at a global scale. It is likely that the two World Wars sped up technological advance, and took it in a particular direction. It seems probable that there would have been an internet, but only

The Tiger: "Curious! I seem to hear a child weeping!"

FIGURE I.3. Wil Dyson's cartoon in the *Daily Herald* in 1919, where the French prime minister notices the Treaty of Versailles has set up WWII for 1940. It predicts WWII to the year.

if computing arrived in time to save the industrial revolution as it hit its fuel limitations.

To understand the nature of prediction it is useful to follow Ken Boulding,[5] who said all predictions are that nothing happens. What did not happen in the past will generally continue not to happen in the future. Scientific predictions are based on constraint that limits events to exclude what did not happen. Of course further into the future there is more opportunity for things to occur that did not happen recently. However, one simply goes further into the past that is commensurate with the length of the prediction. The pattern is that only more general predictions can be made further into the future. Note that Dyson stood little chance of predicting the Cold War in the same way he predicted WWII, because the Cold War was too far into his future. But he could have gone back further from his present to lengthen his prediction.

"SAVE ME FROM MY FRIENDS!"

FIGURE I.4. The cartoon is by Sir John Tenniel in *Punch* magazine, November 30, 1878. It shows "The Great Game," which continues to be played out today in Ukraine, Iran, and Syria, so the signal persists. Compared to Figure I.3, going back further to 1880 here gives prediction into a longer future.

Perhaps going back further to make reference to "The Great Game" Dyson might have been allowed to see the rest of the twentieth century, and even into this millennium. While Russia was allied with Britain and the United States in WWII, that was a detail in the longer history that went back to Victorian imperialism, into the Crimea and the Charge of the Light Brigade. Sir John Tenniel in *Punch* magazine, drew a cartoon, some forty years earlier than Dyson, on November 30, 1878. As we regard it (Figure I.4) we see the Cold War, and the present political situation in Ukraine, Syria, Iran, and A fghanistan all predicted in general terms. Dyson only refers to events in 1919, and that permits him to predict only the 1940s. To reach forward further than he did we need to go back to the late nineteenth century. It is still all in the Great Game.

The more general prediction is that imperial powers with territorial disputes will eventually go to war, but who and where in particular is up for grabs, only guided in a general way by reference to the longer past. Disputes will tend to resonate from earlier battle lines. Our future now may be understandable from reference to Early European imperialism, as the Dutch and Portuguese inserted themselves in the Far East. The immediate unrest in the Near East goes back further when barbarous Normans were set against the civilized world of Arabia in the Crusades. The Arabs had optics and kept the wisdom of Classical Europe. We must emphasize that there is no determinism here, but the long past does appear to allow us to understand the big changes in our present.

Historical systems in detail are medium number, so the particulars are unpredictable, even if what is going on now would be predictable as another chapter in the Great Game. In medium number systems, scientific experimentation does not give a way out because repeated experiments give different results (Figure I.5). Social systems commonly take this form, but some physical systems like bubbling Champagne can also be cast in medium number terms. In Champagne bubbling there may be trillions and trillions of particles, but that is not enough for reliable ag-

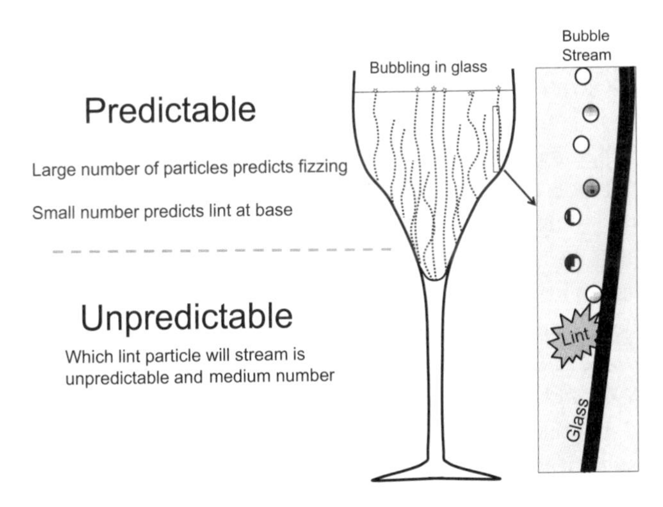

FIGURE I.5. Pour a glass of champagne and some aspects of the fizzing are predictable, but others are not. Large number specifications of champagne involve partial pressure of carbon dioxide, and surrogates can predict fizzy or flat. Small number specifications focus in to address individual extant streams of bubbles, and they consistently find a piece of lint at the nucleation site (seen in the inset). But which nucleation site actually does have a stream on a given pour is consistently unpredictable. In such a medium number specification a large number of constraints play and switch unpredictably, even under tight experimental control.

gregations for some phenomena. Champagne bubbling can be stated as large number, such as the partial pressure of carbon dioxide in the bottle. There are well over Avogadro's number of particles of carbon dioxide, so the wine reliably fizzes or reliably does not as a large number system. On May 22, 1897, Edward, Prince of Wales, opened the Blackwall Tunnel under the Thames in East London. It had been constructed using air pressure and a tunneling shield that moved forward at the digging face. The pressure kept the Thames water from coming into to the tunnel before it was sealed air and watertight. The ceremony was scheduled earlier than when the tunnel project was finished, so there was still pressure in the tunnel for the opening ceremony. Predictably the Champagne was flat because pressurized air did not let the bubbles form. One imagines dignitaries coming up from the ceremony to be met with uncomfortable eructation, as the wine did bubble in their royal tummies now at ambient air pressure.

A medium number specification of Champagne comes with a desire to know the position of nucleation sites that occur on a given pour. Each stream of bubbles usually has a piece of lint that sets the stream going, and once started the stream thus formed tends to keep going. There are predictions to be made. We can give a small number specification and address each extant nucleation site one at a time. Then we will fairly consistently find, and so be able to predict, lint at the base of almost every stream. What makes the question medium number in this situation is to ask which pieces of lint will achieve nucleation. The answer to that question would tell us where the set of streams of bubbles will arise in an instant, but we cannot answer that question. The combinations of streams of bubbles will be different every time wine is poured again. This is a good pedagogical example, because lint and bubbles are fairly homogeneous, so the problem is not overtly that there are so many types in the system. We can know that bubbling is due to lint and the turbulence of the liquid, but that still does not allow prediction because it is in medium number specification. We cannot deal with all the local partial pressures that will permit this versus that stream to emerge.

Some complain that hierarchy theory is either trivial or obvious. This would be a good time to interject that the above discussion of Champagne does give a simple explanation, but one that for the most part readers would not have teased out for themselves. Scientists specify situations so often that we take it for granted, not recognizing what specification involves. Empiricists are clever and often get their process right enough. Even so they are not conscious of what they are doing in principle. They would not dissect the problem with Champagne. And almost certainly they would not know how to change the question to get the solution. Hierarchy theory is neither trivial nor obvious.

The fact that fizzing and lint at nucleation sites are reliably predictable says two things. One is that the problem of medium number is the way the question is asked. The other is that the solution to medium number unpredictability is to change the question to one where we are still interested in the answer. To get a prediction about Champagne we can change the resolution downscale to inquire about individual lint pieces for individual bubble streams. We could also get a prediction by a change of type from streams of bubbles as distinctive individuals to just bubbles in general. Bubbles and lint are homogeneous classes and yet there is a medium number problem in the neighborhood. This means that it may well not be the heterogeneity of the material system that denies prediction.

Medium number systems are not pathological in nature; they rather arise from asking questions that have no meaningful answers. It is often not easy to see that the question is the issue. A mistake is to presume that with enough effort and reduction any question is answerable. No! Some are simply unanswerable because there is no regularity to be found. For some questions the answer is "It will be different every time." The reason is that certain structures posed by a question are unstable. Things may exist in an external world, but they do not exist there as things. Things arbitrarily exist with an identity frozen by the observer not nature. Some assertions about things are more helpful than others, but all are arbitrary. George Box is raising the same issue, but in terms of models and the structures they invoke, when he insisted, "Essentially, all models are wrong, but some are useful."[6] If the things that are posited in a question are unstable over the time span implied by the question, then the question invokes a medium number system. Many will take Box's point, but will not be able to dissect out its meaning. Yes, models are limited, and that might make them wrong. But Box says they are always wrong, and in principle. What is the principle? And one needs to understand that "useful" is a loaded term, making it all deeply subjective.

Some schools of ecology artificially isolated simple ecological entities and use classical Newtonian mathematics to describe and explain their behavior. Ground here is hard-won. The necessary strong simplifying assumptions make one worry. They might weaken faith in the general relevance of results to empirical field or laboratory studies. If one reduces down to too low a level, the question that was asked originally slips like sand between the investigator's fingers. Pattee[7] notes that if you cannot understand what we are saying, the chemistry of the ink on this page is not going to help.

With the coming of workable computing power in the 1970s there flowered the International Biological Program (IBP). The approach there was one of reductionism, but on a large scale: it worked for the atom bomb so why not for the eco-crisis? Unfortunately, ecosystems often invoke specifications that are medium number and so the massive reductionist simulations seem to have offered all the insightful summary that they easily can. An early critic of the IBP was Scott Overton. "Current mathematical models of ecosystems are so complex and large it is extremely difficult to understand how the model behaves, much less to master the details of the coupling and interactions."[8] Large explicit models cannot be coupled with much confidence to the structures being modeled. David Coleman, who was active at the time in the effort for IBP, recently wrote about the *Big Ecology* of those days.[9] At the center of IBP was a set of massive simulation models. Coleman (p. 34) states George Innis' model had "4,400 lines of code, 180 state variables, 500 parameters. It required roughly seven minutes to compile and run a two-year simulation with a two day step." Seven minutes may not seem so long, but it is big enough so that bigger models, or longer predictions, soon multiply up to an unworkable run time. Coleman reports that the central organizers of the effort wanted models that could be used "in place of . . . field experimentation." Yes, they really believed that if an army of ecologists could get everything they knew into a simulation model they could substitute the model for nature.

The most important message learned from abundant failure across IBP was that models need to be conceived in the light of questions from the outset. Scott Overton sought an alternative to the overly general models and proposed a method called FLEX/REFLEX. It is most unfortunate that the leading modelers in his Coniferous Biome would not support its implementation. They did not realize that the repeat of the modeling process either would confirm previous findings, or would show that something was wrong. FLEX was the upper level unit that treated the situation continuously and in a holistic manner. REFLEX modeled the lower level units that were mechanistic and were updated only discretely. Thus there were two levels of scaling. Overton realized that one rigid time between steps of simulation was going to be inept. The PWNEE model was of the grassland biome.[10] It was reported in literature only so far as it offered reasonable results. But Overton (personal communication) heard an oral presentation that reported longer simulations. There was only one unit time-step to update the whole model. It imposed that short time-frame even to parts of the model with long turnover to which short updates would not apply. As a result, due to rounding error alone, PWNEE covered the Western Plains with feet of buffalo dung. Overton knew each process at each hierarchical level needed to be uniquely scaled. FLEX/REFLEX did exactly that. In another biome the general model of Innis was not focused. Criticism of IBP has been overstated, forgetting we all learned a lot in a classic normal science refutation of the biome model approach.

In response to the problem of huge models with no explicit questions, smaller simulation models that ask very specific nontrivial questions were later constructed. For example, Bartell[11] modeled predation to investigate only its influence on phosphorus cycling. The failure of the IBP came from its naïve belief that with enough data we could make big enough models that would then answer post hoc questions. No, models need to be built for a particular question, which was Bartell's strategy. But do not imagine that the naïveté that Bartell countered is no longer expressed. The cautionary tale of the big models appears forgotten in contemporary large scale reductionism, for instance in the big diversity experiments of Hector et al. 1999.[12] The lines of attack on medium number specification offered above were suggested in the first edition of this book in 1982; it is remarkable that medium number models are still a principle device of ecological normal science today. But we have made some progress. New approaches coming out of the generic area of complexity science are making inroads, but a more coherent and focused use of new methods and philosophies would find merit.

With a medium number system, every time there is a new test, the system gives a different result. Science cannot investigate the unique, it can only deal with things that happen lots of times. So science is impotent in the face of a system specified so as to be unique. Should we simply give up? In one sense yes we should give up, but in another sense hierarchy theory is the antidote. The critical point is that medium number lack of pattern is not a material issue. Basically, everything can be made impossibly unmanageable if you look at it in the wrong way, and most ways are wrong, which is why science puts so much effort into looking at things profitably. Discrete structures blur if the lens is out of focus, without there being any change in what had appeared discrete at another focal length. In a random selection of lenses, most images will be out of focus. Medium number systems emerge from inept questions that are unhelpful system specifications. The only thing to do in the face of a properly medium number systems is to change your question to one that can be answered. Hierarchy theory first indicates something is medium number, and then it goes on to guide the observer/questioner to look at things in a better way, with a question that science can answer. We seek a question that gives an answer that might be different from the original inept question, but is a question that is still pertinent and helpful. The reason hierarchy theory studies complexity is to get rid of it by telling when a new question is needed and suggesting what that question might be. We did that with Champagne.

When the different conventional assaults fail they are running into strategic problems more than tactical setbacks. Rosen[13] identifies that strategic errors in an approach to an investigation and analysis are particularly painful when revealed. The solution points out that all the tactical maneuvers were simply pointless. When an empiricist is caught out, all that clever and skilled practice simply goes out the window. We should not decry empiricism, we should rejoice that empiricists deliver as often as they do. When they miss the target, they are not even wrong, they are just irrelevant. Rosen shows how as he refers to "The Purloined Letter" in his 1981 Presidential address to the Society for General Systems Research.

The situation in the story is that the letter to be used for blackmail must be in the blackmailer's quarters so it is available for use. The French Sûreté were all over the house looking for secret cavities under carpets and wallpaper where the letter might be hidden. Auguste Dupin (the French Sherlock Holmes) arrives and briefly ponders the situation. He then goes back into the blackmailer's quarters and switches the letter for a copy, taking the original (Figure I.6). The blackmailer's hiding place foiled regular police methods because of a strategic error. The logic of that error says the hiding place must be deeply inaccessible so it is secure. In fact the letter was in the open. It was sitting in a letter rack, but folded in the reverse direction, addressed to someone else and sealed with blackmailer's seal. There are always assumptions that lie behind any approach to a problem. Dupin and hierarchy theory both approach problem solving in a strategic fashion. A difficulty for hierarchy theorists

FIGURE I.6. In "The Purloined Letter," Dupin is depicted switching the original letter for the substitute. The blackmailer is distracted by a happening outside the window, and will not notice that there has been a substitution. The French Sûreté were meanwhile making strategic mistakes based on the presumption that the letter must be deeply concealed physically. It was in fact concealed in the open.

is that clients whose problems are solved by hierarchy theory claim they already knew (you didn't know the bubble fizzing problem until we took you through it). One supposes they are unable to face the pain of having invested so heavily in previous tactical wastes of time. In fairness, it is very hard to remember what it was like not to know, which is why most teaching is mediocre.

An Example of Ecology in the Medium Number Range

There is a medium number issue in ecology in the tension between water that runs off soil, on the one hand, and water that infiltrates into soil, on the other. Both are connected to the same source, inundation. Some aspects of soil and water are well described and well modeled. For instance, the way water penetrates the soil is reasonably well understood by experts who use various forms of infiltration theory. But there is another whole discourse of how water runs off soil. A given rain will be split into one or the other fate for the water. Although infiltration and runoff interact over a landscape, traditional runoff theory involves scaling up to watershed dimensions while traditional infiltration theory involves scaling down to soil pore dimensions. Large scale differences between models make for difficult translations between models, and lead to separate scientific cultures.

> Hydrologists, who are most interested in surface runoff, tend to reside in engineering disciplines and are motivated by solving a compelling societal problem associated with periodic flooding of rivers; whereas, soil physicists, who are most interested in infiltration, tend to reside in soil science or physics disciplines and may be interested in practical applications but likely are motivated more strongly by curiosity. (Norman 2013)

Associated with scale difference are the different ways scientists and engineers build their models. So part of the problem in parsing out water falling on a landscape is a clash of cultures as to how to solve problems. In fact many points made in this book are cultural, not just in most disparate fields like science versus humanities, but also inside local scientific issues like predicting runoff from precipitation and infiltration. Thus even hard-nosed soil physicists have something as soft as a culture, and one ignores that at one's peril.

Runoff is an engineering problem, while infiltration resides in the open discourse of physical science. Very few investigators span the divide of engineer and scientist, although as a counterexample Roydon Fraser of the University of Waterloo is well aware he has two hats. He self-consciously takes one off when he is wearing the other. As an engineer he gets it done, whereas as a physicist he has to get it right. Engineers are given an endpoint that is fixed at the outset, but those who are not engineers often do not realize how many creative options are available to get to the goal. Engineers try to create something open in a closed system. Meanwhile scientists try to create something closed in an open system. The "open" that is sought by engineers is a process that captures creative and elegant solutions in reaching the prescribed goal. The "closed" that is sought by scientist is the internally consistent model, which is more closed because distracting alternatives are pruned away, leaving only the closed intellectual structure in an unambiguous model that leaves few alternatives. Scientists do not necessarily know what

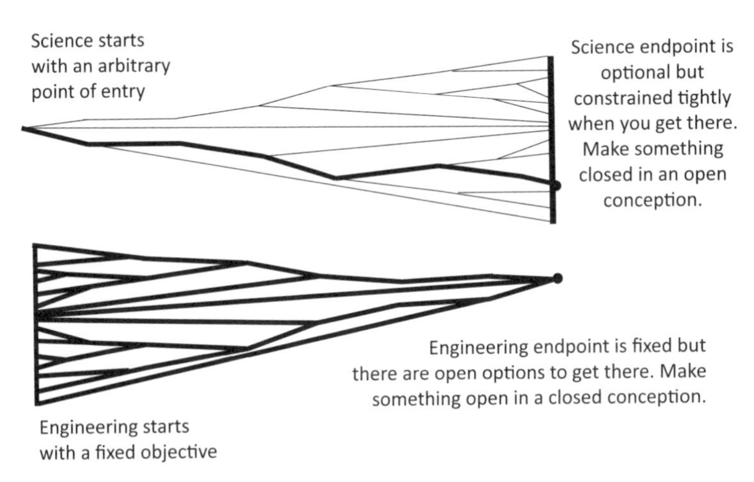

FIGURE I.7. Showing how scientists are open early in their investigations, and move to prune down to a tight, internally consistent model as an endpoint. Engineers are given a certain goal to achieve at the outset. But there are many ways to cross a divide: hot air balloon, bridge of many types, tunnels. So engineers are open to many solutions, and work out which one to use. The focus achieved in a scientific model tends to take the model downscale. The open path to engineering solutions allows engineering models to drift upscale.

their model is going to be, but as they move toward the goal, the model becomes more constrained (Figure I.7). Scientists are open, early in their investigations, and prune down to a tight, internally consistent model as an endpoint. Engineers are given a certain goal to achieve at the outset. But there are many ways to cross a divide: hot air balloon; bridges of many types; tunnels. So engineers are open to many solutions, and work out which one to use. The focus achieved at the end of creating a scientific model tends to take the discussion downscale. The open path to engineering solutions allows engineering models to drift upscale.

With the fundamental difference between the scientist and the engineer, it is no surprise that there is a cultural divide between hydrologists studying runoff and soil physicists who address soil infiltration. Scale differences hide things readily seen on the respective other side of a difference in scale. The separation of cultures is maintained and reinforced by circles of acquaintances that do not overlap. As a result those two sets of experts hardly ever talk to each other in serious terms, although their phenomena are necessarily linked. Investigators tend to cite papers mainly from their own disciplines and rarely attend meetings of other disciplines. Like the investigators who make them, predictions as to how water runs off soil are not often linked to prediction of infiltration. The one essential problem is that insufficient attention is paid to the respective edges of the investigators' purview. Water flows over soil in runoff, and into and below the soil surface. At the edge of runoff and at the edge of infiltration is the first centimeter of soil, seen from the respective other side. The exquisite details of the first centimeter of soil appear to make all the difference to infiltration and ultimately to runoff.

There must be a reason why so few dare to go to that first centimeter. After the first centimeter of soil the science is more easily studied, and many scientists go down those avenues. Idealized infiltration theories are based on treating soils like they are laboratory-packed beds of tiny glass beads. They are well behaved with explicit thresholds that are predictable. But the application of such idealized concepts to soil has failed to obtain even marginal predictability because the top centimeter is the context that determines the input to the body of the soil where the idealized concepts do work. On the other side of the collective issue, once the run-off gets established, that too is well behaved and reasonably predicted by hydrologists. Good behavior promises predictions and many scientists have moved to obtain those rewards as they go down paths that make for easy going. Much science is done because it can be done easily. The breakthroughs in science occur when particularly necessary but challenging measurements are achieved. As Richard Bellman said in 1963, "If one wishes to obtain significant results, it is better to study significant questions from the very beginning." Do not so much do what you can, do what you must.

The first centimeter of soil can act as a constraint on the ultimate disposition of rainwater in a downpour. But that first centimeter will be changed, perhaps in the destruction of rills, which will change how the next downpour may behave with regard to runoff or infiltration on that same piece of soil surface. It is difficult to treat the first centimeter of soil as a constraint, because its properties keep changing. Remember Kenneth Boulding saying all predictions are that nothing happens; something does happen on and in the first centimeter of soil. The first centimeter in fact represents a complex of constraints, in that it can influence infiltration in a large number of ways. Which of those ways gains ascendency is uncertain, and that scuppers prediction. The constraints that link runoff to infiltration into soil in the first centimeter of soil shift often. Treating the first centimeter of soil as a constraint on whether or not water infiltrates or runs off, makes for an unpredictable medium number system, because the prevailing constraint keeps changing. In that situation we need to respecify the system so to as employ a stable set of constraints.

The many turns of the worm can be seen as we tell the narrative of a downpour. This may seem technical, but the point of the telling is to show how convoluted is the causal chain. Do not be burdened by the details, just note there are many of them. No wonder phalanxes of devoted scientists are largely defeated. The story starts when rain begins to fall. It first infiltrates the porous soil surface at rates that depend on the size and arrangement of pores (more sand > larger pores > rapid infiltration; more clay > smaller pores > slower infiltration) and the soil wetness at the time of rain, which is itself dependent on pore size. If the intensity of the rain is great enough for the particular soil, excess water is stored in micro-depressions of roughness on the surface. There is also some infiltration. As rainfall continues at sufficiently high rates, the soil wets up and infiltration slows so that runoff begins with water dribbling from micro-depression to micro-depression downslope. Micro-depressions are highly variable in size and connectivity. They may be randomly placed in nature, but can be systematically organized like tillage furrows or small ridges. In any event, runoff amount and direction is determined by slope steepness and aspect as well as tillage type and direction relative to that slope. The runoff water moves downslope cascading from smaller depressions to larger depressions and sometimes to man-made structures or large natural depressions where large amounts of water can be ponded temporarily. On the way, the moving water creates its own channels (called rills), capturing surface soil in the flow (erosion). At the same time the falling rainwater begins smoothing the surface roughness, depending on storm intensity.

Continued high-intensity rainfall ultimately leads to fast-moving water with much suspended sediment that creates gullies, which can vary in size from something a person can step across to ravines that could swallow a house. The pounding of the rain disintegrates the small soil clods (aggregates) on the surface. This releases the basic soil particles (sand, silt, and clay), and these particles become suspended in the fast moving water. They are carried downslope. As the slope decreases and speed of water movement slows, the sediment-holding capacity of the water decreases. Much suspended sediment is deposited on lower landscape areas. The largest sand particles deposit first, the smallest clay particles usually remain suspended and are most likely carried off the field into streams. The medium-sized particles (silt) may experience deposition or be carried to streams, depending on the rate of runoff flow. The eroded particles that deposit on the lower slopes tend to fill in the channels (rills and gullies) previously created by earlier flows. When an intense storm ends, lower regions of the slope may be covered with a thick layer of sediment devoid of structure. The entire field may have a thin layer of sediment (soil particles) that forms a seal on the surface as drying occurs. Thus a relatively impermeable crust forms so that succeeding rainfalls encounter a drastically altered surface.

Clearly infiltration rate is a key factor in the onset of runoff, and infiltration is determined by a host of factors: 1) soil texture (amount of sand, silt, and clay), 2) soil structure (arrangement of elementary soil particles into aggregates), 3) soil moisture content at the time of rain, 4) presence of a surface seal or soil crust, 5) surface roughness, 6) man-made modifications, 7) slope steepness, and 8) surface features left by previous rainfalls. In addition, some of these factors depend on soil chemistry (surface sealing) and the presence of soil biota (soil structure), and most factors change in time between and during rainfall events. What happens in the top centimeter of soil is crucial for determining infiltration—namely, soil texture, soil structure and its robustness, surface roughness, and surface sealing that depends on all the other factors. Runoff depends on all these factors plus overland flow characteristics associated with moving water. To top it off, the time course of the precipitation intensity, as well as the precipitation duration and total amount, are critical in determining the evolution of these landscape processes. And we have not even discussed the impact of winter processes of freezing/thawing and runoff generated by spring snow melt. The reader should be impressed with how many separate issue collide in a given rainfall event.

This runoff-infiltration process presents an amazingly challenging task for prediction, and a myriad of questions are associated with various aspects of this issue. Hydrologists may be interested in what gets to the stream by direct runoff as well as groundwater flow. Meanwhile the soil scientist may be interested in what is stored in the soil for future plant use by evapotranspiration as well as drainage to groundwater or tile drains when water tables are high. In their respective ken, the two groups of investigators are able to specify their systems so that constraints are stable and reliable. There they can predict. The problem arises when the two lines of investigation are forced together because a prediction is needed that applies to the interaction across lines of investigation. Then constraints for the joint phenomenon are mixed between discourses. The separate sets of constraints jockey for position, with uncertain outcomes. The switching constraints make the models medium number. Here we see how unfortunate is Weinberg's labeling of systems by number of constraints, because the trouble is not number of constraints per se, but the labile nature of constraints, not strictly a matter of number of particles.

In landscapes dominated by agriculture, the role of management can overwhelm many natural processes and radically alter how water infiltrates and runs off. This is particularly unfortunate because management of the soil with agriculture is often the situation that pertains with the need to predict runoff. Runoff tends to be much larger for most agricultural systems compared to the natural systems they replaced. Norman (2013) reports on chisel plowed land on a slope. One treatment was plowed up and down the slope, while the other was plowed with the same plow across the contour of the slope (Figure I.8).

There is a simulation model for soil runoff called PALMS. It is the

FIGURE I.8. Erosion in a field at Arlington Research Station in Wisconsin on May 26, 2004. On the left plowed rows are up and down a 5% slope, and on the right plowing gives rows across the slope. Deposition is only below rows that are up and down the slope. Insert shows 11 cm (4.5 inches) of deposition at a location.

most realistic and well behaved model we have for runoff. An illustration of the medium number nature of runoff from agricultural soils is demonstrated by an observation from Bonilla et al.[14] The field was a sandy loam, with a 10% slope that was tilled on the contour. The random and systematic roughness were set at 2 mm everywhere in the field for this tillage. These modeling decisions were based on observations published in the literature. A single rain event of about 30 mm a few days after tillage with a disk and planting resulted in one of the largest differences of runoff between PALMS model predictions and measurements based on 6 site-years of measurements. The runoff from PALMS was predicted to be 8 mm but measurements indicated 0.09 mm (very close to zero). By keeping the surface roughness at 2 mm on steeper slopes and changing it to 4 mm on gentler slopes, PALMS predicted no runoff. Clearly the difference of a few mm of roughness from farm tillage machines is hardly predictable and indicated the limits of predictability for any runoff model. This narrative is an indication of the medium number nature of agricultural runoff. Fortunately most runoff events do not come this close to the threshold between initiating runoff and not initiating runoff. Runoff depends strongly on the intensity of rainfall as well as many other soil and management features. Of course, once runoff begins, the rills generated by the runoff itself can modify the surface to provide the pathways for rain water to move down the slope regardless of tillage. However, the initiation of this rill generation process itself is not a predictable deterministic process, in general.

One answer to problems of soil runoff is calibration. On a particular piece of landscape, models can be developed by iteration so as to predict runoff and infiltration fairly well. The down side to calibrated models is that they work only on that piece of ground, with failure any time the model is applied to runoff on any other piece of landscape. So sometimes prediction in a medium number situation appears possible, by tightening constraints in a way that gives a specific and local solution. The downside is that the predictions that are achieved are of little use if the situation is to be generalized. The Industrial Revolution turned not so much on power to drive shafts as it did on tightening the constraints so as to generalize the outcome in a highly contrived production process.

Engineers use calibration to solve particular problems and get around the generalization issue by simply not generalizing. Some engineering solutions are achieved by deep and tight specification of the model. In the heyday of the Industrial Revolution there was a radical shift in gunsmithing. Before the Smith and Wesson six shooter in the middle of the nineteenth century, guns were made one at a time with one model for each gun. One gun was like one landscape; the issue was constrained to a highly calibrated situation. The model was for making that particular gun, not guns of that type in general. Unable to machine parts with sufficient precision, workers made each part for each gun and then adjusted relative to the other parts on that particular weapon; a blow with a hammer here, a bit of filing there. Thus parts were not generalizable but if the model was restricted to one gun at a time, it worked well enough. Before the industrial age, making guns was a medium number problem that was solved by something like calibration. So how did they manage to generalize in what was a medium number world of gunsmithing? Medium number failure comes from parts (particles) being so individual that the particularity of some part reliably causes failure. The economies of scale of industrial production were not possible until there was machine tooling to a thousandth of an inch. But with that precision (tight constraints) the parts would fit together in a streamlined process. Control of industrialized manufacture was thus achieve by forcing deep uniformity. Under such uniformity individual particles in the gun (each trigger or barrel) had no functional differences. Once you could say "when you have seen one Smith and Wesson barrel you have seen them all," an individual barrel would not create failure in a medium number emergence. In this way medium number gunsmithing was translated into a large number process. Like one gas particle being functionally identical to all others so that the gas laws work, each trigger or barrel was made uniform enough to allow something like a gas law of gun manufacture. So there are various ways to escape the failure of medium number to predict. One is to change the question. Another is to calibrate. Yet again one could impose tight constraints, as in the barrels of Smith and Wesson all machined to be functionally identical. It helps to have failure generalized in the concept of medium number because then one has a generalized understanding of failure, and can model past it.

Reliability can be manufactured or not as in planned obsolescence. This is a way of using the phenomenon of medium number models to gain control. One often cannot tell which part of a washing machine will fail, but the fact that one of them will do so in the end is reliable. So we are dealing with prediction in both success and failure, the difference being in level of analysis. In the contemporary problem solving, level of analysis is everything.

Post-normal Science

Hierarchy theory becomes pertinent more often as computational power allows society to address big social issues. "It is all so complex," was the excuse before, but now society appears to have some sort of shot at ameliorating large scale complexity. In the 1990s Funtowicz and Ravetz[15] addressed the same dilemmas of medium number and hierarchical solutions, but in a remarkably new way. They refer to post-normal science. In a sense post-normality is the context for using hierarchy theory. In their terms, difficulties cited above arise ever more often, particularly when science attempts to solve practical problems. Not all applied science is medium number, but applied science is often the place where medium number comes into play. In those practical situations the stakes are high, time is short, values are in conflict, and variability is large and intrinsic. In some cases we discover that the challenging situations are not after all medium number. But what to do if medium number is still in your face? That is when one assumes Funtowicz and Ravetz's posture.

A case in point would be a medical doctor trying to diagnose and treat a patient. Certainly some science helps, but if the patient is expected to die in a few days if nothing is done, there is a certain urgency. A "few days" would arrive before the scientist in the physician can get the p values down to a scientifically acceptable .05. Therefore, the physician must act in good faith with the best scientific advice available. The stakes are certainly high, and there are conflicting values all over the place as to how aggressive to be; how large should be the risks that are taken? The bravest internists often demonstrate nerves of steel by simply waiting while poised for action. Experienced old doctors know that in general patients do seem to get better on their own. Medicine, mathematics, and the law are taught the same way: teach by example and repetition. The method is to teach with cases until in the end the student simply gets why calculus or constitutional law or waiting to act on a patient has certain advantages.

Ravetz[16] accepts that science as an institution in the service of society has great potential to be corrupt, taking more money to do more of the same research. He says the system is simply broken. What to do? After procrastination has run its course, in the end pressing resource issues will become so urgent that remedial action will be taken. That action might be to find a serious substitute for fossil fuels. But until that time, political forces will not allow the expense of prudent action. Jimmy Carter, a former nuclear engineer, was one of the most intelligent Presidents. He did try to act prudently. In the oil shortage of his day he proposed to increase taxes on gasoline until use declined. Had he achieved his goal, the world of fuel would be much brighter now, but Congress would not pass it into law. There is no real remedy to be offered ahead of imminently approaching disaster; Congress, or some equivalent body, will always veto prudence. Ravetz suggests that the only thing to be done ahead of an actual good faith effort at remedy is to clear the decks ready for the crisis. Experts are not allowed to make good faith efforts at solving problems like climate change and greenhouse gases; the cost is too big to make it politically feasible. Any politician who tried to do so, would simply be voted out of office. Honest brokers cannot get elected in the first place. But tidying up is not expensive, and is politically allowed. With Ravetz's protocol, when political players are forced to enact a remedy because of scarcity, the path of the overdue correction is at least cleared of debris. Then there is a shot at not going over the cliff. Going over the cliff means running out of resource sufficient to buy increased efficiency.

The deterministic approaches use exact results of running simulations of some sort. They are hampered by their strong simplifying assumptions, which keep the meaning local or time specific. Meanwhile the stochastic, statistical approaches are less and less justified in their use of averages. For instance, the statistics behind the use of statin drugs for lowering cholesterol are questionable. Say that if sixty people are put on stain drugs for four years, on average just one heart attack will be avoided. Meanwhile all sixty will be compromised both physically and mentally. Statins work by poisoning mitochondria, a drastic thing to do. With insulted mitochondria muscles work harder and so produce more lactic acid. Acidity is good for heart function. Stephanie Seneff suggests simply eating more yogurt. Her home page includes several pertinent articles.[17] She comments about the seriousness of the situation. "Should the disinformation [generated by the medical establishment and drug companies] be overcome, there may be massive, chronic liability, akin to the mess left after asbestos." The simulation approach is often limited financially and technologically by its need for more data to set unreasonably exact initial and boundary conditions. There is also the problem of exponential requirement for computational power. There is a bottom line in genuinely medium number systems: experimental approaches and the usual devices of science simply fail. So back off and start working hard at finding a new system specification. The issue is one of poor specification not number of particles per se. The usefulness of the medium number concept is it often serves as a guide as to how to respecify.

Seneff also addresses the issue of increases in asthma, diabetes, and autism in children. These are all diseases with more than a touch of post-normality and medium number systems. The critical political differences are the public attitudes in the United States and the European Union with regard to safety in food. There is much resistance in Europe to genetically modified organisms (GMOs) in the food supply. Americans do not much care about it, and it is almost impossible to eat in America without consuming GMOs. Our position is that the European populace raises bogus arguments with regard to genetic modification. Objections that say we are "playing God," are unwarranted. If God or nature were not doing it already, human technology of gene transfer would not be possible. It emerges that genes are promiscuous and jump between organism and species all the time. GMO technology is not that special, it is just a new way of giving the same genetic screw one more turn. There is nothing in principle that is particularly dangerous about most genetic modifications. One can argue that GMOs are one more turn of the screw, and that increased pressure is undesirable in principle, but that is a different discourse. Of course stupid tinkering with disease genetics is a special case, but it is not genetic engineering as a general issue that is at fault. The irony is that European negativity on GMOs in the food supply has been beneficial, but not for the reasons that activists say. Let us explain.

Genetic modification is not often used to improve the food supply, but is principally used as a device to wrest control from farmers, so as to give the biggest seed producers increased control over the money, which gives farmers fewer options. The commonest use of GMOs is to introduce tolerance of the weed-killer *Roundup*® so that all weeds in the field are killed without hurting the crop. *Roundup*® is glyphosate. It was noticed that it appeared to kill all plants by interfering with a pathway universal to plants but not animals. As a result, glyphosate does not harm humans directly because we do not have the pathway that it targets. On that point the big agricultural companies are right, but there is damage to humans indirectly, and post-normality can find that out.

Now the worm turns. We are not directly poisoned by glyphosate but bacteria do have the critical target pathway and so are affected. It is a relatively new understanding that bacteria are crucial for human health. There is an argument with data to support it that peanut allergy is best treated with peanuts. Infants fed peanuts get the allergy later at lower rates. It is the same argument that children in too clean an environment (e.g., living with sanitizing dishwashing machines) have weakened dysfunctional immune systems, because of the absence of bacteria in their environment to keep the immune system tuned up. The benefit of bacteria arises in particular with regard to the gut flora. A particularly debilitating bacterium is called *Clostridium difficile*. It can come to dominate the human gut flora when antibiotics are used to treat some other bacterial disease. The antibiotics incidentally remove the healthy gut flora, leaving an opportunity for *C. difficile*. A second round of antibiotics sometimes works, but often enough this only keeps the door open for *C. difficile* reinfection; so what to do? Patients thus infected and close to death from chronic incurable diarrhea have been cured with fecal implants from a healthy person's gut.[18] Gut flora are very helpful in digestion and the like.

Returning to Seneff's analysis, the trouble with glyphosate is that residues in food grown with *Roundup®* could reasonably be expected to damage the gut flora. Suspicion about damaged gut flora as a cause of chronic physical and mental disease is not new, but it is hard to prove. Symptoms are vague and distinctively acute cases are unusual. Of course, all complicated epidemiology is difficult to pinpoint, but here the unwarranted attitude of Europeans with regard to GMOs comes to help. While almost all corn and wheat products in the US diet will have in it genetically altered material, such is not the case in the European Union. If glyphosate residue is damaging the gut flora to increase asthma, diabetes, and autism, we have a test case. All those diseases have greatly increased incidence in recent decades. So we can compare their epidemiology in the United States and the European Union. And yes, Europe has not had the same increase. Seneff suggests it is because there is not glyphosate in the EU food supply.

We might be viewed askance because of our suggestion that GMOs are not in principal worse than other genetic manipulation, such as haploid genetics in regular breeding programs. We will probably be vilified for suggesting that an organic diet is not much healthier in general than food that does not qualify for the organic label. Most chemical fertilizer and the like do little harm. That being said, eating organic bread in the United States would allow an end run around *Roundup®* residue. So we have the irony of two wrongs actually making a right. A wrong idea about GMOs in Europe combined with the weak-to-wrong arguments for the benefits of an organic diet actually allow us to see that organic bread can make a difference. Context is everything. Post-normal science is explicit about context and one of the solutions to medium number lack of control is to change the context. One does wish Weinberg had not used medium "number," because he is in general terms right, albeit not about number per se.

Conclusion

So having set out the problem of scaling, hierarchy, prediction, and levels of analysis, we are in a position to explore ramifications in the rest of this book. First a history of the ideas herein is a good starting point. The ideas here have their origins in the early twentieth century when Bertalanffy developed ideas for General Systems Theory.[19] We move them into the context of scale and hierarchy. Another point of origin of these ideas comes from Rashevsky[20] and his research in the 1930s into the concept of organization. We owe a huge debt to Robert Rosen,[21] a student of Rashevsky, who developed the master's ideas into Rosen's own masterpieces. All of this is summed up in what is called relational biology. Relational biology stands against reductionism in that it addresses what is lost when biological systems are taken apart. Louie[22] starts his book defining relational biology as that which is left in biology after all the materiality of life has been removed. Hierarchy theory addresses organization. In this book we do consider material organisms and the like, but in the end the nub of it is organization, which is not in itself material. Ahl and Allen say that hierarchy theory is "a theory of the observer's role in any formal study of complex systems" (p. 29).

We see this book as something of a sister to Margalef's (1968) *Perspectives in Ecological Theory*. Like his ideas, ours have an origin in the study of aquatic ecosystems and in the tension and unity that come about when aquatic and terrestrial systems are compared. Although we do not emphasize information theory per se, we share his interest in a signal-and-message approach to ecology. The flowing of water seems to engender thoughts on information flow. We are also pleased to share with him the University of Chicago Press.

We divide the book into three parts. Part 1 has four chapters where the basic idea of hierarchy is laid out. The second part deals with general biological issues such as evolution, reproduction, and boundaries in living systems. The third part moves on to address generally ecological hierarchies and the scaling that is used there. We finish with a set of scaling strategies for ecology. In the first edition we anticipated metabolic ecology, and so can present it now as a mature view. There is much recent work that is mistaken by the mainstream as a fad, but we show it has roots that go back to biological orthodoxy. The first edition was written at the very front of the new version of complexity theory. It was revolutionary then, but is likely to be more acceptable now. This edition is to get biology and ecology over the hump, so complexity can be understood in terms that are more familiar to mainstream practitioners. The intuitive appeal of hierarchy as a concept is a likely vehicle to move biology and its social extensions into a more comfortable position in a world that demands competent treatment of complexity. It is time to move beyond the reductionist excesses of the preceding decades. Maybe we will make it this time.

PART I

A Theory for Medium Number Systems

CHAPTER ONE

Hierarchies

Many disciplines use hierarchy theory. Herbert Simon wrote the seminal paper in 1962 with ideas stemming from organization theory in business administration and economics. Another early hierarchist is Ilya Prigogine, who addresses emergence of new levels of order in physical chemical systems.[1] Simon and Prigogine were both Nobel Laureates. Ahl and Allen (1996) point to the central titanic figure in psychology, Jean Piaget, as an early constructivist. Constructivism supersedes both empiricism and nativism (nature versus nurture) by suggesting that learning is a process of interaction. Piaget's constructivism arose with a theory called genetic epistemology, which explicitly uses levels of observation and analysis as well as psychological hierarchies.[2] Piaget late in life worked collaboratively with the chemist Prigogine, which indicates that hierarchical thinking is an endeavor in itself that transcends disciplines. Koestler (1967) worked on social and linguistic hierarchies. O'Neill et al. (1986) were explicit even in the title of their book, that ecosystem analysis needs a hierarchical accounting.

In all these areas of discourse, technical improvements allowed the practitioners to achieve finer grain data over wider extents. It follows from coarsening of grain and widening extent that at some point hierarchies will invite themselves into the discourse. With the gradual encroachment of ecology into larger and smaller realms (global ecology and microbial ecology) ecological insights are to be won outside commonplace human experience. There are changes of perceptual scale taking place that allow the discipline to transcend the limits imposed by the naked senses. Rich direct human sense experience is held in a fairly narrow range between large and small: we cannot literally see molecules, and we lose track of individual organisms across vast tracts of remotely sensed landscape. Technology widens our experience such that the environmental movement of the second half of the twentieth century is rooted in the first view of our whole planet in space. The width of the extent of observations increases in ways that change the sense of where we think we live. Across the wider extent of experience we have technology that lets us detect and remember remarkably fine grained detail, perhaps of individual trees across county-wide swaths of land. With that enriched set of senses, the science of ecology can perform at levels of analysis as never before.

We appear to face greater complexity with the new technology, but that is only a matter of appearances. The common assertion that complexity increases with the size of the unit is a holdover from naïve realist positions of modernism in the middle of the past century. For the realist some lower level grain is given privilege as being the real level, and it is held constant as the lower reference when the discourse expands in scope to address a larger situation. The chosen level is an observer position not a material necessity. If the extent is then widened the increased size of the unit (the extent) does appear to increase complexity because of the wider difference between grain and extent. Complexity arises from the gap between grain and extent, not extent by itself. There is no grain that exists without human decision, and referencing external reality ontologically only offers confusion. The privilege of the grain taken as a primitive in the discourse is a decision, not a reality. As a rule increase in extent demands a coarsening of grain, otherwise we exceed our memory and analytic capacity (although technology helps). As grain and extent move upscale together complicatedness might actually stay the same or even decrease.

The error in assuming increased size increases complexity is to imagine that complexity is a material issue as opposed to a matter of how the system is observed. In biology many discussions that claim to address complexity in fact speak only of complicatedness. Complexity is something normative, based on decisions about how to observe. It is a special sort of scaling issue where bigger does not mean more but rather means different. Hegel anchors his philosophy to qualitative distinctions arising out of increased quantity, and we tie complexity to it as well.

Let us justify that complexity is a normative issue, not a matter of material distinction. The brain is particularly complex if it is considered as a mass of neurons. The questions there might be about how thoughts have a basis in neurons. By contrast the brain is simple if the question is,

"How did the reptilian brain become mammalian?" The answer to that question invokes a grain that distinguishes only three parts (hind-, mid-, and forebrain), and one trend of the forebrain getting bigger. That is a simple situation. There is no real grain that exists without human decision. Complexity comes from the question and is not an intrinsic material property. Complicatedness, by contrast, may reflect a material situation, but only one that is tied specifically to some announced observation protocol of a set of parts, whose identity is only asserted.

Our complaint about realism may have set off the reflex response of modernists that science gets closer to reality. But do notice here we have not denied a material external reality in principle. We are not so anti-realist as to deny that Allen, the writer here, is in his real office typing on a real keyboard; we would hasten to add that the meaning of such a belief, for that is all it is, is far from transparent. When we object to realism, all we are saying is that realism often muddles the material and the normative in the conduct of science. When we object to realism in the conduct of science, we are not at odds with a belief that science, when it is all over or is at least at a stage for reflection, appears to approach some sort of reality or at least a biologically, culturally shared understanding of what is real. The two positions on reality address either practice, on the one hand, or the larger meaning, on the other hand.[3] As such they operate at different levels, and so are only orthogonal not opposite. The conduct of normal science includes a narrowing of discourse, so the universe is highly contrived. In a narrow discourse we can be more confident about the reliability of what emerges with that exact protocol but will be less confident as to general application. There are fewer checks through alternatives. When colleagues agree in the light of the full study there have been many alternatives introduced, the agreement of which leads us to have confidence in generality and be less fearful of local artifact. The apparent contradiction about local and general verity is only superficial. Understanding the muddle of material and normative issues helps so we can know what comes from us as opposed to what comes from elsewhere. That distinction gives us control in problem solving, and gets us unstuck from an epistemological mire.

So complexity is significantly a matter of relative scaling. It is the aim of our contribution to deal with the scale problem in its own right. That lets us work toward tools and general models that may encourage ecology and other scientific disciplines to escape from limitations that have historically prevailed. Rosen's modeling relation indicates when scaling

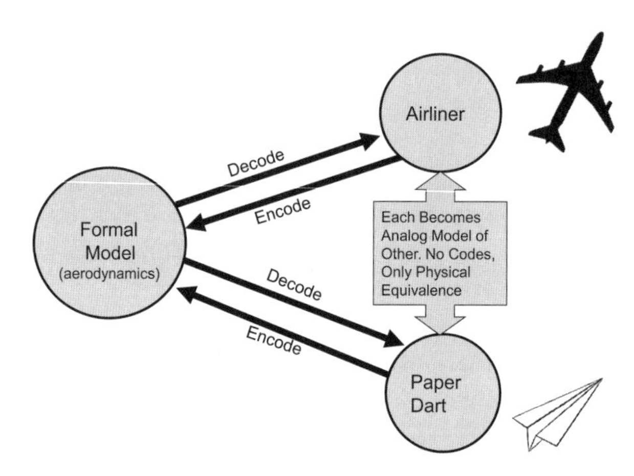

FIGURE 1.1. The Rosen modeling relation. The formal model is scale-relative and so scale-independent. If two material systems can be encoded and decoded into and out of the formal model, they become analog models of each other. The material objects specified here are flying winged planes: a passenger jet and a paper dart.

applies, and when something is scale independent. The modeling relation suggests that when two different material systems can be encoded and decoded into and out of a formal model, the two systems become analog models of each other. All experimentation depends on that relationship. An unspoken indication of analogs is that everything is the same except scale; that is not true, but is one of those assumptions in science that cleans things up for an orderly scientific treatment. A formal model might be the laws of aerodynamics (Figure 1.1). The power of a formal model, such as aerodynamics, is that it captures scaling relationships, such as speed and lift, through the Reynold's number, a dimensionless number involving drag. Therefore formal models are scale-relative and so are scale-independent. Experimentation uses two material systems that can fit into and be derived from a single formal model that applies across scales. Experimentation turns on analog models, which are not coded, but rather look at the compression of the two material systems into a relationship.

A Historical Perspective

Biologists and ecologists have long been aware of scale as a central issue, for example in data collection in ecology. Even so some thirty five years ago there was a ground swell of increased concern about problems of scale in ecology.[4] This scale orientation could be viewed as just the result of the meeting of field biologists, mathematicians, and computer technologists. Perhaps that made the time right for the first edition of this book. Even so there were theorists focusing on scale in biology even a century ago. That size matters has a long history in biology. Even as early as the end of the eighteenth century, Goethe identified problems of shape and relative size in biology. Works on homology in organisms of different size were well advanced before their evolutionary significance was realized. In fact these studies on comparative form were part of the intellectual climate that spawned evolutionary concepts. Theories of evolution were essentially attempts to provide a mechanism for the unity Goethe saw across organism of different size. Evolution came from a scaling issue.

By 1917 D'Arcy Thompson had written his treatise *On Growth and Form*. His appreciation for quantifiable aspects of scale is still very relevant today. In a foreword to a later edition of D'Arcy Thompson, Bonner (1961) comments, "The most conspicuous attitude in the book is the analysis of biological processes from their mathematical and physical aspects." Thompson's quantification of scale problems is all the more impressive in that it antedates both the revolution in applied statistics that began only in the 1920s and the advent in World War II of computers large enough and fast enough to make a critical difference to strategies of problem solving. We see expressed in Thompson the biologists' instinct that tells them that biological systems are complex because they involve the interaction of differently scaled processes. Biologists know that their material must be considered as importantly multilevel. What is astonishing is that Thompson saw this as a quantitative problem before the tools were at hand to carry the program of analysis through. If he could see the necessity of quantification at that time, it is little wonder that inquiry into problems of scale arises more generally now that the methods and machines for numerically probing such questions are generally accessible.

Computers and notions of hierarchy have a reciprocal relationship. Computing machines are at once child of and parent to the development of hierarchy theory. On the one hand evolution of computer technology has certainly incorporated hierarchical form into the construction of both hardware and software. On the other hand the development of a general theory about hierarchies has in turn been facilitated by computer-assisted summary and modeling. For example, fast machine logic has liberated for general use the powerful techniques of Fisher[5] and Hotelling[6] of the 1930s, and these methods of data reduction have allowed the display of hierarchical structure in ecological and other data. Using such multivariate methods, masses of fine-grain data may be summarized so as to reveal large-scale structure. Data transformation takes a matrix of datum values and treats them all consistently, often with a relativization. The binary transformation leaves all the zero entries alone, and converts all other entries to one. With judicious use of data transformation before analysis, we can change the level of organization that is apparent, achieving summary to greater or lesser degrees. Without the assistance of human value judgments, using only simple, objective, numerical criteria, the computer is able to display structures at various levels in the hierarchies associated with given data sets. Since the objective criteria can apply broadly and since a large number of data sets based on very different phenomena have yielded hierarchical structure, indications are that a hierarchical approach has general utility.

Since the first edition of this book, computational power has exploded. Around the time of the first edition the new science of complexity was emerging but was not sufficiently generally known for it to have received our full attention back then. Over the past forty years complexity and emergence have become hot topics. A general treatment of fractals (patterns that are robust across scales) had to wait until the computer power of the 1970s. Chaos theory, a related body of knowledge, emerged in ecology (May 1974) and elsewhere (Lorenz 1968) for those same reasons of sufficient numerical capacity. In the intervening decades, chaos theory became part of the mainstream culture, enough to turn up in the movie *Jurassic Park* (not very well explained by a know-it-all character therein), to the same extent as relativity theory is known widely although not properly understood by the lay public.

Hierarchies invoked as change in level of analysis are now everywhere. Even so we do not mean to imply that reality, independent of our cognizance, is in its nature hierarchical; in fact we are not sure what that could mean let alone what it does mean. What we are trying to say is that somewhere between the world behind our observations and human understanding, hierarchies enter into the scheme of things. The perception of hierarchy is something like a Kantian a priori or the human propensity for a dualistic thinking. Simon (1962) makes a point that echoes the issues of medium number systems:

> If there are important systems in the world that are complex without being hierarchic, they may to a considerable extent escape our observation and our understanding. Analysis of their behavior would involve such detailed knowledge and calculation of the interactions of their elementary parts that it would be beyond our capacities of memory or computation.

The medium number issue in that quotation is that "important systems" refer to important enough to demand attention to its complexity. "Complex" for Simon[7] means that the implied question cannot ignore the individuality of the parts. So many parts cannot be averaged and they then can be expected to overwhelm "our capacities of memory or computation." Hierarchical treatment, Simon's central contribution, gets us out of a medium number specification, although Weinberg did not raise the issue of medium number until later. Simon could be interpreted as suggesting that hierarchical structure is a consequence of human observations. Even if this is seen as limiting the significance of hierarchical conceptions (we take an opposite view), hierarchical approaches are of at least heuristic value.

The classic notion of hierarchy involves discrete levels. Ecology often focuses on the discrete, as when island biogeography works on discrete islands so as to allow invasion and extinction to be cast in self-evident universes. Although they may be both conceptually and pedagogically helpful, the implicit discontinuities between levels that appear singularly real and discrete are significantly arbitrary. The underlying problem in island biogeography is that on the mainland invasion and extinction are continuous situations, in contrast to islands.

Discrete levels need to be recognized as convenience, not truth. Even so, some arbitrary levels of organization are of more general application than others. For example, the level of organization that defines the whole human individual is a helpful level for many models throughout the social sciences. Humans less a bit or with some more to them are generally less useful. Philosophers are careful to identify whether an argument pertains to existence in terms of objective external reality—that is to say, whether an argument is an ontological discussion. The alternative argument is concerned with experience and is restricted to that which is knowable, an epistemological discourse. We use just the latter arguments. Throughout this book we do not much address questions of ontological reality for given levels but prefer to take an epistemological stance in a utilitarian philosophy. We ask, how do we know what we know, leaving metaphysics for the most part out of it. Hierarchies are rich enough a concern without efforts to link them to external reality. The muddle that is almost ubiquitous in ecology often arises from invoking an undefined reality in an arena where only dealing with what we know is hard enough.

Despite the utility of certain levels and the explanatory power that can accrue from an approach that models with discrete levels, sometimes it may be necessary to take account of hierarchical continuity. On these occasions it becomes necessary to invoke a continuously varying function that can describe the continuum of levels and their interactions. That function is called scale and is a central concern in this treatise. We hope to be of service to contemporary scientists who are beginning to come to terms with more than just the states, interrelationships, and interactions that have concerned scholars from classical times through the Enlightenment; we hope to encourage those who find themselves tempted to model dynamic multivariate structures; these structures amount to the interaction between many levels of organization.

CHAPTER TWO

The Janus-Faced Holon

The first extended treatment of the general notion of hierarchical structure was in Koestler's (1967) *The Ghost in the Machine*. Others before had considered hierarchical structure, but the unique contributions of that book were the development of a useful terminology and the presentation of a broad overview of hierarchical communication patterns. We will use Koestler's terminology since it is powerful. It did gain acceptance in ecology some forty years ago.[1] As a pioneer in the field, Koestler very appropriately conducted a detailed argument on the superiority of a hierarchical over mechanistic approaches and warned against the dangers of obligate reductionism. We will support his view with examples of our own. But at the outset we will rely at least initially upon his defense and will devote our efforts to some of the concrete consequences of the use of hierarchies. First let us outline Koestler's scheme.

Doors of Perception

Most of our high school Latin is gone now, but a few words remain. Along with *amo amas amat,* another word from Vocabulary I, *ianua* [a door], sticks in our memory. The janitor was so named when his primary responsibility was opening the door; the doorway to the year is January. In Roman mythology the god who was guardian of portals, and who was patron of beginnings and endings, was the god with two faces, Janus.

In biological taxonomic hierarchies a unit in the scheme is a taxon, such as species and genus, at various levels. For Koestler the generic term for entities in a hierarchy beyond taxonomies is the *holon*. The holon is a difficult concept that is at the heart of hierarchical thinking. Hierarchists carefully unpack its deeper meaning. The term and concept came from Koestler. He uses the image of a doorway between the parts of the structure and the rest of the universe. The holon is an entity that has a duality in that it looks inward at the parts and outward at an integration of its environment; it is at once a whole and a part. At every level in a hierarchy there are these entities, and they all have this dual structure. Koestler chooses to restrict the usage of holons to biological contexts and their social extensions. We, however, prefer to generalize the meaning of holon to include entities of any type.

> Every holon has dual tendency to preserve and assert its individuality as a quasi-autonomous whole; and to function as an integrated part of (an existing or evolving) larger whole. This polarity between the self-assertive and integrative tendencies is inherent in the concept of hierarchic order; and a universal characteristic of life. The self-assertive tendencies are the dynamic expression of holon wholeness, the integrative tendencies of its partness. (Koestler 1967, p. 343)

Koestler is so taken with the self-assertiveness of biological holons that he certainly implies that the holon is not arbitrary. While he may not argue that there must be particular levels at which collections of holons exist ontologically, he does imply that the word *holon*, at least epistemologically, is only applicable at discrete points in the hierarchy. In his explication he has particular holons in mind. In biology they are the classic levels in the biological hierarchy: organelles, cells, tissues, organs, individuals, demes, and species. In other fields he also identifies particular holons: "The ethologist's 'fixed action-patterns' and subroutines of acquired skills are behavioral holons; phonemes, morphemes, words, phrases are linguistic holons; individuals, families, tribes, nations are social holons." Richard Dawkins in his *The Selfish Gene* coined the term *meme* for ideas that pass from human to human. These could be music, fashions, or technological ideas like the arch. Memes are social holons.

Koestler arrives at the notion of holon from a concern for the nature of particular holons that he and everyone else can readily see as "things." Generally his holons are tangible. He is interested in working out a better model for the familiar. If, however, one comes to hierarchies from a series of scale problems, as we did, then there is more generality to the doorway analogy than is given in Koestler's account. The doorway may be stretched or moved. Perhaps the analogy of a lens that is continuously adjustable from strongly fishbowl through planar to telefocal is an image that encompasses the continuum of doorways that we have in mind.

The holon is an integration of its parts; but the definition says nothing about how many parts there should be or indeed what criteria should be applied in order to determine what is and is not a part. This means that while the holon model may relate to something ontogenetically real, nevertheless the holon in its composition and its boundary is essentially arbitrary in our treatment in this book. In a similar vein at the beginning of his book, Margalef[2] is at pains to emphasize that "everywhere in nature we can draw arbitrary surfaces and arbitrarily declare them boundaries separating two subsystems." The holon is an integration of a set, in set theory terms, and sets are a matter of ad hoc definition. As we said in the first chapter, some holons may be more useful as model devices than others, but that does not deny the possibility of definition of an infinite number of similar but very slightly larger or smaller holons. Not all holons are differentiated on size; we use size here as a pedagogical device because the link between different sizes is readily understood. A human being is no more ontologically real than a human being with its skin excluded or its immediate air jacket included; it is just that the human-no-more-no-less holon is more generally useful in model building. The exact human has more general relevance. In this epistemological discussion we wish to deepen the definition of holon so as to emphasize that it is not the parts themselves that become blended, but it is rather information from the parts that is integrated. If a holon is defined only as an integration of information from the parts, then clearly what is included and excluded may be freely chosen by the observer. We suspect that, in part, it is Koestler's failure to exclude explicitly ontological reality from his definitions of holons that has slowed the acceptance of his hierarchical approach. We feel he is mistaken to allow ontological inferences to arise.

As our arguments unfold it will become apparent that our views are in sympathy with dialectical materialism. The parallel begins to emerge at this juncture, and we can quote Levins and Lewontin (1980) in their writing proposing dialectical materialism as a philosophy for biology. In a negation of alternative views they say

> Idealism and reductionism in ecology share a common fault: they see "true causes" as arising at one level only with other levels having only epistemological but not ontological validity.

We share their view that reification of just one level alone is misguided. In fact we generally mistrust reification altogether. A difference between their approach and ours is that they seem prepared to give ontological validity to multiple levels while we avoid this issue. If we were to abandon our resolute epistemological stance then, as we argue now, we think our arguments would lose power.

Peter Checkland's discussion of holons was very kind to our treatment of holons in our first edition. He said, "Allen and Starr are on the side of the angels." The reason he gave was our insistence that what is in a holon is purely the choice of the observer. Holons are concepts, not material entities. Suggesting otherwise is to commit the error of mistaking the map for the area itself. While that idea was first made famous by Alfred Korzybski early in the twentieth century, it was more recently popularized by Gregory Bateson in *Mind and Nature* in a section called, "The map is not the territory, and the name is not the thing named."[3]

There are consequences if we allow holons only in certain positions in the hierarchy while dismissing others as invalid. Problems of defending that particular choice immediately arise. Bastin (1968) asserts

> that two levels are specifiable if and only if the ratio of time constants characterizing the two levels is large (numerically). The ratio rather than either of the separate time constants is the important quantity here, for it is clearly only the ratio that has any experimental meaning in the absence of an independent time reference.

In biology we do not have an independent time reference, for time can be properly measured in units from microseconds to circadian rhythms to generation times, depending on the question at hand. From this Bastin speculated that

> one could never with complete certainty settle the allocation of hierarchy units into levels because complete certainty could only be obtained at the expense of an infinite ratio of time constants of the levels. This reasoning was based upon a sort of engineering guess work.

The argument indicates that assignment of holons to particular ontogenetic levels leaves their relationship to holons of other levels undefined in an epistemological account because of the intrusion of infinity to which Bastin refers. We suggest that it is more profitable to view the discreteness of levels as a product of human perception and conception. If the discontinuity has its source in the human perceptual apparatus, then it is hardly surprising that the levels appear very concrete.

As more is included in the arbitrary boundaries of a system, the cycle time for unperturbed behavior increases (i.e., the reciprocal of the endogenous or natural frequency of the holon; see Figure 2.1). Patterns take longer to repeat the larger is the holon. Therefore from top to bottom of a hierarchy there is a continuum of natural frequencies. English as a language is particularly assertive about being and discreteness. "Is" invites those assertions, President Clinton's statement notwithstanding: "It depends on what the meaning of the word *is* is."[4] Unfortunately there is no vernacular language for describing such a condition of generalized continuity, and so we must, for the moment, retreat to words in the vocabu-

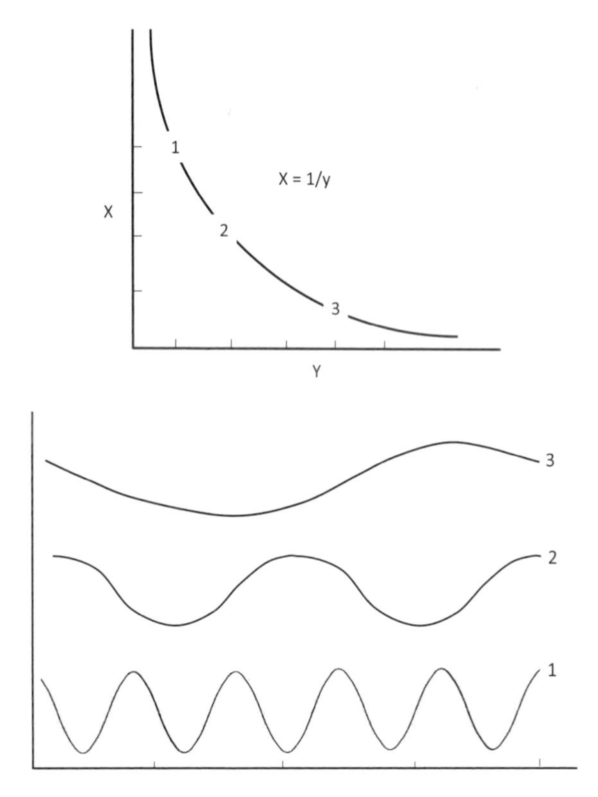

FIGURE 2.1. a) This graphical display shows the inverse relationship between cycle time for characteristic behavior and natural frequency. b) The significant point is that there is oscillation, repeated behavior at different natural frequencies.

lary of the discrete case. We therefore use terms like "net" or "tree" that imply discrete branch points, but only intend these words as expository tools. We hold to our conviction that hierarchies are best conceived for most purposes as being vertically continuous as in bottom to top.

Korzybski, in his discourse of "The map is not the territory," developed a philosophy of General Semantics. This is not to be confused with more usual issues of semantics. His philosophy addressed exactly the issue of "is." He identified different "orders of abstraction." In a "consciousness of abstracting," he notes that commentary on a person is a higher level of abstraction than the person in person. Rather than saying "The painting is good," he would prefer to say, "I liked the painting." This book is in a sense an unpacking of the consequences of Korzybski's philosophy, although we do not avoid the verb "to be" as much as he did. Instead we draw attention to the pitfalls, and then go on to use common parlance.

Codes, Memory, and Holons

Since the first edition there have been advances in complexity science. The notion of emergence has matured. What is new here is an appreciation of emergence held in the context of coded constraints. Emergence is not mystical at all, it simply forces a higher level of analysis, which by definition does not have direct connection to what applied before. Emergence is not simply an explosion driven by a steep gradient that drives positive feedbacks. It arises when those positive feedbacks encounter some limit, perhaps in the form recognizable as a negative feedback. Emergence comes from gradients applied to material such that positive feedbacks generate structure without any plan. A whirlpool would be a case in point. The gradient that drives the positive feedbacks is gravity pulling water down a hole so as to increase the speed of the flow. Positive feedbacks spin the water until there is a vertical water/air interface. Such an arrangement of liquid/gas interfaces does not generally appear in nature. The limit imposed on spinning makes the whirlpool stable. It invites being called a holon. Its stability depends on it being an open system fed by water from above. If you had never seen one, you would not invent it, which makes emergence hard to predict.

In physical systems constraints that make stable structures apply in some region of time and space where there is no other contravening constraint. In physics, structures emerge at such different scales that the one structure can nest unnoticed in another; subatomic particles in themselves do not feel the processes that make an atom, they simply live nested within. The galaxy rotates, but not like you would notice. If the universe of discourse is so small as to span only the diameter of a proton, chemistry going on in the neighborhood cannot be experienced. A wider universe of discourse would allow chemistry to emerge. Physical hierarchies often describe quarks, subatomic particles, and atoms. Such hierarchies are very deep. That is, levels in physics have massive scale differences between them. The constraints at one level see constraints of adjacent levels as simple constants. Constraints that are proximately scaled will see the behavior of each other. Those behaviors will come into conflict, making the hierarchy unstable, inviting a new set of holons. In essence a medium number specification invokes levels that can interfere with each other. Competing constraints either will have a winner or both will disappear. That explains why the gyres in a turbulent fluid appear to alternate in direction of rotation relative to their neighboring gyres. Turbulence is caused by sheering stress between fast and slow flux that is adjacent. A flow must go around an obstruction and so must go farther and faster than fluid that gets a straight passage. Gyres in emerging turbulence that rotate the same way will rub against each other and usually both will be destroyed as their sheering stress is dissipated. That is why we only see adjacent gyres in turbulence that counter-rotate. Such gyres only kiss, they do not rub.

In biology things are different from physics. In biology the patterns are coded by natural selection and other processes that create plans. If there is an emergent in biology, the coding keeps the patterns stable even if there already exist in the vicinity patterns of similar order that will interfere. For instance, individual selection works at one scale, while group selection works at a larger scale. But, close as they are, the two can coexist. For instance sex is a group-selected phenomenon that is at odds with individual selection. It is much better for the individual to avoid sharing its offspring with a sexual partner, and simply reproduce asexually, so as to keep the whole genome to itself. And yet sex is often involved in individual selection with regard to other traits. Coded stability is why there are so many patterns of similar scale in biology. Biological hierarchies have many closely packed holons. The hierarchy of physical systems is deep, with levels well separated. Coding and symbols in biology make for dense flat hierarchies, with levels closely pressed together. Physical hierarchies invite fewer holons.

Physical happenings are rate-dependent—the effect of gravity on falling objects, for instance. Codes, symbols, and meanings are rate-independent. The process of creating a plan, such as natural selection, need not be directed. As useful emergent structures appear in biology and in social systems, codes are selected to stabilize them. Codes give limits not rates. The rate-independence of biological constraints blunts the effects of competing processes. Codes also arise in social systems in a somewhat different manner, but the rate-independence of coding remains critical. There is nothing magic about the coded information, but it is distinct from thermodynamic processes.

Unlike dynamical processes, codes exist independent of scale. Change the size of a thermodynamic process and it works out differently. That is what we mean by scaling effects. Big animals all have to be stocky, because of scaling effects. But codes and symbols are independent of scale. It therefore becomes possible for coded patterns to cross scales with unchanged effects. The result of past selection in real time can be captured in the codes of populations or species. Because codes are scale-independent, coded information of whole species can then be applied to lower level individuals who have not had that experience. This allows biological entities such as organisms to anticipate without it being unnatural or magical. A foal appears to know what to do with its mother even though it has never seen a horse before, let alone its mother. But there is nothing strange in this; the patterns that anticipate have been learned by the upper level of aggregation that makes the species from its past experience—no big deal there. In coded form that experience is handed down. Hierarchy theory deals with scaled processes but is happy to recognize qualitative differences that can appear under scale change. Hierarchy theory unites scaling with structure and meaning, and so is adept at dealing with what appears otherwise mysterious in life and society.

Holons and Causality

One could argue for holons in the physical sciences, but they would, in our opinion, be much less interesting than biological and social holons. The interesting issue is coded information in holons. In physical systems one does still have the problem of observation, which does bring in the purposeful, biological human observer. In physics the systems are sufficiently well behaved for physicists to drive right to the heart of observation. There is no such thing as an observer-free observation. Pressed to the edge, the observation problem manifests itself in quantum mechanics, where observation leads to a change. In the dual slit experiment of Thomas Young in 1803, photons were fired through a slit that appeared to allow the particles through with a central tendency. Photons are detected on a screen behind as discrete particles. More photons appeared on the detector in a direct shot across through the slit. Less and less were detected at wider and wider angles from a straight shot through the slit. Close the first slot and open a second slot, and as expected the distribution is the same except moved over by the new position of the new slot. The puzzle comes when you open both slots so electrons can pass through either. One would expect photons to pass through the two slots in a distribution that was the sum of the previous two experiments: a double peaked distribution that accounted for the two slots. But that is not what happens.

When the second slit is opened the pattern on the detector is a wave interference pattern, with its largest peak situated between the two slots. Once again it is a bell shaped distribution, but this time seen as a series of peaks and valleys (Figure 2.2). The valleys will be places where the waves cancel out each other. Ah, so the electrons are going through not as particles but as waves appearing to interfere with each other. Yes, but the problem is that the photons appear on the detector as individual hits, so they must be particles. But how can individual particles be influenced by the slot through which they do not go? One might even entertain that the particle splits and goes through both slits; somehow that does not seem right.

All right, let us put a device that tells us through which slot the photon did pass. The trouble is then that the interference pattern disappears. It looks as if nature appears to know whether we are looking, or at least that the act of observing changes the situation in a fundamental way. So even at the most basic level, we cannot see what is going on in nature without the situation being changed. In physics the scientist can press the issue so hard that the dilemma cannot be avoided. Pattee (1979) argues that Niels Bohr really did mean that these critical dualities occur in biology at the macrolevel, well above the level of quantum mechanics. In macroexperiments, some of the influence of the observer can be taken

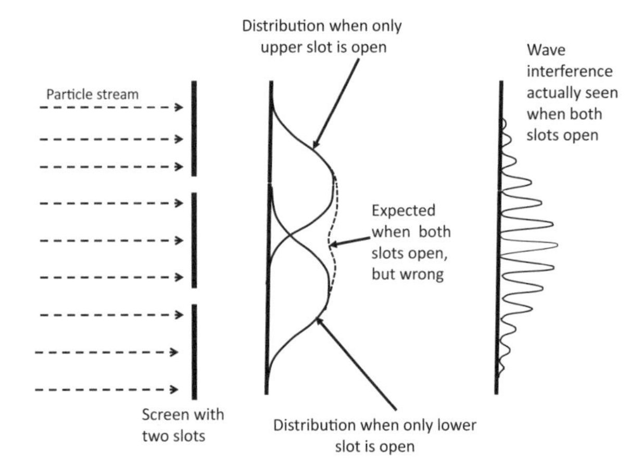

FIGURE 2.2. Wave interference with the double slit experiment. The signal is delivered as individual particles but appears in aggregate as wave interference. The quandary is how one particle contributing to the wave can be influenced by the slit through which it did not go.

into account, and its effect can be corrected. Yes, scientists are good at that, but that is not the intrusion of which we speak. For instance, we can measure photosynthesis by putting a leaf in a bag and measuring gases that come off the leaf. But the bag will of course change things by altering gaseous boundary effects. All right, so let us calculate those effects and subtract them from the measurement so we can see what is really going on. We can do that, so what is the problem? The intrusion is not so much the bag, as it is the concept of photosynthesis in the first place. You cannot measure photosynthesis without first having a model for it, and that is what causes the intrusion for which you cannot correct without quitting the discussion.

So important is the measurement problem that Rosen (2000) uses models to get a something of a free shot. He says that the power of a model is that you get insights into a system without having to look for them. At the San Francisco meeting of the Society for General System Research in 1980, we were told by a speaker from the American Association for the Advancement of Science that these systems thinkers need to understand that science does not begin until you make an observation. It was a rebuke coming from the speaker's perception that systems scientists come to see their models as reality. Rosen told us in our own private meeting that the speaker was wrong. "No," he said, "science does not begin until you make two observations, because then you can get the differential." The great thing about a differential is you do not have to look to get it. You already did your looking to get the two measurements before the subtraction. You do not actually look when you do the subtraction. Differentials are what science is about.

So all this applies to physical holons, but in biology there is a further twist that makes biosocial holons much more interesting. In social science and biology the holon often makes its own observations. And it does so according to its own models set in its own narrative, told in its own semiotic scheme. The implications are that coded memory is different from inertial memory. Inertial memory keeps a whirlpool turning in the same direction. DNA, hormones, legislation, and mating dances all invoke memory and carry meaning, but they do so in a rate-independent way. Meaning has no rate, it just is. Meanwhile there is energy involved in life and society that creates gradients that drive rate-dependent thermodynamic processes. Thus biosocial holons are entities wherein both dynamical and coded meaning need to be represented.

Pattee (1978) makes much of the distinction between rate-dependence and rate-independence. Rosen (2000) heightens the distinction in his piece on mass. Einstein viewed it as most remarkable that mass in gravity is the same as mass in acceleration. The one is an expression of rate, and so is rate-dependent. The other is static, and so is rate-independent. Thermodynamics is rate-dependent. Efficiency is a matter of some preferred state. Preferences and states are rate-independent. Physical systems work without preferences but do have rate-dependent explanations as in the memory of the whirlpool with regard to direction. But there are no codes or symbolism. It is like the distinction between signs and symbols. A footprint is only a sign someone was there, while a symbol carries with it codes and symbolism. Juarrero in her *Dynamics in Action* starts with the insightful question, "What is the difference between a wink and a blink?" Biological systems do have codes that are rate-independent, which makes them particularly interesting in ways not pertinent to physical systems. In biology we are concerned with the way that constraints limit flux. Organization is a negative thing in that it is what the constraints deny with regard to flux. So we see two sides to holons, one is the rate-dependence of their thermodynamics, while the other is the rate-independence of codes and plans.

Figure 2.3 shows the holon as containing two parts. One part is thermodynamic, while the other generates codes. The thermodynamics are driven by an external gradient that might be food, sunlight, money, or imports. The thermodynamics of living and social systems involve consumption of some resource that exists in some concentrated form. The concentration exists at the top of a gradient in the environment. If it is food, then left to itself, food moves down the gradient as it slowly decomposes. But food eaten is quickly degraded to waste, which is then extruded. The degradation sets up gradients inside the holon. Those gradients are what drive system functioning. Petroleum left in its gas can degrades, and becomes "bad gas." But put in a car it is much more quickly degraded. That degradation drives the parts of the drive system to produce work, the movement of the vehicle. The car can be seen as a mere physical holon, but it becomes much more interesting when it includes the transportation of the species that devised it. In that sense an automobile is a living holon.

The other part inside the biosocial holon is the coded and coding elements. True, it takes thermodynamics to make and read DNA, but the coding itself is not thermodynamic. Codes are rate-independent. So the coding is inside the organism or whatever is the holon, but the coding is separate from the thermodynamics. Codes invoke plans. Plans constrain the thermodynamics by imposing constraints. Plans change as the holon becomes something else, such as an adult derived from a child. The plans are executed at a rate, perhaps the rate of protein synthesis. Call that rate d/t. But notice that the rate of becoming an adult is slower, say at rate d/ζ.

But as the planned structure behaves, it does things that were not part of the plan. There are unplanned processes of emergence. All this is seen by other bodies in the environment, such as the system's prey, predator, mate, or scientific observer, all of which have their own models and codes. The information that there are new plans along with unplanned emergence comes out as a narrative. A narrative is a representation of a compression down to only what matters. Notice that the narrative told to the world is told at a rate different from both the rate of development and the rate at which the system becomes something else. The narrative unfolds, perhaps at rate d/θ. Being the storyteller does not mean that the holon fails to hear the story. It does hear, and the implications are, in the light of the tale just told, the holon needs a new plan to deal with the emergent properties that arose last time round the cycle. The plan in the coded half of the holon must be updated. And in this way the holon plays out its role in a process of becoming. Becoming is driven by the story told by the holon. Note that a role comes from the outside. The US president has a role to play, imposed by the US presidency.

If the holon is political and social, the US presidency has a history, which is updated by public reaction to the action of the most recent incumbent. So the dual nature of the holon is at the core of its functioning. It communicates to its world. That changes the holon's context, which then steers the way that the role of the holon is played out.

Rosen argues for holons as realizations of essences.[5] His position is not idealist, in that the essence is not pertinent until there is a realization that pertains. Essences are not out there as ideals; instead they exist within the epistemological not the ontological realm. A species might be an essence that then realizes individuals of that species. The bizarre way that life appears to anticipate comes from the passage of information from the larger essence. The species horse has seen what happens to mares and foals. While changes of scale change the effect of thermodynamic processes (with all else equal, bigger bridges fall down), information can travel freely across scale change unaltered. Changes in the scale of thermodynamics force changes in scaling effects because the larger spaces influence flux. While information can change with the application of a filter, change in information is not forced by shifts in scale as it is in thermodynamics. The reason a foal knows what to do with its mother, when it has never even seen any horse before, is that its essence has seen it before and tells the foal what to do. The functioning foal holon may have received the information from the essence, but the information is now embodied within the foal holon. The foal knows even though it has no experience. The holon reacts to two separate streams of information. The environment of the holon offers one stream of information, and the essence has provided another stream that tells what to do about the environmental information. The holon updates its plan from how the world reacts to its story. An organism does not know it is an item of prey until the world treats it that way. Evolution uses the responses of the realized holons to update the essence. That is how natural selection works, without having to invoke purpose or direction.

If the essence knows something, some might argue that we are saying the essence exists in an idealist way. It is best to think of the essence as a device to know and understand about the classes we erect. The essence is well treated as a dummy variable, as is used in calculus, something we would not assert as real and external. We do not mean for essence to be a particular thing, but rather as an intellectual device for capturing what evolution does. Significance, perhaps, in evolution cannot exist independent of a decision-making observer. If we deny that holons are material things, we can also deny that evolution is something material. No: evolution is an intellectual device to understand what we observe in realized entities. We suspect it is the reification of evolution that leads to the strident pronouncements of a Darwin Police, as it asserts its particular correctness. After Noble's recent work[6] the central dogma and neo-Darwinism are now in tatters, but as a paradigm change the Darwin Police will not notice the problem. The old guard in a paradigm change will only complain that the old important questions are not being asked. The old important questions do not matter anymore.

We have shown the two parts of the holon, the thermodynamics and the coding elements. These parts pertain to several different aspects of the working of the holon. So the holon has several aspects that are separated by different reaction rates. While information is rate-independent, there is rate to the change of codes. We can dissect the holon to see wherein are the four Aristotelian causes. Ulanowicz[7] breaks out the four causes by considering a house. The material cause is the stuff of the house. The house is material because its bricks are in this and that relationship to each other in space. The material cause invokes thermodynamics. The material cause is the left hand circle of Figure 2.3 at level N.

The final cause for the house is found in the role the house plays, captured in "We need housing." Role invokes meaning and is rate-independent. The information for the role is found in the response of the world to the holon. The final cause is thus found at level N + 1 and it is linguistic and meaningful. There are values at play, and values are rate-independent. Thermodynamics has no reference to values, it just happens. Note that efficiency of a motor car is not thermodynamic, because it invokes preferred states, values which have no intrinsic materiality.

The efficient cause of the house is that some external force came to bear. The holon is set up a certain way. In the case of the house it is the force brought to bear by the builders, who put the bricks in place. In the diagram of the holon it is the gradient in the environment. Thus the efficient cause exists at level N + 1 and is thermodynamic.

The last cause for the house is that the plans said to build it that way. The plans are a set of constraints put on the parts of the house as they are moved into place. In that they work at the level of the parts, the plans work at level N − 1. They are clearly linguistic in that they amount to a

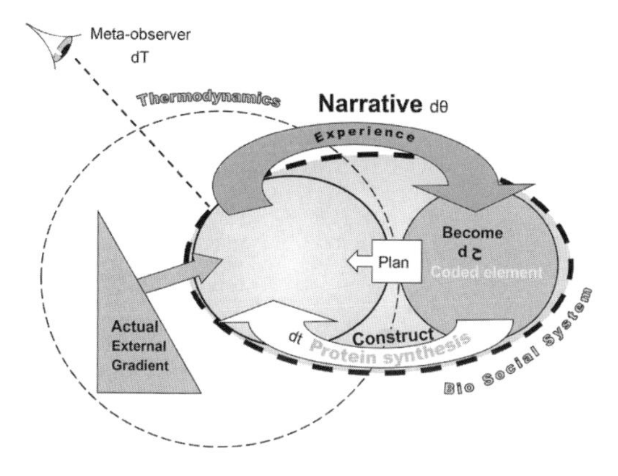

FIGURE 2.3. The holon seen with its various types of part and context. The holon is bounded by the strong dotted line. Inside the holon are two separate considerations: thermodynamics and information. The thermodynamic substance of the holon is the inner oval on the left. The thermodynamics of biosocial systems include the gradient of food, or money, or whatever drives the metabolic system (see the fine dotted line as containing the strictly thermodynamic parts of the functioning of the holon). The internal dynamics are integrated in the whole holon. The oval on the right is inside the holon, but separate from its thermodynamics. This is the coded information (perhaps its DNA or mating dances). The whole holon is the integration of the two ovals. Note the external gradient, which might be food or trade. There is change everywhere but it happens at different rates. The information sector is responsible for the way the system becomes something else. Becoming occurs at the rate d/ƨ that is slower than creating the whole at a rate of d/t. The narrative told as output from the whole moves at a different rate, d/θ. The meta-observer rules and decides all at an evolutionary rate of a meta-narrative d/T. When that story ends the project stops (the meta-observer or holon dies or loses interest). The plan is executed and makes the stuff of the holon. In the thermodynamic real time change of the stuff there are unplanned patterns of emergence. These are communicated to the external world as a narrative to mates, predators, conspecifics, competitors, and even the meta-narrator. The changes require an updated plan that drives the becoming.

series of instructions. The plans are labeled on Figure 2.3 as relating the planning element to the thermodynamics. The Figure 2.4 of the holon is updated now to add the position of the four Aristotelian causes. Stacey, Griffin, and Shaw write in *Complexity and Management* of using various teleologies as causes. Teleology is frowned upon in biological science because, in unskilled hands, it can lead to crude life forces and reification of imagined anthropomorphic intentions, like "The plant wants to . . . ." However, even when one cannot clearly identify a reason for action in biology, it is not possible to understand living systems except as they play out roles and fulfill goals. One cannot discuss disease without reference

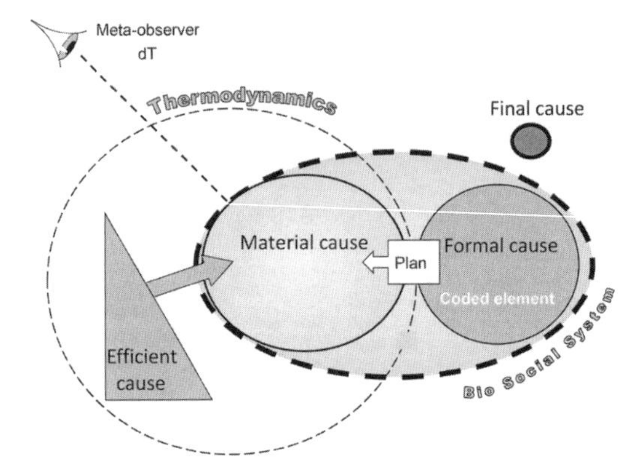

FIGURE 2.4. The holon in figure 2.3 with Aristotle's causes superimposed: material cause, formal cause, efficient or immediate cause, final cause.

to some preferred state of wellness. Rosen closes his *Anticipatory Systems* (1985) with this:

> The study of anticipatory systems thus involves in an essential way the subjective notions of good and ill, as they manifest themselves in the models which shape our behavior. For in a profound sense, the study of models is the study of man; and if we can agree about our models, we can agree about everything else.[8]

Aristotle's four causes do need to be given account.

There are two distinctions to be made in all these discussions: Thermodynamics versus linguistic meaningful aspects of the system; and level relative to the holon itself. It makes sense if we start with the final cause, in a sense the reason for the thing in the first place. Why would such a thing even arise? That relates to an upper level N + 1, that defines the role and meaning of the holon in its context. That is on the linguistic side of Figures 2.5 and 2.6. Two levels down at level N − 1 we find the formal cause. Organization is achieved by putting constraints on the parts of the holon. Planning takes us down to level N − 1, where lies the formal cause. The formal cause exerts control on the lower levels of the thermodynamics. Controlled parts lead to mechanisms, restricted pathways of effect. Mechanisms cannot do anything by themselves, so we need the upper level N + 1 to provide the drive. That is done by the efficient cause, which applies force. When mechanism meets a force to drive it, structure is produced, the material stuff of the holon configured. That is the material cause, which of course exists on the thermodynamic side. The holon tells narratives to its context. That moves us through meaning diagonally back up to formal cause again. Figure 2.6 lays out that pathway.

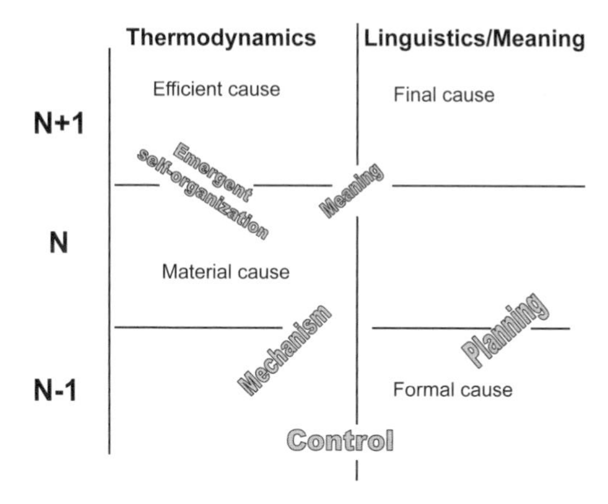

FIGURE 2.5. The Aristotelian explanations linked. They are presented at three levels of discourse expressed in terms of thermodynamic process and linguistic meaningful structure. Linking events and processes connects the parts of the holon.

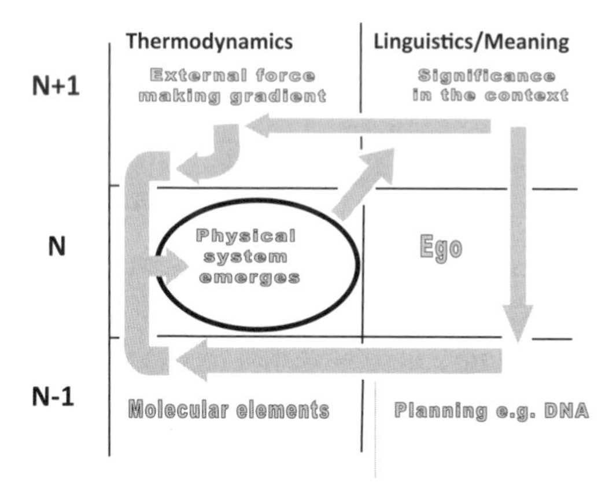

FIGURE 2.6. Showing the path whereby one causality is linked to the next. Like Figure 2.5, the three levels of discourse are expressed in terms of thermodynamic process and linguistic meaningful structure at different levels. The figure shows dynamic connections between Aristotle's causes: 1) final to immediate or efficient; 2) final to formal; 3) formal joins immediate to give material; and 4) material to final cause. The means of the linkages correspond to their arrangement on Figure 2.5.

Constraint and Order

So impressive is organization in biological systems that conventional biological wisdom always views organization in positive terms. It is sometimes advantageous, however, to view organization not positively as a series of connections, but rather negatively as a series of constraints and boundaries. In fact as a rule of thumb, anytime one has a deep insight, it is always a good idea to turn it on its head. For instance, bilateral symmetry in animals can be well explained by simply recognizing it as a bipolar departure from radial symmetry. In that spirit we note that ordered systems are so, not because of what the components do, but rather because of what they are not allowed to do. The emergent properties of nerves are so full of positive achievement that it is hard to remember that they work only because of restrictions placed on the position and movement of sodium and potassium ions. It is what sodium and potassium in the nerve cannot do that supports the emergent property of a nerve firing. A nerve fires through disorder not ordering. The constraints on the potassium and sodium are relaxed as membranes become permeable, and the ions simply move to a more likely position under the probabilities of the second law of thermodynamics.

Hierarchies can be profitably viewed as systems of constraint. Any holon higher in the hierarchy exerts some constraint on all lower holons with which it communicates. The proviso of communication here is important, for there are two ways in which constraint on a lower holon is not imposed by a higher holon in the same hierarchy. Both failures to constrain arise from lack of communication. One failure comes from the higher holon being in a different stem of hierarchy; an example would be a supervisor but one in a different department. There the parts of one holon are not influenced much by the parts of another; things are separate. The other failure comes from too much vertical distance. Here the control exerted by the higher holon is so general that the lower holon lives out its entire existence without encountering any direct control from far above. The lower holon does not exist long enough, have memory long enough, or exert an influence wide enough to have behaviorally significant communication with the higher holon.

An example here is the almost complete irrelevance of any behavior of our galaxy to the processes of life on this planet. The Milky Way does provide an inertial frame that is a constraint in a certain general sense, but there are few galactic constraints that have relevance for the existence of life. Any or even no galaxy would for the most part do just as well, and so there is no functional galactic constraint of behavioral import to living systems. Occasionally the rotation of the Milky Way has an effect in that meteoric activity appears to have a sixty million year periodicity, but life would go on with or without that effect. The extinction of the dinosaurs and the attendant rise of the mammals appears related to the impact of a large meteor some sixty-five million years ago. As mammals ourselves we may care, but life as a whole simply goes on independent of the Milky Way. A different galaxy, or even no galaxy, might serve as a context of life just as well.

The larger stems of a hierarchical tree could be said to constrain the finer stems below. A more robust model, however, is one of an n-dimensional hierarchical reticulum in which several superior holons will simultaneously and to different degrees exert constraint over a lower holon. There is, then, the potential for a lateral continuum in our conception of hierarchy as well as the vertical continuum that we emphasized earlier. A hierarchy can be considered as a suite of complex fuzzy sets.[9] According to Gold (1977), "The concept of *fuzzy* sets recognizes degrees of membership, ranging from 0 (not a member) up to 1 (is a member). The set 'tall people' might be an example" (his italics).

The higher holons provide, to various extents, the environment in which the lower holons operate. This may amount to a nesting of parts inside wholes, but sometimes the upper level being contextual to the lower level is sufficient. A holon's environment can be defined as being made up of all things with slower behavior with which it interacts. The holon itself can be seen as part of the environment of all faster behaving holons with which it interacts. We define sister holons as having no asymmetric effect on each other. Holons may be sisters in one discourse while failing to meet the criteria for sorority in another discourse. Therefore constraint is seen as relative to defined information exchanges between holons. There is no direct constraint between sister holons at the same level in the hierarchy because there is no asymmetry in the exchange. Meanwhile at the same time there is no functional constraint of a very small holon (low) by a very large holon (high). A meaningful measure of constraint must be zero between the reader and another human of equal standing—that is, sister holons or holons at the same level. Con-

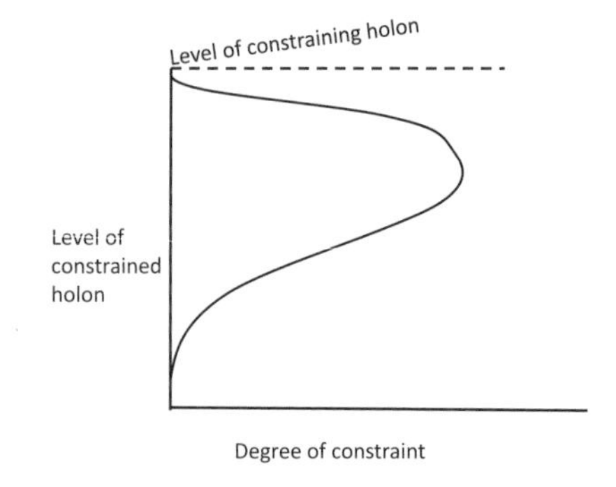

FIGURE 2.7. Showing the degree to which a constraining holon constrains a constrained holon, relative to their respective levels.

straint will be strong between local geography and the reader (one enters a room through the door, not the wall); yet it must again be low between our readers and the galaxy in which they happen to sit (see Figure 2.7).

Our model for constraint has two parts. One is the amplitude coming from physical happenings. Middling amplitudes will usually have more significance as information is exchanged. Here information and action are used interchangeably (the equations for information in information theory and thermodynamics are interchangeable). Essentially amplitude is the quantity of signal received by the lower holon. The other part of constraint is the difference between reciprocal signal exchange—that is, the asymmetry of information exchange between lower to higher holon as opposed to higher to lower holon. The asymmetry results from differences in the cycle times for the endogenous behavior of the two holons. In vernacular terms, how easily can the higher holon play a waiting game in any challenge to the conditions it imposes on lower holons? A measure of constraint can be viewed as some sort of product of its two parts: the amplitude of signal on one hand, and the asymmetry of signal exchange on the other. The asymmetry of constraining versus constrained causes hierarchies to be, in technical set theoretic terms, partially ordered sets.

If two holons are at the same level in a hierarchy, then there is no asymmetry in information exchange, for neither holon can outwait the other. No matter how forceful the information exchange, that large quantity is multiplied by zero asymmetry, and therefore constraint is zero. Newton would say that two identical billiard balls can still influence movement, but that is not what we mean by constraint.

Colleagues can have a vociferous disagreement and both can walk away leaving the other unconvinced. The situation can remain the same as it was before the argument, so long as one of the colleagues is not the department chairman. On the other hand there can be enormous asymmetry between holons as in the Milky Way and life comparison. Here, however, the significant information exchange is close to zero, for life evolved with the Milky Way only as a simple backdrop, so in its functioning, life finds the Milky Way fairly much untextured and ignores it. The Milky Way behaves so slowly and over such a large space that the life we know will be gone before the galaxy can offer much significant for it. The enormous galactobiotic asymmetry is multiplied by zero significance to the exchange, so again giving zero constraint.

In between these cases are many others where quite intransigent slow holons offer very significant signals. Large asymmetry multiplied by great significance produces high degrees of constraint. The walls around the reader constrain movement between rooms to doors. The only exception could be if the reader were to make the wall behave faster than the reader by taking a sledgehammer to it.

The positive aspects of organization emanate from the freedom that comes with constraint. The constraint gives freedom from an infinite and unmanageable set of choices; regulation gives freedom within the law. Since constraint comes from environmental inertia and intransigence, the fast reacting constrained holon is free to do its will within the constrained region. In the powerful constraints in the rules of the fugue form, Bach became free to write any fugue of his choosing.

The power to constrain gives the burden of responsibility, whereas being constrained gives freedom from those pressures. The self-assertiveness of the constraining environmental holon in its struggles with its own constraining super-environment essentially protects the subholons that the constrained environment itself controls. Cold-blooded animals may lie on rocks when too cold or put their tails in water when too hot. This keeps the organs inside, the lower level subholons, at a working temperature. Warm-blooded animals go one better in that they are even more organized. They consume so much food that they can maintain body temperature in cold by increasing their metabolism. Humans generate so much heat that even at 120°F in Arizona we still on balance lose heat. Cold-blooded animals with their heat exchanges significantly going both ways have a nonlinear temperature problem and can be whipsawed. The warm-blooded linear temperature regulation of mammals allows greater control of body temperature so that the enzymes in the body are evolved to be maximally efficient at normal human body temperatures. By protecting their enzyme systems, warm-blooded animals are better served by their metabolism.

In summary, entities (holons) in a hierarchy may be viewed as the interface between the parts and the rest of the universe. On its journey to the outside, signal from the parts is integrated through the whole (holon). The same applies to signals reaching the parts from the rest of the universe. The holon itself is the surface between the parts and the whole. Stan Salthe refers to the triadic aspects of holons: the level of the holon itself; the level of the parts; and the level of the holon seen from the environment. Giampietro[10] goes further in identifying a level above the environment that keeps it stable. He also addresses the level below the parts that needs to be a constant presence that supports the stability of the parts. Giampietro addresses five levels.

The holon itself is simply the surface around its parts. Not all surfaces are tangible; consider the boundary of an ecosystem. Surfaces separate by filtering information in both directions. Connections occur mostly if the connected entities work at roughly the same rate. A cell membrane separates the inside from the environment. However, in the evolution of life, membranes only appeared after it was all over but for the shouting. The separation before then was the different reaction rates of organic chemistry as opposed to biochemistry. Biochemical cycles could well have existed in a diffuse state before cells. Notice that inside cells are diffuse structures like the Krebs Cycle[11]—strong connections within and weak reactions between. The Krebs Cycle has a surface, but one that separates in time not space.

All holons can be considered the skin, either in spatial or temporal terms, through which inward and outward integration occurs. Allen once attended in 1982 a Master of Fine Arts thesis show at the University of Wisconsin. The artist, Pier Gustafson,[12] had created holons, and in doing so he makes our point. He sculpted objects small and large out of paper with ink drawing on it. He had made a paper clarinet. Larger objects like drill presses were also created out of paper. The pièce de résistance was a life size paper household basement. It was so convincing that before Allen went into it, someone had tried to climb basement stairs and had crushed the bottom step. The integration of the drill press was displayed to Allen as holons in his environment, with him outside the press. By contrast, when Allen was in the basement he was a part inside the holon. He could hear muffled sounds of other attendees, as the basement filtered the noise from the ambient gallery, which is what holons do for their parts. The clutter of other holons in the basement is seen in Figure 2.8; a broken piano with a chair on top of it, a boiler and ducts. These items appear as holons in the basement along with the viewer. It worked so well that when it was time to leave Allen pressed the button to call the "real" elevator gingerly for fear of punching a hole in it. The whole show was about holons, although it is not at all clear that the artist even knew the word or explicit concept. Jack Damer was a professor in Education at Wisconsin who taught Gustafson in those days. He has a Gustafson surveyor's tripod made of inked paper. Damer says Gustafson would work fast, creating things like a movie projector out of paper in about one hour.

More apparently integrated wholes, such as commonplace discrete objects, happen to have for the observer a greater influence upon signals received than do more diffuse wholes, such as ecosystems. However, things that appear as integrated may emerge as less so when viewed more closely. The arbitrariness of qualities of wholeness is more apparent in less integrated wholes, but such arbitrariness is present even in the most apparently particular and discrete wholes, such as chairs and tables. Observation that is in tune with the patterns of integration of objects of certain spatial and temporal size will easily lead to definition of a class of objects that represents a level of organization. Although Koestler works with discrete levels of organization, the arbitrariness of levels may be derived from the arbitrariness of holons. Hierarchies may be viewed profitably as being vertically continuous with levels emerging as devices of the observer, helpful in conception, communication, and calculation. Higher holons in a hierarchy constrain lower holons and provide the context in which lower holons function.

By expressing complexity in terms of context and content, hierarchies offer a particularly general organizational frame. The order in hierarchy is often fully apparent, but unlike more restrictive ordering systems, such as seriality, hierarchical order is more emergent and less imposed. Koestler shows this by the example of sentence construction and interpretation. A serial model for a sentence using stimulus response insists on a particular word placement. In a hierarchical model, nouns and their articles are grouped together and then subjects with verbs. Ideas

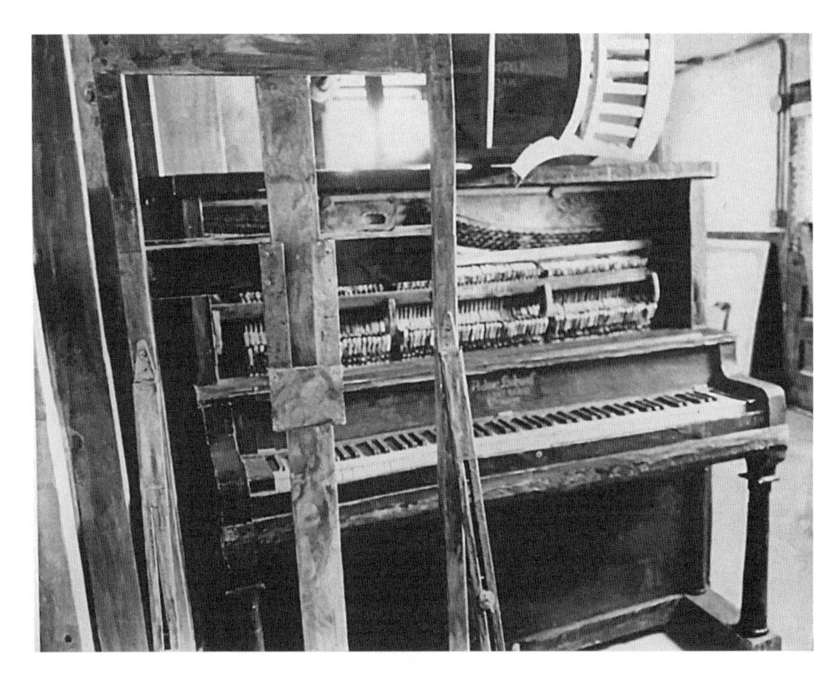

FIGURE 2.8. Panels a, b, and c show a sculpted environment (household basement) made by Pier Gustafson out of paper with ink on it. The basement is a holon viewed from inside such that one cannot see the studio in which it is displayed. The basement has a piano stored in it and a chair stacked up, all made of paper. The piano and chair are exactly freestanding holons; the thing is actually only its surface. (Courtesy Pier Gustafson.)

are contained in phrases and so on. This allows for the construction of many sentences all with the same meaning, merely by rearrangement within the rules of association set down by the allowed classes: "The dog bites the man" as opposed to "The man was bitten by the dog." Hierarchy in sentences also allows predictions as to the changes in meaning that would follow from certain changes in word order: "The dog bites the man" and "The man bites the dog." The rules of the hierarchy give a very general although perfectly accurate account of the sentence and its intended meaning.

CHAPTER THREE

Scales and Filters

In the previous chapter, hierarchies are seen in terms of constraint described by quality and quantity of information flow and its consequences. Therefore the scale of a structure can be defined by the time and space constants whereby it receives and transmits information. That information may be carried by any number of vehicles in either energy or a matter. In German the distinction between the study of electric motors as opposed to electronic circuitry can be literally translated into English as "strong current" versus "weak current." The information associated with scale may be transmitted by either strong or weak currents. It is the significance of information, not the voltage, that determines the amplitude of the signal. The impact on the recipient through a telephone may carry high amplitude message. Meanwhile the working of a motor may have less significance, although the amps in driving it may well be less than the current driving a telephone.

The distinction between strong and weak current is more important these days, in the information age. Although computing since 1965 on mainframes and mini-computers, Allen got his first microcomputer in the early 1980s. That let him communicate much more easily with letters to family back in England. Allen's father, Frank Allen, was a holdover from the industrial age, before information was all. He congratulated his son on the effectiveness of his new "engine." Machines were all just machines as far as Frank Allen was concerned.

A Definition of Scale

A signal starts with a stream of energy or matter in transit between communicating entities. We need to make the distinction between a sign and a symbol. It is instructive to look at Denise Schmandt-Besserat's *How Writing Came About.*[1] She shows how writing arose, and in that she makes the distinction between signs and symbols. With agriculture came tokens for bookkeeping. For instance the token for dog was a banana-shaped piece of clay. Barley was a wedge shape. These persisted for some 4,000 years before writing. A problematic series of these tokens were dated just before writing came into existence. They were poorly made, relative to the majority of tokens found. The surprise and puzzle is that they were always found in good condition, relative to the better-made tokens that had been knocked about by much handling by their users. The explanation was found in hollow clay spheres. The spheres were containers called envelopes by archeologists. Their contents appeared to be some sort of Neolithic IOU, or some other incomplete transaction. In modern trading, the futures market consists of incomplete transactions. One can imagine a Neolithic version of a futures market using containers and tokens for future reference. The poorly made tokens were in good condition because they spent their working life inside the clay spheres, protected from wear and tear. They were poorly made because the makers knew the tokens would not be much worn to do their job.

A difficulty arose for the users of the envelopes in that it takes breaking open the clay spheres to read the information in the tokens. So it did not take long before the users of the sealed tokens began to make an impression on the soft clay of the sphere with the tokens about to be entombed. The impressions were signs, a mark as a clue to what was inside; they were not symbols, as one might take the original clay tokens to be. The marks were only signs. With the imprint of the tokens on the outside, then it is an easy step to see that what is inside becomes superfluous. But making many impressions on envelopes with over forty tokens inside will have been tiresome and time consuming. Instead of leaving a mark of an impression, one could more quickly fashion something like the mark with a stylus on a clay tablet. Those inscribed marks would indeed be symbols, not signs. Voila, writing. The hard part about inventing writing is the move from 3D tokens to 2D symbols, a considerable abstraction. It took 4,000 years of using tokens and eventually hiding the tokens for tokens to generate symbols in writing.

The critical difference between the indentations left by the token and the symbols in writing is that the indentations are only marks left behind. Written symbols are active abstractions. When Sherlock Holmes sees footprints outside a window, it is only a sign of an intruder. A footprint on the moon is in a sense only a sign that we were there (Figure 3.1). Of course, the image of Buzz Aldrin's footprint in the dust has been circulated around humanity, and in this way the sign turns into a symbol of human achievement. Notice that human achievement is a higher level conception with more generality that enriches the significance of the

FIGURE 3.1. On July 21, 1969, Apollo 11 astronaut Edwin Aldrin photographed this footprint to study the nature of lunar dust and the effects of pressure on the surface. The dust was found to compact easily under the weight of the astronauts, leaving a shallow but clear impression of the boot. It is at once a sign that humanity was there, but unusually has also become a symbol for the achievements of all humanity. Photograph provided by NASA.

mark on the moon's surface. The footprint as only a mark carries much more specific and inflexible significance. The critical difference between the image of the footprint as a sign as opposed to a symbol is the focus of the rich significance of the mark. It is less a footprint and more *the* footprint. Meaning is always found at the next level up. The meaning of "humans were there," is a lesser consideration. Neil Armstrong has captured the contrast between the footprint as a sign versus a symbol. His poetic statement "That's one small step for a man, one giant leap for mankind" has two levels of analysis juxtaposed.

In her book, *Dynamics in Action*, Alicia Juarrero shows how two actions can be either a sign or a symbol, the latter releasing a cascade of larger meaning. The first lines of her treatise are "What is the difference between a wink and a blink? Knowing one from the other is important."[2] One is a sign, the other is a symbol.

A signal string can be considered to be infinitely fine grained, although a digital string would consist of only small, not infinitely fine grained, bits. Any attempt to retrieve all possible information about the signal would involve both scanning with an infinitely fine observation point passing over the signal, and all of the infinite number of possible integrations of that entire scan. As information leaves a body as a signal string, the string amounts to an integration of all the parts of the body with respect to the particular mode of energy or matter of which the signal string is composed. At departure, the signal represents a freezing of the infinitely rich dynamics of the transmitting holon. Although meanings can change, a single meaning has no dynamics of its own, for meaning is frozen in time. Meaning is independent of any rate. This is in contrast to, say, a process or system behavior, which does flow or behave at a rate. If there are not in fact transmitters giving signal meaning, then there must be something very like them; however, since we have access only to received signals, the above is beyond knowledge but is a necessary assertion for the argument in our framework. Receivers of signal cannot know the meaning of the signal for the transmitter, and different receivers may read common signal in very different ways. Moreover, receivers can assign meaning to transmissions that are not signals from the perspective of the transmitter, but they become signs to the receiver.

In a communication (as opposed to a signal stream) the information content is ambiguous between two bodies, for the meaning of a communication is both in the eye of the beholder and in the visage of the transmitter. These two parts to the communication are different. A communication has meaning for the transmitter and for the receiver, but between them the signal is undefined as to meaning. A signal by itself is undefined in terms of scale, for it is potentially infinitely fine grained and may be smoothed in an infinite number of ways. A signal has no intrinsic time or space constants. All this is reminiscent of Mandlebrot's argument that the coastline of Britain is infinite.[3] There must always be some sort of measuring stick. If the meter is long, Britain is outlined close to a triangle (Figure 3.2). But shorter segments pick out increasing detail. The triangle appears to have a rectangle on its western edge, which is Wales. Shorter segments again begin to show Cardigan Bay, the hollow that separates North Wales from South Wales. Yet shorter segments will show the Wash, a shallow bay in the crotch between East Anglia and the North. King John lost the crown jewels trying to cross the Wash at low tide. Segments shorter still will detect the Humber, a slash of an inlet on the side of Yorkshire. In the end, as one measures the irregularities around the sand grains on Brighton Beach, the coastline is so long as to be functionally infinite. The same applies to signals divided up over time.

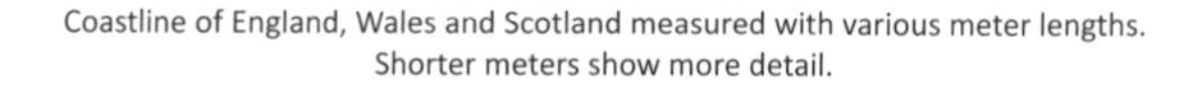

Coastline of England, Wales and Scotland measured with various meter lengths.
Shorter meters show more detail.

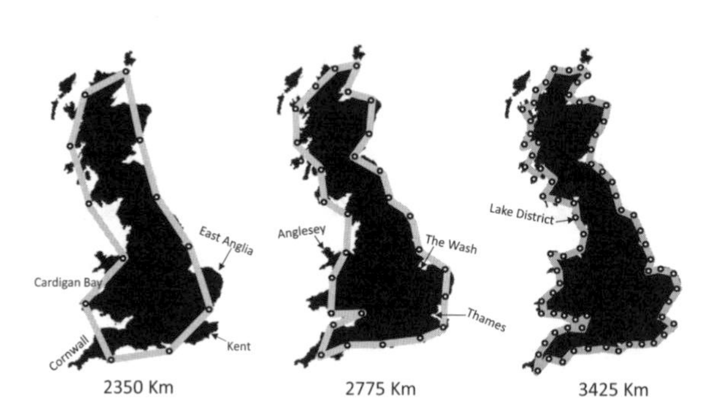

FIGURE 3.2. As the coastline of Britain is measured with various linear meters, the respective calculated perimeter lengths are about 2350 km, 2775 km, and 3425 km, left to right. The shorter the meter, the longer is a measured coastline. This means that the coastline of Britain is infinitely long if it is measured with an infinitely short meter. The longest meter does not even detect major regions like Cornwall, North Wales, Kent, or East Anglia. Shorter meters show more local landmarks like the Wash, Anglesey, the Lake District, or the Thames. The images were recast, and perimeter calculations were taken from "Britain-fractal-coastline-combined." ("Britain-fractal-coastline-combined" by Avsataken is licensed under CC BY-SA 3.0 via Wikimedia Commons.)

A scale is the period of time or space over which signals are integrated or smoothed to give message. Transmitted messages have particular meanings; that is, they carry particular information for the transmitters, as do received messages for receivers. However, between message transmission and message reception a message becomes a signal. Information in a message transmission is not usually the same as information at reception. There are, of course, two messages involved: party whispering games depend on the difference between them. A message in either transmission or reception mode has defined time and space constants and so has scale. The time and space constants are related to the patterns of integration in the conversion of either transmitter dynamics into signal stream or signal stream into received message.

Patten and Auble[4] identify the term *genon* with the patterns of integration used to convert internal holon functioning to external signal transmission. They use the term *creaon* for the patterns associated with signal reception. Patten and Auble also split up the environment as that which transmits inputs to the holon as opposed to that which receives the signal from the holon. The extra terminology is useful to the point that it dissects the process of movement between levels. Network theory derives from box and arrow diagrams of Forrester models.[5] Patten and Auble show how network theory carefully steps upscale, while hierarchy theory reaches farther over greater differences in scale and type. Network and hierarchy theory are twins. We do not find a sufficient return for the effort of expanding our vocabulary to include theirs, and so we do not use their vocabulary, but this is only a matter of taste.

In both transmission and reception, the integration may be seen as involving a window in time or its spatial corollary. Integration is across the window, which averages either the internal behavior of holons in transmission or the signal in reception. A signal stream includes details well below the capacity of a receiver to distinguish details. An instance is human males usually cannot hear the squeaks of bats. The first part of scaling is to assert the grain, the finest distinction between the smallest units of integration. For instance, a photographic plate receives light at the size of photons, but the plate cannot distinguish one photon from another. Different films, from the days before digital photography, captured images over particular time frames. Fast versus slow film works with units of time. The grain of the recoded signal is determined by the literal grain of the crystals on the film. Fast and slow film can be distinguished by the size of the grain. On a silver nitrate film, the grain is the size of the individual grains that turn from light-colored to dark when exposed to light. On remotely sensed images there is a certain grain that corresponds to an area of landscape, commonly ten meters on a side. Finer grain can allow detection of smaller scale phenomena. There is some prohibition of grain on images at less than thirty meters on a side because below that grain one can tell the difference between water and water with a submarine in it. That is one of the few national secrets that actually needs to be secret. The second part of scaling is the extent of the universe that is included. As extent increases, grain usually has to be chunked up so as to meet limits on memory and the capacity to handle data. Scaling is a practical matter of handling experience in a workable fashion.

In temporal scaling the extent is the longest time associated with the signal string. The grain of temporal scaling is the smallest recognized instant between bits of the signal stream. Temporal scale is determined by integration of signal over a window (time period) that moves continuously down the signal with time. At first one might prefer to apply equal weight or significance to all parts of the signal stream within the window. In the simplest case, the weight applied to the signal is not only constant through the window but also constant as the window as a whole moves to a new portion of signal with the passage of time. That sort of window is akin to a moving average in statistics. Spikes in the signal are smoothed out by integrating them with adjacent parts of the raw signal.

As scaling becomes more sophisticated, the even weight applied to the signal across the window is allowed to vary. That is to say, some parts of the signal are given greater account because they are deemed to be more significant. This is akin to a moving average where a greater contribution to the average is made by some values in the window, perhaps the most recent entry. Weighting the most recent always gives more significance to the front end of the signal as it is received. But we do not have to always use the same pattern of weights across the window. The pattern in the window can itself change over time, for instance not always giving high weight to the most recent entry. If one wished to emphasize some particular event, as time goes on the event moves back in time relative to the leading edge of the window. Emphasizing an event means that as time moves on the largest weight moves back in time with the progress of time. Accordingly the higher weights applied to the event move back in the window with the passage of time. Perhaps the leading edge of the moving window, the most recent entries in the signal, may be given heavy weights at some time-zeros but light weights at others. It takes some ingenuity, but an echo effect is just an elaborate filter that references itself.

There is nothing to stop the window from becoming larger or smaller by integrating over short segments of signal at some times and long segments of signal at others. An example in biology here might be the significance of what a bear has eaten. As the bear moves into hibernation the time of the last meal recedes deeper into the past, but that deeper past remains important to surviving hibernation. Eventually the bear begins to arouse, at which point the meal before hibernation becomes less important; it has already done its work. The important part of the signal becomes the anticipation of the next meal, the first food consumed upon awaking. The bear's window ceases to reach far back into the past and moves into the future. And we do mean "moves into the future." Organisms do anticipate—that is, they read ahead of time zero. Margalef put it as signals traveling faster than the speed of light; the signal is read before it has been given. But the future does not have a direct effect, it is anticipated through a model, which might be wrong. This opening and closing of the window gives significance in message construction to respectively longer or shorter segments of the signal string. Making the window longer is achieved by changing zero weights outside the window to nonzero values. Shortening the window is achieved by changing to zero weights in the window at its leading or trailing edges. May[6] uses a similar integration period in what he calls a "weighted average time delay." Barclay and Van Den Driessche[7] employ the integrated time delay in a multispecies ecosystem model.

May's concern is for the influence (messages) that past populations (signal) have upon resources at time t. He integrates the signal N using a particular weighting function Q. "The function Q(t) specifies how much weight to attach to the populations at various past times, in order to arrive at their present effect on resource availability." Q(t) is dragged across the window that trails from the present, time t. As Q(t) passes across the window, Q asserts its weights as they are specified by the respective differences in time away from the present, time t. All this common sense activity, say of the bear hibernating and waking up or of past populations consuming food, can be captured in an equation. Some readers may find the specification of all this in an equation helpful, particularly if creating a quantitative model is the goal. Others may prefer simply to intuit from the word-model of the hibernating bear that things in the past vary in significance, and which part of the past matters can itself change in significance as time passes. In formal terms, the total effect of past populations on resources at time t is as follows:

$$\int_{-\infty}^{t} N\,(t')Q(t-t')\,dt'$$

where N = the number of individuals in the past populations.

But May's weighted average time delay is a special case of scale. While he uses different weights for populations from various past times (t−t′), he keeps the same form for Q(t) as t changes; he maintains the same weighting structure as Q is applied at different present times. In our usage the form of Q(t) itself may change as present time passes; the weights applied to populations from any particular lag (t−t′) vary as time t passes. Q therefore becomes a two-dimensional function dependent on both lag from the time of the application of Q and the time of the application itself as t_0 moves forward.

The position of a holon in a hierarchy is determined by patterns of constraint. Constraint may be translated into terms found in the weighted average time delay. Patterns of constraint are determined by the amplitude and asymmetry of information exchange between constrained and constraining holons. A discrete lag involves delay of the signal for a particular time period as it is translated to message, although the form of the converted signal is unaltered. Continuous lags delay the signal, but they smooth it as well. The distinction between continuous and discrete is well captured by regarding the move to the digital age with digital cameras and CDs as opposed to the old film in cameras and vinyl LP recordings. The signal on a vinyl record, an LP, is played with a needle in a wavy groove. The groove was cut by the sound waves captured by the recording microphone. We refer to this sort of recording and playing as an analog system. Most recordings today are digital, as on CDs, where the sound is not captured in a direct mechanical fashion, but is rather on the disc as a set of digital numbers. The recording is in a code where there are symbolized numbers between the recording and playing of the music. The same contrast arises in the analog capture of images on film, as opposed to digital camera technology. At the beginning of modern computing, there was a distinction rather like film as opposed to digital cameras, and LPs versus CDs. In analog computers electrons raced around a circuit to carry signal. In digital computers there were discrete bits of information. In simple analog computers, discrete lags cannot be readily achieved but continuous lags can be incorporated by using a capacitor to integrate or average the signal. Remember the analogy of water pressure for voltage in a water/electricity analog. A capacitor in an electrical circuit is like a bathtub that accumulates water that arrived in the past.

Constraining holons integrate the signal of constrained holons so that the signal is averaged or lagged such that it exhibits negligible behavior for the receiver. The small fast variations in the signal of the constrained holon are smoothed out to become close to a flat line. In the reversed situation, constrained holons also integrate information, but over what is effectively so short a period that the signal from the governor is read also as a constant. The constrained holon is influenced by the governor while itself exerting little influence over the governor. Thus the scale (i.e., the position in the hierarchy) of both the governing and the governed holons is determined by their information-handling attributes. The way a holon converts unscaled signal into scaled message determines its functional environment and its scale.

Thus the scale S is determined by the weighted integration of the signal N, where Q is the two-dimensional weighting function. One dimension of Q gives the weights through time, and the other dimension changes the form of the weights as the time of application of Q changes. With anticipation put into the scheme, the integration is from minus infinity (the past) to plus infinity (the future). Thus

$$S = \int_{-\infty}^{+\infty} N(t')Q(t-t',\infty)\, dt'$$

(see Figure 3.3). Electrical engineers will recognize the above equation as a filter, and they will probably be familiar with filters whose behavior is very akin to the adaptive response of a biological system reading the environment. Organisms read their circumstances, and then may adapt the weights in the scale to accommodate to where the organism finds itself. A mammal that reads "too hot" may sweat and so function as if it were in a cooler environment. The same can occur in an electronic filter. For example, a signal from a vinyl record contains music and noise. The noise may be read by the amplifier in a way that distinguishes music from surface noise of the recording. Modern electronic filters can identify the general patterns of order (music) just played as being those parts of the whole signal with high autocorrelation, that is with lots of internal structure capturing loops of recurrence. Autocorrelation indicates struc-

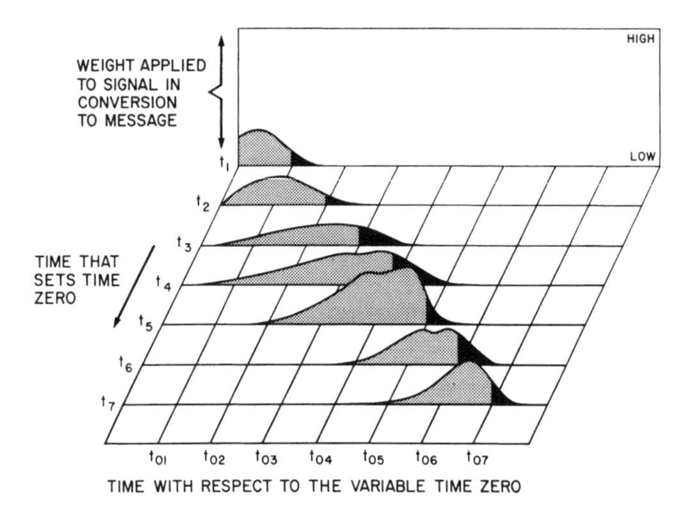

FIGURE 3.3. The two dimensions in this figure capture two relative times: first, time of the application of the filter (t1–t7), and second in the progress through the filter as it is applied at a respective time (t01–t07). The intersection of the lines t and t_0 diagonally across the plane are the positions of the respective time zeros as the filter is applied. On any given application there are the weights to the left of time zero that capture the influence of the past in that instant. Those weights are shaded gray. Notice how the past is taken into account varies with the time of application of the filter. That is the gray regions change as t1 to t7 respectively come to pass. The black profiles occur ahead of time zero and are our expression of the influence of anticipation. The message of the whole figure is that as time passes the degree to which the past and future affect the signal itself changes over time. Sometimes the deep past matters, and sometimes it does not. Sometimes anticipation is further into the future than at other times.

ture in the signal, such as a note repeated at a certain frequency or notes in a certain key. The filter uses past structure to anticipate what is music and what is scratchy background noise. It uses that identity to give heavy weight to the music and zero weight to the surface and static noise. The message from the noisy signal is clean stereophonic music.[8] In a discussion of information flow through a hierarchy, Koestler uses the idea of filters although he does not develop a formula. Our development of the filter formula lends independent support to Koestler. Allen here showed his hard-won new formula for scale to an associate, Robert Friedman, who said, "Oh, that's a filter"—disappointing it was so commonplace but reassuring it was not far out. Later Allen was surprised to find that Koestler had arrived at the same point; signal in hierarchies is filtered.

Koestler uses the image of military command where an instruction from a general is the beginning of a chain of trigger reactions going down the hierarchy. An unspecified request from the general for reconnaissance eventually becomes a specific instruction of time and place from a platoon leader to the private. As the soldier makes his report to his platoon leader, information begins its return passage up the hierarchy. This time, instead of a chain of trigger reactions to "do this," the information passes up through a series of successively coarser filters, becoming a broader account of the enemy position until the general receives the intelligence that he requested. While the trigger-filter duality is a helpful way to emphasize the asymmetry in hierarchical control, it also has the unfortunate property of masking the essential similarity of filter and trigger. In both cases the information passes through what an electrical engineer would call a filter. We then prefer to talk of the scale of a holon in terms of its transmitting and receiving filters (genons and creaons for Patten and Auble).

Simultaneous signals between two holons may, of course, have very different subject matter and may be delivered by quite dissimilar vehicles. Even if the signals are in the same direction, one might be a "strong-current" signal amounting to a mechanical force while the other could be a "weak-current" signal like a telephone message. It is, then, necessary to recognize that signals and messages pass between holons multidimensionally with differently scaled filters for each dimension. In organizing governance Mario Giampietro[9] has captured these distinctions as he developed a method he calls Multi-Scale Integrated Analysis of Societal and Ecosystem Metabolism (MuSIASEM).

Scales of Transmission and Reception

The transference of information between two objects involves two scales where the message of the transmitter becomes a signal to the receiver. That is to say, the signal may be integrated by the transmitter over periods that are different from those employed by the receiver. In such a case the communicants are differently scaled and in some ways the signal is distorted, but the distorted signal is all that the receiver has and becomes the firm context for any responses the receiver might subsequently make. It was a telling title that McLuhan and Fiore[10] chose for their startling book *The Medium Is the Massage*. It was massage, not message, but we think they meant message too.

In showing a film strip, the projector is the transmitter, and the cinemagoer is the receiver. By changing Q through time, the projector lags the continuous strip twenty-four times a second to produce a flickering message on the screen; that message is taken as signal by the human eye, which, being limited in the speed at which it can erase impinging light signals, weighs light heavier than dark. Thus Q for the eye changes through time relative to the present and has positive nonzero weight over a period longer than a twenty-fourth of a second. The eye integrates a light-dark sequence to receive a message that is continuous, not flickering. Thus doubly scaled information transfer from celluloid to human experience gives Humphrey Bogart smoothly raising an eyebrow as Sam plays it again. The essential asymmetry of the scaling process could be displayed by running this process backward. Read the physical film strip like the human eye sees, and it all looks very different. A film strip transmitted at the scale of a human eye would be not a smoothly changing sharp image, but the blur of a film passing continuously through the gate (see Figure 3.4). That is what happens when a filmstrip jumps out of the gate.

A simple reverse is not the same as passing signal through mirrored filters. If filters are mirror images of each other, the message can be coded one way and decoded the other. This is different from simply reversing filters, as when the film strip jumps out of the gate to give a blurred image. The movie camera creates the movie strip with a filter that is a mirror image of the filter of the movie projector. It is no accident that a movie camera and projector make roughly the same chattering sound. It is also no coincidence that when the film jumps out of the gate the projector changes the sound it makes. This is because the film moves through continuously because the sprocket is disengaged from the film strip and so cannot correctly constrain the movement of the film strip. The relationship of the film strip to the proper cinematographic image is the same in both directions as the camera takes the film and the projector projects it. The one creates the physical film itself and the other reads the signal on the image frames in the way they were created. Humphry Bogart looks more or less the same to the audience as he does to the camera. The trick is the perfectly mirror-imaged filters of filming and projecting. The message from the film set is the same as that on the screen. The receiver and sender of the massage simply reverse the process of transmission to copy the process for receiving. Coevolution is the evolution of two species together, as when the length of the spur containing nectar corre-

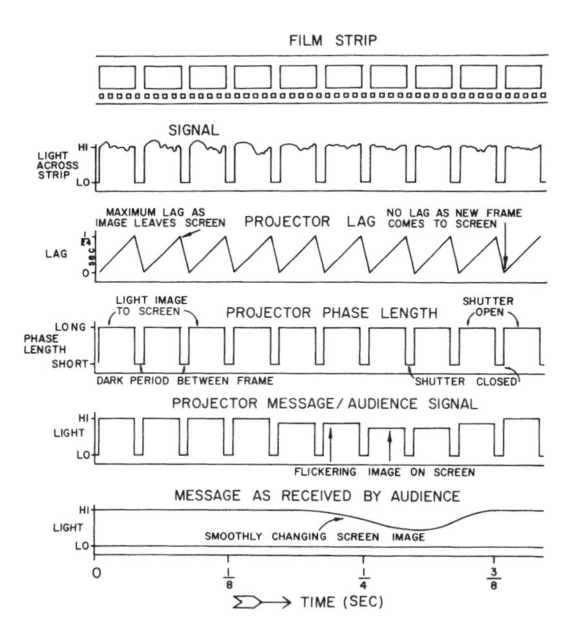

FIGURE 3.4. Film strip to audience. The top row is the film strip. The second row, the signal, shows the infinitely fine-grained scan of the film strip. The third row shows how the projector modifies its signal-weighting function so as to lag the strip twenty-four times a second. The particular image comes into the frame and stays there for 1/24th of a second, after which the frame jumps forward. The fourth row shows how the weighting function integrates short and long segments so as to show whole light frames and whole dark gaps in between. The fifth row shows the projector's flickering dark-light image. In the bottom row the audience has smoothed the flickering images by not being able to see dark quickly enough to notice it. There is a gradually changing picture over the dark periods between frames. That happens both in the movie theater and in life in general. For instance, you cannot see the 50 to 60 cycles of light and dark as light bulbs use alternating current.

sponds to length of the tongue of the moth that eats the nectar. Coevolution can be seen as tuning the respective output filters to the input filters of the coevolved partners. That is so highly contrived that the relationships stay stable across long periods of evolution. This is evidenced by the way that paired relationships occur across genera, such that moths of a given genus, are paired with flowers of the plant genus. Across the genera of the plant and the animal partners, different species are respectively paired, as the general relationship survives evolution of both species in a given partnership. Every time you see *Casablanca*, Sam does indeed play it again.

Bill Wimsatt uses the notion of flicker fusion frequency in his discussion of environmental heterogeneity perceived by animals. He notes that a checkerboard of environmental patches of ten meters on a side, varying between patches discretely from 0°C to 40°C, would kill a *Drosophila* by either freezing or overheating but would not even activate thermoregulatory systems in a cow or human walking through such an environment.

CHAPTER FOUR

Sirens of Certainty

The complex of scaling phenomena makes it difficult to be sure of much at all. Merely asserting we know from observation does not make it so. Scales playing off against each other deny certainty. In this chapter we unpack that issue. Since human filters have evolved to make present human experience, the effect of these filters feels more concrete than it is. The Siren says it seems so real it must be so. No; it is not!

Information may pass between two holons without alteration only if they both use the same filters—that is, only if they work at the same scale (like movie camera and projector). Since the human perceptual scale is different from that of any nonhuman observed holon, information received through experience must have an important individual human component. In some ways we can read other humans because we all have roughly the same filters. Even so, it is commonplace for humans to misread each other. Clues as to the form of that human filter are found in the consistency of the observation under what are apparently the same conditions. The sameness of conditions is sometimes hard to engineer, which is one of the things that requires science to be exacting. Psychological abnormalities can be used as a contrast to gain insight into the general human condition.

The scale of a holon has concrete consequences for everything else with which it interacts. It determines not just what is known about the entity, but also what is knowable. What is knowable is further modified by the scale of the observing holon.

> All our evidence (*of external reality*) deals with transactions . . . entities have meaning only in encounters. . . . The only constants are functions—exchanges

> as Heraclitus said, connections as E. Mach expressed it. Nothing exists by itself or in itself. Everything exists through reciprocity, *Wechselwirkung*. (Boodin 1943; our italics in parentheses only)

And this reciprocity is scale mediated.

Arts and Science

Both science and art depend on signals from holons; these signals are scaled and filtered to give human experience. Contrasting the scaling strategies of the artist and scientist is instructive in identifying the role of scale in scientific investigation.

Usually there are three scales in any observation: the scale of the transmitting holon, the scale of the receiving holon, and the scale assumed by the observing scientist. The first two scales determine the structure of the models. The third (human) scale will be different from the scale of either of the observed holons. That difference in scale between the human and that which is modeled changes the messages from those transmitted by the object of study to those received by the observer. It is possible to map the positions of a pair of holons from observations of each. Then a series of reasonable expectations may be recorded for the inter-holon relationship. Because the human perceptual scale will be different from the scale of both holons, the recorded relationship contains a human component that is absent from the observer-independent relationship. Up to this point the artist and scientist travel together.

The Artist

No matter how carefully the original observations were made, if the human scale is different from that which is observed, then the calculated relationship between the holons will not live up to expectations under all conditions. The artist focuses attention on the inconsistencies between expectations and that which unfolds. The inconsistencies then draw attention to what has been mistaken for granted. It draws attention to what it is to be human; the apparent contradictions focus on the human scale.

The optical illusions of 1960s "Op Art" fit easily into the mold, as do the Impressionists. Impressionism carefully distills the feeling of a warm summer day, but with a minimum of cold fine grain signal that a camera might perceive. The savage eye of the snapshot captures little of the warmth of summer scenes. As we look at summer photographs, we might wonder where summer went. It probably went into Impressionist paintings or the what-it-is-to-be-human. The comment on the human scale is that there is a time dimension of experience that blurs the sharp images the eye holds for each instant. Fine-art photography, as opposed to holiday snapshots, can capture summer, but there the photographic artist takes time (the dimension missing in the snapshot) to bring that blurred summer experience to a carefully sharpened image. Time taken in careful framing and printing may be substituted for the time in experiencing summertime. Impressionism had to wait for photography (the other holon) before it could properly develop.

The Scientist

Both scientist and artist are conscious of the human scale, but the artist celebrates it while the scientist tries to eliminate its effect. Scientists make every effort to measure the holon interaction at the scales of the holons themselves as directly as possible. They try to avoid distortions of human perception; the scientist tries to look at what is modeled the way it looks at itself and exists in itself. In order to make measurements the scientist must be acutely aware of the human scale and the effect it has on incoming signals. Since the only information is the messages received, and they are defined by the scientists' own input filters as much as by the signal coming into the filters, the scientist must guess what is the signal coming from the observed holons. It is not even the signal that is of primary interest, for what the scientist really wants to know is the meaning of the message for the holon that gave rise to the signal. The special use here of the term *meaning* has been considered in chapter 3. The relationship between the signal and the infinitely rich dynamics it reflects is the primary interest. However, the observer only has access to the signal stream, and that through personal input filters. The elaborate subjectivity of observation in science is one of the reasons for our reservations in relying on objective reality as a benchmark for science.

Good science deals with its signals not by trying to extract all their information (which, given the infinities involved, is worse than cumbersome, it is impossible), but rather it tries to derive signals by making informed guesses as to their parent messages. Unlike the signal, the message is finite and defined in terms of scale. The informed guesses are the hypotheses of elegant science that are missing in ill-conceived or brute force approaches. From observations and subsequent calculations, an interrelationship between holons with its attendant filters is hypothesized. Now fresh signals reach the input filters of the observed interacting holons; these signals may emanate from the scientist (e.g., the administration of radioactive CO_2 in a metabolism experiment), or they may be derived from the observed holon's environment, not importantly including the scientist (e.g., storms and natural lake nutrient studies). Given the hypothesized filters, a resulting input message to the scientist is predicted.

If predictions hold, further tests are conducted to find out more about the observed holons' filters. These tests may involve a different signal, say radioactive CO_2 but with an enzyme inhibitor to alter uptake of the gas. Alternatively the input may be left the same but with the scientist predicting different results because of knowingly changed observation filters. The change in filter may be the same administration of radioactive CO_2 without the inhibitor. The filter may be widened by using a longer delay in fixing (killing) the material. That way radioactivity can pass further down the hypothesized metabolic pathway. If the original test fails and the predicted message is not found, then scientists have two options: 1) change the hypothesized observed holon filters (erect a new model) or 2) change their own observation filters with a new experimental design, having presumed that the signal was there but they missed it. The scientist then goes on to build some apparatus and experiments designed to collect signals at the hypothesized external scale. When scientists have achieved an aspect of the filter associated with the scale of their observed holons, the hypothesized relationship matches the observed and the model is presumed to be confirmatory. The filter is never perfectly correct, and even if it were, scientists have no way of knowing that to be the case. Later generations observe more carefully and find inconsistency or ambiguity, and the process is repeated as the model is rejected. One of the hard parts of physics research is building an apparatus that is scaled so as not to ignore the messages from the studied phenomenon. In holistic science there is an attempt to capture the complex consequences of simultaneous communication at different scales. In reductionist science each scale is applied one at a time in a causal chain: transmitter message to receiver signal, through receiver scale to received message, to response that is seen as a message transmitted by the receiver, and so on.

Observer Intrusion

Careful interpretation of the inconsistencies between observation and expectation suggests alternative scales for observation. Nevertheless, no matter how brilliantly the models are conceived, nor how carefully the technical tasks of observation are executed, the scientist is greatly influenced by scale problems that limit what can be known. If scientists are to observe their subjects completely and accurately, they must observe at close proximity with the same scale as the observed holon. The problem then becomes one of intrusion. Since the identity of a holon is determined by its scale, when they apply an identical scale at close quarters, the scientists become part of the phenomenon. R. D. Laing in *Politics of Experience*, an early postmodern work in psychiatry, notes that psychiatrists cannot observe the patient; they can only observe the patient being observed by the doctor, and that is different.

Thus the scientist is double bound; either there is distortion in the message by applying a different scale in order to avoid becoming part of the system, or scientists may intrude by changing the components of the holon so as to include the scientists themselves. Some sort of compromise is the usual way out, but even then scientists must remember their humanity or suffer delusions of objectivity. The technologist aims to intrude into the world and modify it, so the more applied is the study, the less important are epistemological problems of intrusion.

In the final analysis the scientist must translate back to a human understanding, and it is there that the scientist and artist share moments of creative insight. Like many, Rindler (1969) points out that "a physical theory, being an amalgam of invention, definition and laws is regarded as a model for a certain part of nature asserting not so much what nature is but rather what it is like." The etymology of the word *like* puts Rindler' s remark in perspective.

"Lich gate, a roofed gate at the entrance to a churchyard, where a coffin can be set down to await the arrival of the clergyman."[1] Thus Lich and body are associated. Hence which, what body; and each, one body. Our reports in science are emphatically in-the-body-of ourselves. The scientist, while accepting that limitation, tries to model at the scale of that which behaves to produce the data; the artist, on the other hand, points out what it is like to be in here, inside a human body. There is an ironic echo here. “So *God created man* in His own image; He created him in the image of God.”[2]

In the end the bottom line in science is narrative. Narratives are neither true nor untrue, they are simply announcements of a point of view. There are more and less compelling stories, but that is a matter of subjective utility not truth. There is some subtlety involved here introduced by N. Kathryn Hayles in her ideas on constrained constructivism and the semiotic square.

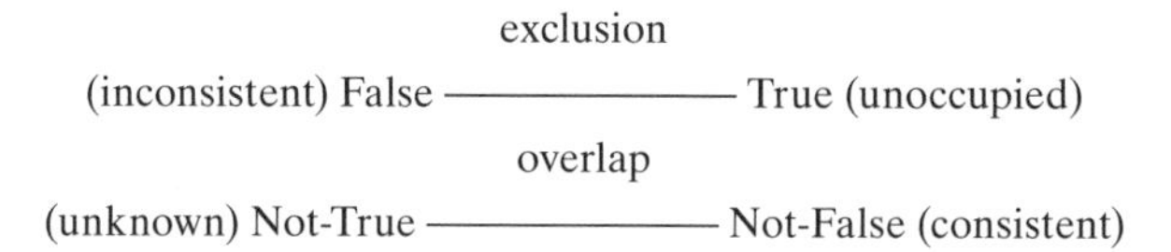

> Realism tends to elide the differences indicated by these markings, assimilating not-false into true and not-true into false. When a scientific textbook states “All the matter in the universe was once contracted to a very small area,” the difference between the model and the reality tends to disappear, as do the position and processing of the observer for whom the statement makes sense. Far from eliding markings, the semiotic square displays them along the vertical axis. (Hayles 1991)

Primate Knowledge

For ecology the problems pertaining to Heisenberg’s uncertainty principle might appear a trifle esoteric. Ecologists deal with more commonplace experience than electrons. Trees and flowers do not overtly appear to present epistemological difficulties of import. But there is a danger; success at a habitual human scale may lead to a conviction that what is not seen may not be seen.

Ecology, like many other disciplines, finds itself often limited by the habitual size and time scale of its primate practitioners. Ecological structure abounds, but ecologists are usually only cognizant of that which can be readily perceived. We detect nonrandomness. When we study holons scaled very differently from ourselves, something is lost in translation to the human scale. To identify the bias in the pursuit of ecology, Hoekstra, Allen, and Flather looked for patterns in which organisms were used more often than expected to perform research in broad ecological sectors. When using a microscope, it is hard to remember that a *Paramecium* does not see a bacterium with light-sensitive senses any better than do microscopists. Bacteria are blurred by the wavelength of light. Hoekstra and his colleagues saw distinctive patterns in where ecologists put their effort. The patterns confirmed that we detect readily only ecological phenomena that work at scales commensurate, and consonant, with commonplace human experience. There is of course scaling in enhanced technology, but that was not seen in their results. Ecologists appear to work at time scales bracketed by naked human observation, and less so influenced by fancy machines. Ecologists work with a lower limit of seconds and an upper limit of about three score years and ten, and at most the generation time of long-lived trees. Imposing their experience of time, ecologists often forget that natural phenomena are usually paced by time scales that are not linearly related to time measured by chronometers. On the other hand we can rejoice that ecologists have amazing perceptual powers if a phenomenon is in tune with innately human perceptual patterns.

An example of a complex high-order phenomenon that does strike harmony with human perception is the convergent evolution of a set of northern Wisconsin bog flowers, all a pink mauve and formed in bilaterally symmetric stars. Although they are from different families, they flower in an overlapping temporal sequence, and they are pollinated by one species of bee that is kept busy, but not too busy, throughout the season. The flower community is tightly integrated through time although members are chronologically separated.[3] The point here is that the information transmitted to give the community its structure is, by chance, at a scale and in language that runs deep in human phylogeny. It uses color code (red, although bees see the blue part of mauve) in an annual cycle, with morphological mimicry at a scale of about one inch. Being primates, ecologists detect the pattern easily. Ernst Mayr noted that both scientists and people living in tropical forests see just about the same number of species. The quotation from Mayr is in his obituary.

> "But I discovered that the very same aggregations or groupings of individuals that the trained zoologist called separate species were called species by the New Guinea natives," Dr. Mayr said. "I collected 137 species of birds. The natives had 136 names for these birds—they confused only two of them.

> The coincidence of what Western scientists called species and what the natives called species was so total that I realized the species was a very real thing in nature." (Yoon 2005)

Mayr worked in the middle of modernist realism, so he would easily reify his system. We now know better. It is not that the species are materially real, it is that our input and output streams are of the same types and combinations as birds. Primates and birds both feed on colored fruits and signal pleasure at finding them. We address the world orally, and perceive the world visually. We can readily experience the same signals that the birds use in breeding strategies. There are mechanical reasons that humans can recognize information that serves to separate related species. Birds need to separate bird species, and we can read that with our human filters. But that does not indicate that bird species are more ontologically real than bird genera or families, which taxonomic levels are generally considered as abstractions more than material.

It seems very likely that not only in ecology but also in such diverse fields as bacteriology and political science similarly high-order and complex natural systems are commonplace. But such structures go undetected because the salient naturally selected information is not readily observable through human filters shared with other primates. Common sense and general experience are of less broad utility than might be expected, for they only perceive in the manner of the consciousness of highly evolved hominids. The problem is particularly acute in the higher-order sociobiological disciplines (from ecology to history). There the investigators are limited by the essential tangibility of their material; tangibility narrows human vision because it encourages prejudice. Biological and human sociologists have not been pressed as far as they might into probing counterintuitive relationships. Pristine areas of study exist at nonhuman scales, which involve information that is not a part of normal field experience.

The greater is the direct involvement of human perceptual apparatus in scientific observation, the more revealing is the observation. Practically speaking, no pattern can be observed by a human scientist until signals are translated into the range of human physical or cognitive perception. Hence, the prevalence of visualization, graphics, and even ordered tables in scientific communication. Rare indeed is the scientist who can perceive pattern in numbers alone. Despite all the elegant environment sampling systems used by Viking 1 on Mars, the most powerful and subtle observations were made by scientists looking at the visual images of the Martian surface. The slightest shadow was used to assess intricacies of geomorphology on that planet. The assessment is that the images from the rover Curiosity suggest free-flowing water was once present on Mars. Were we not human, such a plane of light and dark patches would be very difficult to interpret even grossly, let alone with the precision and insight that was achieved. Looking at a photograph is certainly a strange and indirect method of geophysical data collection, but it works surprisingly well for us humans.

The general point made here is that observations may be made more revealing by translation of phenomena into signal streams that can enter directly and easily through primary senses. In another example, chess games are hard for the moderate player to follow if they are expressed only as moves in formal notation (P–K4, etc.). Although all the information is there in black and white script for the master, it is not until black and white pieces are diagrammed or actually moved that understanding comes to the novice. Similarly, computer programmers can couple their program to a speaker that then twitters and squeaks. It is not that the noises are interpretable as such, but it becomes immediately apparent when the program is behaving in an unusual way, because it does not "sound right." The change in the pattern may be subtle, but the human ear can tell the difference. Facility for interpretation of phenomena may be enhanced by conversion into a signal stream that flows readily through sensory portals.

Thus, of the many signals from the observed, some can be chosen such that the signal stream is particularly compatible with human senses. Alternatively, incompatible signal streams may be converted, say from electricity to sound waves as in the computer example above, so that they become compatible. Both these aspects of observation are under the control of the observer: there is, however, an aspect of revealing observation that cannot be so readily controlled by the observer. If the inputs are instinctive we cannot so readily control them, but they give us insight and recognition. We make the richest observations of other humans or humanistic objects, because as holons they filter the world as we do. In some cases this may reflect the singular importance of group selection in our evolution; humans readily reading each other leads to bonding that can enhance selection of groups over individuals.

Readily reading human faces can lead to coevolution. Human faces are not completely symmetric. Duplicate the left side of the face in a mirror image, and then put the original left with the inverted left to make a full face. The whole face will have a different expression made of two left sides instead of two right sides. We look at faces to capture unity and asymmetry in that a human face is addressed with a move up the nose and then a diagonally to the left eye. Critically, dogs also look at humans that way. But dogs do not look at each other's faces that way.[4] Tail and posture matters more to them. It is telling that wolves, which of course have little human influence in their heritage, do not look at human faces in the human way. Domesticated dogs have evolved a human mode of addressing humans so that it is now instinctual. Deep insight comes from familiarity.

Science often works by achieving an original insight by seeing the phenomenon in a particularly human fashion. The softer end of a science achieves deep insight because of a resonance with ready human interpretation. The new insight may then be generalized by taking the notion toward the harder end of the discipline. The insights of harder science are often more generalizable. That happened with Malthus and Darwin, as Malthus saw deeply, while Darwin generalized social science insights into the huge and general notion of natural selection.

Not until Malthus, working from a social class with vested interest, was it realized that important events occur in the second derivative of populations (of the poor about to preempt a gentleman's resources). Here intrusion associated with observation returns to the discussion. It was the very same class consciousness that ultimately made Malthus intrude selfishly into his observations so as to blind him to their powerful generality. Not until half a century later did Darwin and Wallace, both under Malthusian influence, derive the generalization to evolution by natural selection. Their insight came from a concern, not with human beings, but with wild exotic populations to which Victorians did not have an emotional attachment. The cutting edge of scientific perception is beaten and honed by the intrusion of humanity, while general models with powerful implications come only from the less familiar and the nonhuman. Data for a physicist are often hard-won, but their implications are often very general. For most purposes when you have seen one electron you have seen them all, although this may also be a bias of human perception and scale of observation; electrons may themselves be able to distinguish differences. By contrast information about other humans is gained easily, but the data are dirty, they are full of subjectivity. Because of similarity in scale of observer and observed humans find it very difficult to ignore small differences. Seeing them we interpret them subjectively. If we are less personally involved, we can see the forest despite the trees. So generality is hard-won in data at the so-called soft end of disciplines, but deep insights come easily there.

Darwin predates postmodernism. His opinion is generally viewed as privileged by biologists, while in a postmodern world privilege is more fluid. In postmodernity, once Darwin has announced what he said, it is in the open and anyone can give privilege in any way they please. In a remarkable postmodern take on Darwin in 2009, Kayla Grove[5] created a collage to make an image of Darwin taken from the 1881 portrait by John Collier. The trick is that she used text from the *Origin of Species* printed in fonts of different size and compression. The print she used was computer generated. The original of Figure 4.1 was a collage where

A

FIGURE 4.1. Parts a-f show a postmodern image of Darwin, a collage of text of different sizes from *On the Origin of Species*. So important were the ideas of Darwin under the influence of Malthus that the *Origin* takes on an iconic form in itself. In postmodernism the original loses its privilege. The collage was handcrafted for a project in Allen's class by the artist Kayla Grove, at a scale of about 2 ft. by 1 ft. 6 inches. Pieces of text of different font sizes were applied at different angles. The image used as the template was the 1881 oil painting on canvas of John Collier now in the National Portrait Gallery. In a postmodern vein the collage shows a fire in the old man's eyes not present in the oil painting.

(*continued*)

FIGURE 4.1. *(continued)*

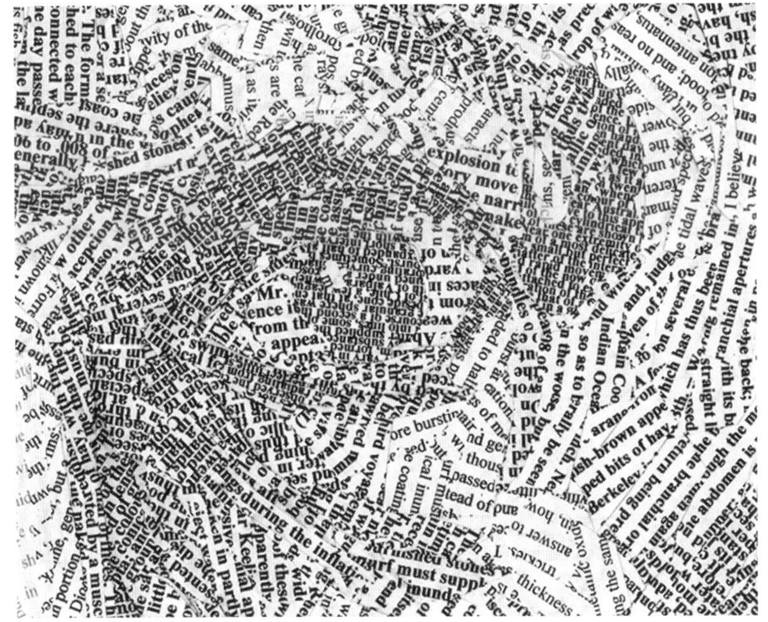

FIGURE 4.1. *(continued)*

FIGURE 4.1. (*continued*)

paper was affixed to create the image. Telling in this postmodern image is that Grove herself appears to be taken with Darwin, the great man, and his great creation, one of the most important books of all time. In her portrait the old man has a fire in his eyes that is not present in Collier's original. In Grove's image his ideas are not just what a great man wrote, they become immortal.

PART II

Origins of Life as a Complex Medium Number System

CHAPTER FIVE

A Wrinkle in Time

Evolution by Preadaptation

Nesting and Non-nesting

We have defined a hierarchy to be a system of communication where holons have an asymmetric relationship up the hierarchy as opposed to down. Often the asymmetry is in speed of cycling time. There slow behavior would be at the top while holons behaving successively faster occur lower in the hierarchy. The characteristic time for behavior of a holon is determined by the cycle time of endogenously driven behavior, also called the natural frequency.[1] More or less everything cycles, even things that appear stationary, like rocks. In the end almost all rocks are recycled into the Earth's mantle. It just takes many millions of years. Meanwhile unstable isotopes decay in very small fractions of a second. The endogenous cycle time is related to the time taken for a holon to return to its equilibrium behavior after being influenced by an external signal. This is sometimes called the relaxation time. We can see it in the familiar population biology example, the logistic growth equation. The equation is:

$$dN/dt = rN(1-N/K),$$

where N is the number of individuals in the population, K is the carrying capacity of the environment, and r is the growth rate term that determines the response time of the population to displacement from K. The expression 1/r is identified as a good approximation to the relaxation time by May.[2] The bigger is r, the lower is the population in a hierarchy of populations defined by equivalent equations. The bigger is r, the shorter is the relaxation time. The shorter the relaxation time the lower is the holon in a scaled hierarchy.

Ahl and Allen called scale-based hierarchies observational hierarchies. Observational hierarchies are ordered by spatio-temporal scaling of observed entities. The levels are scale-defined. But the same authors highlight an alternative sort of hierarchy ordered instead by interlocking definitions. Definitional holons are composed of levels of organization, which can be ordered in any way the definitions dictate. For instance, the concept of organism as a level of organization has a relation to the population level of organization not based on size but rather by definition of what it takes to be a given holon at a given level. Intuitively it would appear that population is larger scale than organism, but there is the inconvenient fact that some organisms are larger than many populations. Intuition appears mistaken. The distinctions between organism and population and their order are a matter of definition, not scale. The definition of a population subsumes the definition of organism, but only within the domain of the specific definition of population. The definition of a population is an aggregation of organisms but with a special proviso that the organisms are in some way equivalent. Non-equivalent organisms cannot form a population, by definition. One can change the definition, but not willy-nilly, and not without announcing that fact.

A collection of organisms with nothing in common is not a population. The host of a population of mites is not an organism equivalent to the mites living upon it. An aggregation of equivalent organisms is clearly by definition larger than the members of the aggregate, but only because the members are equivalent. The hook is equivalence. Organisms that are larger than the population simply do not meet the criterion of equivalence, and so do not meet the definition of what it takes to be part of the population. Yes, aggregations are larger than the things aggregated. That scaling rule is itself a shallow nested hierarchy addressing aggregates in general. But the relationship between organisms and populations is a separate hierarchy of a more specific sort. Some organisms in the vicinity are not part of the aggregation, and so the scaling rules of aggregation do not apply to them. Thus the host is a landscape for the mites living on it; for the mites the host is not an organism. Meanwhile an organism will by definition be smaller than any population to which it belongs. It is precisely because of subtle distinctions between definitions that biology has such rich terminology. Words like *predator*, *par-*

asite, *epiphyte*, and *commensal* are necessary so as to put flags on the potential for confusion between scaling and definition. Organism as a general principle is lower than population as a principle only by virtue of the specific definition of population, not because of the size of particular organisms or populations. One distinction is specific, while the other is general, and different rules apply to the specific as opposed to the general. Order by interlocking definitions is not the same as order by size. Order by definition gives not scalar hierarchies, but hierarchies of levels of organization. These are not to be confused with observational levels. Observational and definitional hierarchies need to be distinguished, a point that ought to be trivial. But the ubiquitous muddled presentations of ecological situations suggest such a distinction is generally ignored, and swept under the rug.

An example of blatant confusion arises in the hierarchy of life presented in many biology text books, so the error is commonplace.[3] The hierarchy of life generally goes from cell to organism, through population, ecological community, ecosystem, landscape, biome, and biosphere. The authors including the hierarchy of life seem to think that it is a general and overarching scheme; they are simply wrong on that. But the grand hierarchy of life is not wrong, it is merely not general. Most facets of the functioning of life do not work with those categories. The problem is that the hierarchy in textbooks mixes scalar and definitional criteria. It is not that such an ordering is wrong, it is simply not the general case that textbook authors think they are offering. It muddles scale with type. So if textbooks are universally slovenly about relating scale and type, and if they remain stubborn in the face of earlier statements about the problem, it is a general problem surrounding biological thinking.

The hierarchy of life applies only to particularly narrow situations; it not any sort of general condition. There is a general appearance of a scaled hierarchy, when in fact the levels are linked by interlocking definitions not relative scale. The concepts of population over organism, or ecosystem over biotic community, are not scalar. The different definitions demand different treatments that are not scale dependent or even scale-general. It is possible to consider an ecosystem as a community with its physical environment rolled in. That would indeed make for something more inclusive than just the collection of organisms in the community. But just as in the scalar muddle of the organism/population level relationship, many ecosystems are smaller than many communities. We cannot be talking of a scaled relationship in the general distinction between communities and ecosystems. Some ecosystems are so small that they can even be parts of organisms. For instance, a cow's rumen can be seen as an ecosystem. A rumen takes in and expels material and energy, and functions through flow and flux. There is some recycling. The idea of small ecosystems appears in Robert Bosserman's PhD dissertation where he reports how a small floating plant, the bladderwort, functions exactly on the normal criteria for ecosystems. It captures water fleas and digests them. That food ends up as nutrients that are released by the plant. Nutrients feed periphyton living on the bladderwort. Periphyton are a significant part of the bladderwort ecosystems in that they feed the water fleas.[4] The idea here is that the concept of ecosystem is not scale or in any way size dependent, which was an important new insight at the time.

Ecosystem and community are defined very differently, in ways that change an effective manner for addressing the respective types. Organisms in communities matter as individual units belonging to a species. In ecosystem function they work not as organisms in species per se, but as physical connecters between ecosystem compartments. In an ecosystem view, animals simply melt into pathways. Animals in animal communities do function as discrete organismal structures. Once the environment is folded in with the organisms in a community, the guiding principles turn from evolution to mass balance. Apt description switches from a set of biota in a place to become a series of relationships based on flux and connection. Evolution does not apply to organisms seen in an ecosystem, because the organizing principle in ecosystems is mass balance and the first law of thermodynamics (a conservation principle). As a rule some other species can often be substituted without affecting ecosystem function.[5] The mass balance essential to ecosystem calculations does not apply as an explanatory principle to evolution. An evolving population will have members that eat and drink in balance with how much they sweat, respire, urinate, and defecate, but that mass balance does not address the evolution that may be occurring. The bookkeeping in evolution is of animals largely as individual units that are vessels for genetic material. The bookkeeping in ecosystems is not of animals as units, but as crosswalks through which matter and energy flow. Communities are held together by processes that surround evolution. Community and ecosystem are types of biological entity, not entities that are distinguished on scale.

In ecosystems organisms as organisms are not generally the functional parts. True, one cannot have an ecosystem without biota. Also, under other observational criteria biota should be seen in organismal terms. Science as it practices does not deal with all of reality, all at once. As science focuses on ways of knowing, observing, and modeling, it narrows the investigation explicitly to sets of rules of functioning. Just because other rules may also apply under some other criterion, changing the rules in midstream in the scientific process causes a muddle. Changing the rules often changes the subject unwittingly. It says, "I am going to change the subject by shifting the level of analysis, without admitting I am doing that." Animals as discrete organisms change the rules from mass balance in a way that takes the ecosystem out of the discourse. There is baggage that comes with ecosystem research that has to ignore organisms as organisms to do its bookkeeping. In a grazing ecosystem, the identity of a cow as a cow is not the point; rather the cow connects green primary production with detritus. The cow melts away to become something that connects its mouth at one point in time and space to its anus at a later time, and probably in a different place. Cows as organisms is a different discussion from cows as connectors. We cannot discuss everything all at once, and if we try we simply create a shambles. Biology is often confused. The value of the conceptual precision in hierarchy theory is it avoids messy semantic argument or at least exposes what is and is not semantic. It is wise to speak only of relationships that fit the discussion. The value of the ecosystem concept is that it takes a point of view at a level of analysis and makes all that consistent. A chronic problem in biology is the new entrant to a discourse who says from the back of the lecture room, "But what about . . . ?" That question will often be outside the discourse as it introduces something simply unrelated. We can avoid confusion by including only pertinent points of view. Pertinence not verity is the object here. Otherwise we lose the *what* of what we are talking about.

Allen and Hoekstra organize their treatment of ecology by distinguishing between scale and type, between level of observation versus level of organization.[6] Scale sets a protocol for observation perhaps with the decision to use binoculars instead of a microscope. Scale also applies to the things once they are found; organism as a concept has no size but a particular observed elephant is indeed a certain size. Type is an entirely separate issue. Type identifies what is seen in the foreground. Before observation, type may indicate what sort of class member is being sought. After observation, type is the class into which the observer assigns that which is in the foreground. Types are simply classes that define what is equivalent across members of a class. Allen and Hoekstra find that there is a much clearer view of what is a particular type if scale is treated as a separate issue. In ecology, the types identify what are the parts that make the whole entity in the foreground. Type goes on to define the relationships between parts such that the whole emerges. Different types invite different ways of looking at entities. The various ways of seeing are associated with the different methodologies that define the practices of the different subdisciplines in ecology. Types in other disciplines define the different methods in, say, micro- versus macroeconomics, or family as opposed to individual therapy in clinical psychology.

Separate from the issue of levels of organization versus scalar levels is nested and non-nested hierarchies. Either type of hierarchy discussed above can be nested or non-nested. The notion of hierarchical arrangement is central to biology and even has an Aristotelean origin. The two standard ways of organizing biological systems are into taxonomic units (subspecies to families) and structural relationship (cell to organism), and both represent nested hierarchies. A nested hierarchy is one where the holon at the apex of the hierarchy contains and is composed of all lower holons. The apical holon consists of the sum of the substance and the interactions of all its daughter holons and is, in that sense, derivable from them. Put all the parts together and above the collection of the parts there will be emergent properties of the whole that are predictable from the parts, even in nested systems. Individuals may be seen as nested within populations, organs within organisms, tissues within organs, and tissues are composed of cells. In taxonomy a family consists of its constituent genera and their component species. These two hierarchical arrangements (physical and taxonomic) are ubiquitous in modern biology to such an extent that these arrangements are commonly used to order the material in introductory textbooks.

Biologists encounters this pair of hierarchies so frequently that they are wont to consider nested hierarchies as being more properly hierarchical than their non-nested counterparts. Set against that convention, our more general approach to hierarchies says it is profitable to view the nested hierarchical condition as only a special and restricted case rather than a perfect hierarchy. Nested hierarchies meet all the criteria of the more general hierarchical condition. The higher holons in the system are associated with slower time constants of behavior (i.e., longer cycling times and relaxation times), and, if manifested in space, they occupy a larger volume. Higher holons in a nested system constrain the behavior of the lower holons to be within the higher holons; where in the world a heart shall beat is determined by the movement of the whole organism. Non-nested hierarchies relax the requirement for containment of lower by higher holons and also do not insist that higher holons are derivable from collected lower holons. An example of a non-nested hierarchy might be a food chain or a pecking order. There is constraint between levels in a pecking order but the top dog does not contain or consist of the others. In a non-nested hierarchy the criteria for moving between levels generally remains the same, such as upper level eats lower level or upper level bosses the lower level. There is an essential structure to a food chain and eating creates that structure.

Nested hierarchies are very useful in the exploratory phase of an investigation because if criteria change, the nesting keeps order. For instance, cell surfaces cohere to make a tissue, but organs relate to each other often through hydrostatic relationships, as in the heart and the kidneys. Heart failure commonly is associated with kidney failure. Meanwhile the whole human organism might relate to another in some social way that has nothing to do with hydrostatic connections. But in all this variety of connections, the nesting keeps things straight. It would be possible to change criteria in a non-nested system, but that mixes scale and significance and invites confusion. Non-nesting is an opportunity to expose a unified set of relationships between many levels, and so keeping just one way of connecting levels top to bottom makes sense. Non-nesting is a more mature part of an investigation that occurs once general principles are understood.

With scale-based relationships out of the way, non-nested systems can be reversed or reorganized in other ways depending on the criteria for relationship. There is a political correctness that objects to power hierarchies, and that objection can lead to a premature rejection of the whole notion of hierarchy. Power relationships are generally non-nested, and it is easy to reverse them. For instance, we can reverse a politically incorrect "Rules and enforces" criterion with a criterion of "depends." The ruling classes very much depend on the rank and file working material things. This was the hierarchy that Gandhi used against British rule over India. A few hundred thousand British could not rule without compliance of many of the Indian populace. Once that compliance was withdrawn, India could not be governed by the British. The workers are dignified by work. At that point the peasant becomes the top of the hierarchy. In a food chain a switch from "eats" to "depends on" puts the wolf lower in the hierarchy than the caribou prey. That switch also lines up with periodicity of behavior. An aerial view would see the wolf pack as a satellite of the migrating herd. In a study of Isle Royal wolf and caribou, the researchers were interested in flux of sodium, whereupon the wolves and the caribou appear as sister holons at the same level.[7]

Koestler's example of a hierarchical military command is a useful non-nested example here. It is particularly illustrative because a nested equivalent may be developed with ease. The general in command of an army is the apex to a non-nested hierarchy in that the person at the top does not consist of, nor contain the lower levels in the army. The general is not the sum of the soldiers. The material nested hierarchy is the army in a place consisting of all its soldiers, including the general. The leader in that nested view is just another soldier. At that point the army consists of and contains all the soldiers. The stability of nested hierarchies is nicely illustrated by the army in the way that the general may not be the aggregate embodiment of all the soldiers lower in the command, but the command structure is nested. There is a rigid chain of command. The general's command is the entire army. The commands of the officers are variously parts of the whole army. At the bottom the platoon leader has a command that has nested in it the platoon. This is a nice example of how the form of a hierarchy depends on the purpose of the hierarchical conception, and the hierarchy changes dramatically with a change of conception.

Nesting does imply a certain determinability of the top holons from the lower level, since the top holon can be created by a simple aggregation of the parts. Even so there will be emergence of properties of the whole that cannot be determined, in that we cannot go into water and pick out a wet molecule. Wetness is a property of the nested whole. Pattee referred to the emergence in nested wholes when he said:

> In fact, if we should actually achieve a microscopic rate-equation description of the measurement constraints for the system we are "explaining", we would find that not only the measurement but the system we originally had in mind would disappear, only to be replaced by a new system with an immense number of new initial conditions requiring new measurements. One can recognize this process as related to what we call reductionist explanation where one has explained away the original system. The essence of the measurement problem at the quantum mechanical level is that reductionism apparently "explains away" the measurement itself. (Pattee 1979)

There is an inescapable presence of the observer that is seen particularly clearly in non-nested hierarchies. In every measurement there is present a choosing to make the measurement in the first place, and in that contextual choosing the observer defines the form of the non-nested hierarchy.

In the non-nested condition, the behavior of the highest holon, say a general, is not derivable from even the most complete account of the lower levels. In non-nested hierarchies emergent properties must remain uncertain until the system is actually let run for a while. The nested hierarchical model is the one which generally facilitates a reductionist model. Holistic science, on the other hand, can always operate perfectly well with the non-nested condition. In the non-nested case the individual holons are taken at face value as quasi-independent wholes that are part of a hierarchical system of communication. Conventional thermodynamics requires systems to be nested, otherwise bookkeeping becomes impossible. The incorporation of conservation principles in the construction of complex models, such as those for ecosystems, also tends to make the system nested. This allows large ecosystem models to be built, as well as that task can be performed, in the reductionist mode. Conservation principles are often reasonable, but provide simple cast-iron constraints for the whole model. If, however, the system to be modeled has subtle self-organizing forces, the conservation constraints do not allow the emergence of subtlety in the model. Notice ecosystems depend on conservation of matter and energy as one of their organizing principles as they demonstrate flux and mass balance. Meanwhile biotic communities depend on open notions like competition and evolution, not conservation principles.

Emergence

Higher levels commonly arise because of a process of emergence. Emergence is a favorite property of complex systems theorists to the extent that it sometimes becomes mystical. There is no magic in emergence. Emergence takes several forms, some of which can be disappointing if one wants emergence to be something special. The simplest aspect of emergence arises when the observer looks at things in a different way. Notice immediately our reluctance to give privilege to material changes, but instead our preference for emergence as a consequence of how the observer observes. The size of the universe of discourse affects emergence too. For instance, chemistry cannot be part of the discourse if the universe of the discussion is only the width of a proton. The realist might insist that there is still real chemistry going on, even if it cannot be seen in such a narrow universe; but that is our point—emergence is a matter of how we address the observed situation. The proton example is emergence defined in spatial extent. Emergence over time appears closer to a material change, but even that depends on the observer recognizing the shift. For instance, a whirlpool in a tub might be simply dismissed as flux down a drain. However, one can recognize the appearance of the gyre and identify that it presents the emergence of an unusual vertical water air interface. One can see that such a peculiar arrangement may not be present and then emerge. But there is a parallel to emergence by widening spatial extent in that one simply widens the time horizon until the emergence manifests itself. Chemistry is always a potentiality even when it cannot be seen in a universe so small as to manifest only protons. In the same way, the potentiality of the whirlpool is always there. The emergence of a whirlpool simply needs the presence of a strong gradient of water flow to set positive feedbacks going in place. The feedbacks can drive harder until they press up against the negative feedback that is the structure of the whirlpool. Thus emergence is not simply the appearance of a physically new situation, it demands scaling issues and the process of recognition. With the emergence of animal husbandry one could look at slaughterhouses as simply a rather unsporting and very efficient hunting, but one would be missing the point of agricultural emergence. We wish to disabuse those who become mystical and whose eyes glaze over in the face of emergence; no, emergence is not that special, and is what one might expect anyway. Look at something differently and it will appear to be different.

Complementarity

Non-nested hierarchies are non-determinable devices that go beyond nested hierarchies. As we have suggested, there are ways in which properties of the higher levels in even nested systems are not reducible to the sum of the properties of the lower levels of organization. Helpful here is Pattee's distinction between laws and rules. Pattee points out that the rate-independent dynamics of the lower-levels of organization are constrained by rate-independent rules at the higher constraining level. The rules emerge from the system to which they apply:

> The basic distinction between laws and rules can be made by these criteria: laws are a) inexorable, b) incorporeal, and c) universal; rules are a) arbitrary, b) structure-dependent, c) local. In other words, we can never alter or evade laws of nature; we can always evade and change rules. Laws of nature do not need embodiments of structures to execute them; rules must have a real physical structure or constraint if they are to be executed. (Pattee 1978)

Much in the way that we admit to the arbitrariness of levels but nevertheless find them conceptually useful, Pattee makes much of the distinction between rate-dependence and rate-independence while admitting that rate-dependence is a matter of degree. Parenthetically, nestedness is also a matter of degree.

> This sharp conceptual distinction I have made between rate-dependent and rate-independent processes can, of course, be challenged on the same grounds that the distinction between reversible and irreversible phenomena are challenged. . . . Reversibility and irreversibility are not sharp distinctions, but only a matter of degree, the microscopic reversible description being the more objective (i.e., obeying the laws of motion), and the irreversible description is more subjective (i.e., depending upon the observer's information about the system). (Pattee 1978)

Thus it is possible to gloss over the rule-law distinction and reach some sort of compromise in order to preserve a unified model for complex behavior. There are, however, advantages to avoiding compromise and in heightening the paradox so that the problem is clearly defined. This is a common stratagem in physics, although biologists seem to prefer unity at all costs. With paradox in hand it is easy to identify that it is in the desire for a unified model that the conflict occurs. The unified model is a paradigm, a narrative. Narratives do not have to be consistent. We gain much by rejecting unity and embracing complementarity as a modeling paradigm.

The complementarity principle is quite central to our thesis and should perhaps have been introduced at the very outset so that its importance could have been seen throughout. The reason we have held it back until now is that it is a principle quite out of step with most biological practice and needs a concrete set of biological problems for its exposition in biological terms. As an explanatory tool it is much more aggressive than a mere tolerance of alternative models. The principle of complementarity drives paradox by recognizing that formal incompatibilities are necessary between dual modes of description, both of which are required for a complete account of phenomena. For reasons that will become apparent, the two modes of description are called the dynamic and the linguistic modes of description by Pattee (1978). The requirement for both modes of description should not, however, be seen as a duality in the system itself under investigation; the duality is a matter of observation and conception. The system under observation may be unified, but the requirements for description are dual. The incompatibility of the two modes is not a contradiction, for the modes are formally separate; contradiction can occur only inside a formally unified system of description. Koestler' s duality in the holon, where it is both whole and part at the same time, does not lead to contradiction, for although wholeness and partness are contradictory in a unified description of a unified system, they are disjunct descriptions of the system when set in the complementarity paradigm. The same formal separation is true of boundary conditions (constraint) at a given hierarchical level and the functioning of entities at that level.

Although they do not generally emphasize the incompatibility of the two modes of description (however, see Levins and Lewontin 1980), contributors to the ecological literature did for a while use complementary descriptions. Part/whole dualities are the cornerstone of the systems approach to the concept of niche proposed by Patten.[8] Haefner in his "Two Metaphors of the Niche"[9] actually uses the term "linguistic" for the alternative description of the niche he proposes. Hutchinson[10] cast the niche as a hypervolume with each dimension corresponding to a resource for the species in question. Haefner contrasts the geometric and linguistic descriptions of the niche, respectively, in the simile "a house and a set of instructions for constructing a house."[11] In Ulanowicz's[12] terms, house and instructions for building are two of the four Aristotelian causes (he extends the house building instance to all four causes). As Haefner puts it, the geometric representation

> is most adept at describing community dynamics and stability, niche overlap, and organism response to abiotic environments. The linguistic metaphor represents well organism activities, niche dynamics, and ecosystem assembly.

In a similar use of dual modes of description, Southwood[13] factors pattern from process. "Thus the whole concept of evolution may be seen to have two components: the patterns (of taxonomy, classification and variation, and of biogeography) and the mechanism, Darwin's theory." Southwood's taxonomic patterns are rate-independent and linguistic. His mechanism, Darwin's theory, is rate-dependent. This chapter amplifies Southwood's statement.

As we introduce complementarity here, it may be mistaken as merely the use of helpful supplementary multiple description. We do in fact mean complementarity (not "supplementarity"), and by the end of the chapter we expect to have forced the necessary incompatibilities to the forefront of our argument. The rate-independent rules of constraint are not reducible to the law-dependent dynamics of lower-level functioning. This irreducibility applies as much to nested as to non-nested systems and is responsible for a type of emergence, even in nested systems, which does not derive from incomplete specification. If you can see the dam, you cannot see the river. If you can see the flux of the river, you cannot see the dam. Both pertain, but under separate observation schemes. A wider view could see both, but the process of the river will be lost, as the river appears as a static ribbon.

> It will appear strange to some that complementarity should arise at the atomic level [quantum mechanics], apparently disappear at the classical macroscopic level, and then reappear at the biological and sociological levels of organization. . . . Why are macrophysical systems the only ones that are exempt from functional behavior? I believe the reason lies in the fact that macrophysical systems are the only level where the role of the observer, or the subject in the subject-object duality, can be suppressed and we still obtain reasonable predictions. . . . Because celestial objects are exceptionally well isolated from the control of observers, this suppression succeeded only too well, and now is applied everywhere—even to the extreme of trying to explain cognition and volition by unified, mechanistic models. (Pattee 1978)

The reappearance of complementarity in biological and social systems may also be seen as a consequence of the importance of codes and symbolic representation in the functioning of biological structures. Recognition of the role of biological language, not just of humans but of biological structures in general, is crucial. It prevents the biologist from suppressing the arbitrariness which derives from the heavy rule-dependence of the study material.

> It is a universal property of language (and hence, all description) that the structure of symbol vehicles or signs (i.e. the letters of the alphabet, nucleotides, words, codons, etc.) are related to their referent or their effect by arbitrary rules. These rules are not derived from, or reducible to, the laws of nature. They are perhaps best described as frozen historical accidents—accidents because their antecedent events are unobservable, historical because the crucial events occurred only once and frozen because the result persists as a coherent, stable, hereditary constraint. (Pattee 1978)

These rules are not derivable, even in nested systems, from the law-dependent dynamics of the constrained structure.

> The description of the physical state of the DNA, even to the details of quantum mechanics, would give no more clue to the meaning of that piece of code than the chemistry of this ink would give a clue to the meaning of my words. (Pattee 1978)

The counter-contention, with which we vehemently disagree, would be that the laws of nature are sufficient to explain what we see and that even language systems would yield to an appropriate reduction. Such a reduction not only would have to identify the microdynamics of the system in which occurred the frozen historical accident that set the code, but also would have to find a dynamical explanation for the significance of the code. Such a reduction would run into the problems of explaining away the observation. Pattee (1978) refers to explaining away observations. It is the intrusion of the observer, through understanding of system symbolism, that gives rise to irreducibility of emergent properties of both nested and non-nested systems.

Thus in complex systems the conceptual structure of the observer is important and inescapable. There is a need for not only rate-dependent descriptions of system dynamics, but also complementary rate-independent descriptions of system constraints. In order to amplify this point, Pattee (1978) uses the example of American football, where the individual play involving the movement and coordination of players is strikingly rate-dependent. The critical distinction here is between the parts of the game where there is actual play, as opposed to the huddle, where the plan for the next play is announced to the offensive line players so they know what to expect. The one who passes the football does so at a rate, which to be successful must be compatible with the rate of movement of the receiver of the pass. Nevertheless, so long as the offensive team does not illegally delay the game by taking too long between plays, the rate with which policy is made in the huddle in the times between plays is immaterial. Thus the constraint imposed on subsequent game dynamics is rate-independent. A team in the lead toward the end of the game will generally have longer huddles to run out the clock on the game. A team behind at that time in the game will generally dispense with the huddle altogether, to save time on the clock.

> It is also clear that one team's choice of policy is based on both its model of the game dynamics and its model of the other team's policy, i.e. complementary models are a necessary condition for rational action. (Pattee 1978)

A difficulty commonly experienced by the holistic scientist is that of showing the utility of his point of view in the face of the abundant success of the presently dominant mode in biology, mechanistic reductionism. Pattee (1979) concedes:

> There is no doubt that reductionistic explanation is often needed, especially when the principles of the microscopic system form a more practical or elegant explanation, say in terms of values such as universality, simplicity, and coherence, ease of computation, etc.

We would point out that reduction is performed by both reductionists and holists. It is just that the holist spends a lot of time working out what to reduce and how far, with most of the investigation involving looking for patterns at the macrolevel. Once you have worked out what to do, the problem is almost solved, and self-evidently so too. Mechanistic models are undeniably powerful when the parts of the system to be modeled may be connected together in a fairly simple serial fashion. Such is their success that there is every temptation to reify mechanism and to assert that mechanism as a modeling strategy works because the world is, in an ontological sense, mechanistic. The verity of such assertions is unknowable and therefore falls outside a scientific discourse.

A lever is a mechanism where force is applied relative to some fulcrum. Let us take it apart to expose the essence of mechanism (Figure 5.1).

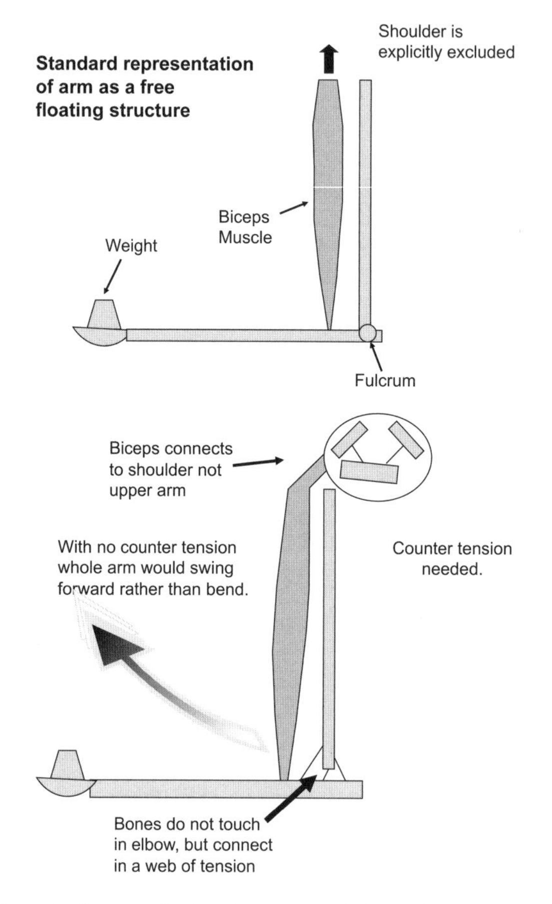

FIGURE 5.1. a) The orthodox arm displayed as a lever. Orthopedic images always remove the shoulder so it cannot complicate the concept of a lever. b) Simple arms with bicep pulling in the absence of the triceps, moving the arm forward and upward, not bending. c) Arm with triceps counter pulling, all connected to the tripartite shoulder. Note the two frictionless joints (like 2 lubricated ball bearings) between the ends of the bicep. This cannot be a lever, but is rather a tension compression structure. Bones do not touch. (Images similar to panels a and c appear in Allen and Hoekstra's *Toward a Unified Ecology*, 2015, Columbia University Press; reproduced with permission.)

Orthopedic surgeons generally use this model for, say, the arm, even though it is demonstrably false to the point that it is misleading.[14] Discussions of the arm regularly exclude the shoulder. Mechanistic models of levers have to do that because the shoulder brings with it impossible complications to the mechanism. The bicep does not attach to the upper arm, it attaches to the shoulder. So between the points of attachment are two,

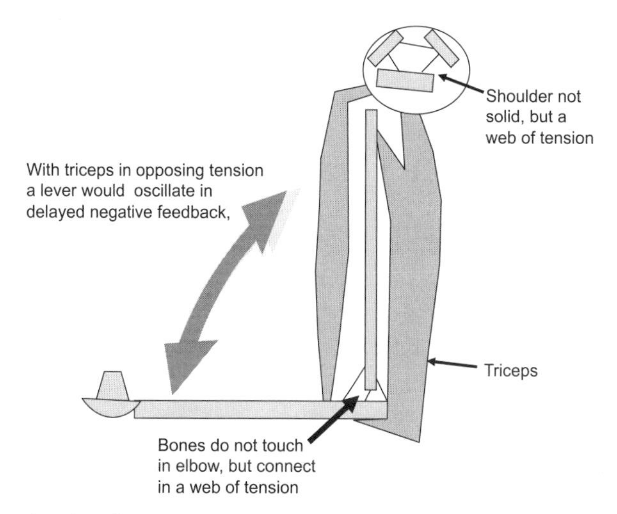
Shoulder not
solid, but a
web of tension
With triceps in opposing tension
a lever would oscillate in
delayed negative feedback,
Triceps
Bones do not touch
in elbow, but connect
in a web of tension

FIGURE 5.1. *(continued)*

not one, frictionless joints. The friction in the elbow is less than in a well-lubricated ball bearing. So the bicep contraction should cause the arm to pivot at the shoulder moving the whole arm in a rude gesture, not bending the arm. To achieve bending at the elbow, there must be the triceps. That attaches also to the upper arm, but also to the shoulder. The shoulder has its own complications in that it is three bones, again with frictionless joints between them. Archimedes said he could move the world, with a big enough lever, if he had a place to stand. Excluding the shoulder from the arm to make it a lever denies him his place to stand. In the end all muscles pull against the feet on the ground. The arm is not a lever, it is part of a web of tension in which float the compression members, the bones. When you fall on your heel from a height, your heel does not break, your fourth lumbar vertebra does. It seems to be the weakest part of a tension compression system. Stacking bones as a set of levers makes the body gravity dependent. Since you can stand on your hands, it is independent of gravity, as are tension compression structures. If the skeletal and muscular system were levers, calculations show that muscles and tendons would tear out, and bones would be crushed, and there is not enough energy in the body to make it work.

So to generalize this discussion to capture and critique mechanism, note that removing the shoulder from calculations on the arm narrows the universe. Any nonlinear system can be linearized by narrowing the discourse. Curved lines on graphs can be made straight by only looking at a short enough segment. The close-to-serial patterns in mechanisms come from narrowing the scope to get rid of nonlinearity. Removing the shoulder allows the effects of the bicep on the arm to appear as a linear structure, a lever. But there are the objections above to that scenario. So the arm as a lever is achieved by narrowing the scope to linearize the system. That means that mechanism is a linear approximation. Linear approximations have their use, but it makes no sense to assert they are ontologically real. So the search for the real mechanism makes no more sense than the search for the real linear approximation, an oxymoron.

Perhaps the main body of predictive biology is readily couched in mechanistic terms, precisely because mechanisms are all that are sought. If there are phenomena that are not susceptible to mechanistic explanation, then the prevalent investigative strategy will not find them. This book is in part a plea for a suspension of judgment on the necessary superiority of mechanism as a model, so that alternative searches may be made. Perhaps it is seriality that not only makes some systems amenable to mechanistic explanation, but also confers on them a degree of stability. Systems break down in the face of strong nonlinearity. Nonlinear variables in an extended range can come to increase or decrease so steeply that the values reported become functionally infinite or zero. At that point what you were discussing becomes undefined, so there is nothing to discuss. That is what happens when a change in level of analysis is forced on the scientist. Stability in a system is conducive to description of the system, and so many descriptions are mechanisms. Stable systems tend to persist and so are more readily observed than unstable systems. But stability, it should be remembered, is not an objective criterion. Therefore it could be mistaken to view the wealth of phenomena that yield to mechanistic explanation as evidence for the ontological verity of a clockwork universe. On the other hand there is, in all likelihood, more to mechanism than its aesthetic appeal or social acceptability in this machine-reliant society.

Certainly a hierarchical approach has something to say about evolution, and a consideration of origins comprises the rest of the chapter. There will be a discussion of the importance of stable holons, and evolutionary advance will be seen to revolve around them. Without a disclaimer at this point in the discussion, a mistaken impression might be given that stable is somehow more real. We hold to our position that hierarchies are conceptual and perceptual, leaving the nature of observer-independent reality out of the argument. Although holons at some levels may have more-general properties than others at other levels, hierarchies are less dependent on arbitrary observer decisions if they are conceived as vertically continuous instead of being seen as composed only of disjunct levels. It is, however, the more general and apparently more discrete levels, the levels composed of commonplace "things" with their attendant stability, that speed evolution on its way. The role of apparent discreteness and its surfaces in the context of stability awaits extended discussion in the following chapter.

Processes of Intensification

Polanyi[15] suggests that explanation of living phenomena cannot always be reduced to physical and chemical laws because

> a boundary condition is always extraneous to the process which it delimits. . . . If we accept, as I do, the view that living beings form a hierarchy in which each higher level represents a distinctive principle that harnesses the level below it (while being itself irreducible to its lower principles), then the evolutionary sequence gains new and deeper significance. We can recognize then a strictly defined progression, rising from the inanimate level to ever higher additional principles of life. That is not to say that the higher levels of life are altogether absent in earlier stages of evolution. They may be present in traces long before they become prominent. Evolution may be seen, then, as a progressive intensification of the higher principles of life.

To continue with Herbert Simon[16]:

> Direction is provided to the scheme by the stability of the complex forms, once these come into existence. But this is nothing more than survival of the fittest—i.e., of the stable. . . . Atoms of high atomic weight and complex inorganic molecules are witnesses to the fact that the evolution of complexity does not imply self-reproduction. . . . Among possible complex forms, hierarchies are the ones that have time to evolve.

Note that Simon talks only of stable systems, not finding it necessary to require that they be living. It is, then, with these thoughts that we begin a discussion of the evolution of the peculiar characteristics of living systems. We conclude in concurrence with Simon and Polanyi that the abiotic world contains the rudiments of all that is necessary.

Life Origins

Seeing laws in physics, biologists ask whether there are laws in biology that are in some sense equivalent, and that can be used to tidy up biology. Dodds[17] is one of the latest attempts. His approach to spontaneous generation is instructive as to the general issue central to this book. On the face of it, it would seem he is justified in making a law out of Pasteur's challenge to spontaneous generation. We are all familiar with the critical experiment in 1864 where Pasteur showed that sterilized broth in a sealed container would not spoil. He also used swan-necked open vessels where curve of the neck did not allow bacteria to fall into the broth. That too remained unspoiled. The broth became infected only when Pasteur rotated the vessel so the curve of the neck allowed bacteria to fall in. Life comes from life, and is not spontaneously generated. It seems like a universal law in biology.

But then Dodds goes on to note what he sees as an exception: the origins of life. And on the face of it, that exception does not deny his law. But that is the problem with laws, particularly in biology. Things get befuddled and arbitrary when the point of view and scale are changed. The original law pertains to very local and recent conditions, relative to the billions of years needed for a window for considering life origins. The origin of life is not an exception, because it is a massively differently scaled phenomenon. Laws appear inviolable only when the scale of the extent is carefully stated. In physics practitioners can get away with ignoring history and narrative. That is how they come to use time zero so often, when biology is hard pressed to find a time that is not encumbered by a history before that time zero. History makes a mess of laws in biology, but not generally in physics.

Over the extended time when it formed first, life did fidget itself into existence, edging into actual viability. So here is the challenge. If it happened then, why does it not happen now? And the painful answer is it does happen now, and more so because there are more of the building blocks around. So why do we not see it? That is because life as a cyclical process of generation is the first full-flowered such cycle of that sort on this planet. Manfred Eigen[18] states a law that holds up pretty well. That is that the first autocatalytic cycle to establish goes on to hold sway indefinitely. That is how come all of life uses ATP as its ready fuel. ATP is one of Pattee's frozen accidents; it got in place first for no reason beyond some local thermodynamic accident. GTP substitutes guanine (G) for adenine (A) for the energetics of moving genetic material as it is read in ribosomes, so ATP did not win hands down, but it took most of the field in that battle. Perhaps the energetics of genetics has some local wrinkle where there is so much guanine around in genetic material that the thermodynamic accident for GTP was the one to win first, if only locally in protein synthesis. Perhaps GTP in moving RNA is an indication that genetics and cell division have a separate origin from thermodynamic parts of physiology. There may be two organisms involved, where now we see only one.

So here is life in place. As the processes that created life de novo continue to work now in the present, the tragedy is that full blown life comes along and eats it. So it depends on how far along the road to life we want to go before we want to call it spontaneous generation. And in fact, the spontaneous generation of life is probably moving along much faster in the contemporary world than it did the first time. The raw materials to make life are far more abundant now than they were four billion years ago. The trouble is that contemporary life sets up road blocks along the path to life by stealing the first products of the emergence. History matters in biology in a way that it does not so much in physics.

There is always the problem for the student of biogenesis that the only material to be seen in the flesh, as it were, is the phylogenetically finished product. The scaling and hierarchical implications of enzymes, cells, and genetic systems are very worthy of consideration, and soon we give them our attention; but they should wait upon an identification of the characteristics of a scale that set the first life apart from its dead matrix. Clearly precursors of vital characteristics must have existed and so probably still exist in chemo-physical systems, at least until life consumes them. We therefore briefly turn our attention first to the development of a physical system to see if there is anything there to be generalized to organic phylogeny.

Early in a star's development the influence of gravity within the cloud of diffuse gases would not be measurable by any discrete event. To adapt Schrödinger's (1967) description of Brownian movement, the gas molecules "are thus knocked about and can only on the average follow the influence of gravity." The probability that any given particle would move toward rather than away from the weak center of the forming body would be almost exactly one-half. Nevertheless, somewhere out in the twentieth, fiftieth, or hundredth decimal place would be a small figure that could just tip the balance in favor of movement to the center. The significant feature here is not that particular movement, but rather the positive feedback of which it is part. As the particles gradually concentrate, their combined gravitational force becomes more powerful, so enhancing the ability to collect more particles.

So it is at the origins of life. The evolution of something as tangible as the first gene, as we would know it, is well along the road; it is analogous to the flash of the emerging stellar properties at the core of the gas cloud. At the very beginning of life we must look for a system as loose as the concentrating cold gas cloud. All that is necessary is an exceedingly weak positive feedback loop. Two organic compounds, the presence of each enhancing the survival characteristics of the other, is quite sufficient at the outset. There were probably many such pairs or larger groups, and so our starting point is arbitrary and imaginary. The significant feature of such systems of positive feedback is the innate propensity for survival of their components. These are "the stable" in Simon's model for the evolution of abiotic complexity. They begin the evolution of hierarchies that are to become life. They are the bricks at the beginning of Polanyi's "progressive intensification of the higher principles of life."

Evolution in Hierarchies

Survival of the stable. In Chicago was a watchmaker, or rather watch-mender, who did very nicely as the proprietor of a jewelry store. He had not, however, always been self-employed, and at one time his services were courted by watch repair shops all over town. His standing among watch-menders was attributable to his being able to mend twenty-five watches on a good day when his most skillful and diligent colleagues could mend seven, and then only by skipping lunch hour, something that Mr. Cittadino would never have to do.

Mr. Cittadino's secret was in a strategy that he developed for dealing with broken watches. The classic school of watch-mending came to Chicago from Germany. It teaches the mending of one watch at a time.

Mr. Cittadino's method was to deal with the watches as a group whenever possible. The first step was to wash them in bulk. That fixed a certain percentage of the watches right here. As the process continued, families of problems were successively solved until the most intransigent cases were left.[19]

In his article at the foundation of hierarchy theory, "The Architecture of Complexity," Simon tells a parallel story of watchmakers. It would seem from Mr. Cittadino's success that Herbert Simon's story of the two watchmakers has a parallel in the nonfictional world of watch-menders. The story has been cited by Koestler and others before, but it makes such a convincing case for the robustness of hierarchies that we tell it again.

There are two watchmakers, Hora and Tempus, whose watches consist of one thousand parts. Tempus' watch is not hierarchical in its construction; should the watchmaker be forced to put his work aside before the last piece is inserted, then the incipient watch disintegrates to the one thousand original components. Hora's watch, however, is hierarchical. The first stage in its construction is to put together ten primary components. Once this is achieved, those ten components as a unit may be set aside while the watchmaker moves on to construct the second ten-part unit. When he has constructed the one hundred ten-part units, he moves on to the next stage. That involves combining ten units of ten parts to make a single hundred-part unit. Having achieved this he puts the hundred-piece unit down and moves to construct another of similar size and so on. Once this is achieved, all ten of the hundred-part units are combined to build one watch.

In the absence of disruption and disturbance either mode of construction works perfectly well and can be completed in about the same amount of time. But the working world is never without distraction and disruption, and the consequences for the nonhierarchical watchmaker are clear. With each interruption Tempus must start again with one thousand separate pieces, a long way from his final goal. At any time, however, the hierarchical watchmaker, Hora, can lose only the work he has put into the ten-entity construction upon which he is presently working. All those ten-entity units completed up to that point are safe and remain intact as do all lower levels of construction.

So it is with the evolution of living systems. Should multicellularity fail, life need not return to the primeval soup, for it can drop back to unicellularity and fight for multicellular existence another day. The example of multicellularity is familiar, but the same would apply to the evolution of life through all its stages. The extinction of photosynthetic forms would only give a temporary return to heterotrophism. The losses due to failure through misadventure are minimized, shortening the time needed for evolution to higher forms to the time at hand.

Natural selection. Evolution by natural selection is a hierarchical model, although it is not usually explicitly identified as such. Some thermodynamic models may also be viewed hierarchically and, as such, can be seen as some sort of inverse of evolution. Layzer[20] uses the image of an open bottle of perfume in a box chamber. As time passes the highly concentrated perfume diffuses out of the bottle, so raising the level of entropy in the box. A hierarchical account of this process notes that the macrostructure of perfume in the bottle is lost. The information, "perfume in the bottle," is transferred to the molecules of perfume and the correlations of their velocities. If all the motions of all the molecules were exactly reversed, the liquid perfume would be reconstituted in the bottle because action and reaction are symmetric. In the thermodynamic process played out in the usual direction, order can be seen to become finer grained as time passes. In evolution, however, events at the lower level have their significance concentrated at the higher level with the passage of time. Organisms of favored structure saturate the next higher level, the population, so that it develops new ordered characteristics. Order runs down through the levels of complex thermodynamic systems, but order runs up evolving hierarchies. Bioaccumulation of fat-soluble poisons in a food chain works like evolution. Organisms scavenge for fat, and so fat in successive acts of eating concentrate the poison in the higher members of the food chain.

Fractal patterns derive from some unifying process that propagates upscale. Fractals are often seen as correspondence and repetition downscale (Figure 3.2). The pattern derived from some fractal equation may be dissected in finer detail with the same general pattern persisting at lower levels of structure. Equally fractal patterns emerge by low level processes that amplify upscale. A line on the ground, in a fractal pattern, forms when the low level process can propagate upscale no more. Then some new process, with a different degree of complication, takes over. The line on the ground marks the transition between causes. A concrete example of that tension between upscale and down, and rate-dependence and rate-independence, is found in the notion of turbulence.

As a moving fluid encounters an obstruction it must take a track that bends around the obstacle. The bent track must therefore go further than that part of the fluid that passes straight and unimpeded to either side of the obstruction. Since all tracks meet on the far side of the obstruction, the longer track must go faster to catch up, and so bending translates to moving faster. There is therefore sheering stress in the fluid as that part going the long way round cannot quite keep up with the linear flow of adjacent fluid. That sheering stress causes wrinkles in the fluid. Increase the speed of the flow and the wrinkles become circulating gyres characteristic of turbulent systems. The turbulent gyres dissipate the stress, passing it down to smaller gyres that hang off the big ones. The pattern repeats in fractal pattern until the last of the stress is dissipated in friction between molecules of the fluid. If turbulence is started a second time under what are apparently identical conditions, the pattern of turbulence will turn out differently. The general pattern of turbulence will stay the same with all else equal, but it will not be identical in detail. Turbulence expressed as differential rates in these terms is a medium number condition.

Even so there are regularities in the gyres. They are always evenly spaced and counter rotational to their nearest neighbor. To explain this regular and predictable pattern we need a different device that cascades upscale not down. As the turbulence starts with many wrinkles in the fluid, those that are close together will share the shearing stress in the vicinity. With sharing of resource the wrinkles get smaller not bigger and eventually disappear. That shared stress then becomes available to more distant but still adjacent wrinkles. With that extra stress, those wrinkles get bigger and start to form gyres. The regularity of the gyres is a pattern formed by competition for sheering stress. Competition usually leads to even separation in space as closer structures are removed. Gyres that are con-rotational rub each other up the wrong way. That friction causes the gyres to slow and disappear. By contrast counter rotational gyres merely kiss each other, and can persist next to their neighbor. Darwin liked Spencer's phrase "survival of the fittest."[21] But it is not survival of the healthiest, it is survival of the "fittest in-est." Those gyres that fit into their environment best survive. The evenness and counter rotation of fully formed turbulence arises from natural selection of structures.

That pattern is opposite to the whole pattern of turbulence, which dynamically dissipates stress. While removal of stress goes downscale, the emergence of larger gyres expresses a relationship that moves upscale. Also the whole turbulent pattern is dynamical and emergent. It comes from the physicality of the gradient that forms the flow. Meanwhile the emerging gyres are not so much seen as dynamical, their circular motion notwithstanding, but are identifiable as separate structures. Evolution selects structures.

As a concession to realists we admit that there probably are things in the external world, but they do not exist there as things; what makes a thing a thing comes from the observer. All structures come from a decision on the part of the observer to recognize this as discrete from that. Structures do correspond to things happening in the observed scheme of things, but their identity as structures comes from the observer taking one last step. It is a choice to freeze what is merely changing slowly, freezing it into some sort of constant. Evolution works on structures that can be taken as constant. While the process of losing in a selection process may be continuous, the final selection of one thing over another creates a singularity that cleaves selection away from the continuity of the process of winning or losing. The distinction of losing arises when epsilon becomes zero, and a known quantity becomes an undefinable number, zero. The relative constancy of the genes means organisms are largely stuck with the genome with which they were endowed at birth (epigenetics is where lies the exception). For evolution to generate structure, it must be able to separate one structure from another. It must be able to select out some structures, while others are left to persist.

Simon suggests evolution as an effective model is not restricted to living systems. If something like evolution occurs in social structures, then it seems unnecessarily restrictive to insist that in biology only the individual is selected. If a population as a whole within a group of populations survives when others disappear then it is the group of populations, not the single population, which evolves. This point is not blunted by the fact that selection, with consequent changes in biological characters, may be simultaneously occurring at other levels, such as selection of individuals.

There is a temptation to reify natural selection and to be so mistaken as to seek the level at which selection is "really" occurring. Michael Wade[22] performed experiments in group selection of flour beetles. Populations were selected as sources of new colonies using the criterion of size of the adult population. Large and small population lines emerged. It misses the point to suggest that the differences arose from selection at the level of the individual. No matter how reasonable, plausible, or economical the model of individual selection might be, it does not apply, because in this experimental circumstance we actually know the criterion for selection to be a group character, not an individual character. If exactly the same individuals had been selected according to some criterion applied to them as individuals and we knew that to be the case, then all group selection models would be inapplicable for the same reason that individual selection was beside the point when selection was based upon groups. Thus the "true" level at which selection occurs does not depend on which individuals survive to reproduce, but rather it is determined by the focus of the selection. The message is this: if both models are consistent with the observations, one cannot tell the difference between individual and group selection without knowing the mind of the Creator, and that is something we will never know, as scientists. A model of group selection as an explanation for a set of changes observed in populations cannot be rejected just because those same observations would be consistent with individual selection. An assertion that the individual selection model is the correct model is a very strong and particular statement. The burden of proof rests not on those who suggest, as we do, that group selection is significant, but on the individual selection partisans who must be able to show an inconsistency of group selection with the data. Certainly individual selection is a powerful model that is often useful, but its historical precedence is not justification for the exclusion of selection models operating at other levels. We suggest, as have others before us, that evolution and selection often occur at higher and lower levels than the classic Darwinian model.

There are also levels issues with Lamarckian versus Darwinian evolution. Paul Kammerer performed work on the midwife toad in the opening decades of the twentieth century. Ferocious attacks from anglophone Darwinians exposed India ink on the last specimen housed in a museum, and proclaimed it a fraud. But new work offers rehabilitation of Kammerer's Lamarckian inheritance work on the midwife toad. The specimen being in a museum, it will have been preserved by a museum curator keeping the specimen looking like it was when it was first displayed. In 2009 Vargas went over Kammerer's notes and found patterns that we now understand. Kammerer reported the pattern but it did not help the case he was making one way or another. He had no vested interest in what he reported. That recent finding was sufficiently high profile that it made *Newsweek* on September 28, 2009 (Begley 2009). True, it is a journalistic piece, but it is based on serious scientific literature.[23] Begley does put a political edge on it when she says Lamarck "drives the self-appointed evolution police crazy." Rigid Darwinians do have it coming. The work used a hot, dry environment. With severe drought Kammerer induced female toads to mate and lay eggs in water instead of on land. The normal condition in the species is to lay eggs on land, whereupon the male carries them (hence the name "midwife") until they are ready to fend for themselves. In Kammerer's manipulation, the survivorship of the fertilized eggs was very low, but there was a next generation. Reintroduced back onto cool land with water available the tadpoles grew up with distinctly aquatic not terrestrial behavior: as they matured they continued to mate in water and abandoned the habit of males caring for eggs. After four generations the toads even developed pads present only in other species of toad that characteristically mate in water (the pads that were found to be full of India ink). It appeared that characteristics acquired in being coaxed to mate in water were persistent over generations and were inherited. The critical new information came from Vargas reporting sex linkage of the acquired characteristics from Kammerer's own notes. We now understand the pattern. At the time sex linkage did not further Kammerer's argument. But now we know that epigenetics are generally inherited sex linked, a fact not known to Kammerer or his anglophone critics. Kammerer had no reason to fake the sex linkage reports in the way he has been accused.

It looks like Kammerer's work was legitimate and shows something like Lamarckian evolution with acquired characteristics changing the performance of genes in a way that is indeed heritable. We now understand that under distinctive environments epigenetic waves of methylation pass over the genome, silencing some genes and turning on others. The silencing of normal behavior is heritable because the signal for silencing is in the DNA. The significance of Kammerer here is that rejecting him stemmed from an error in level of analysis. Level of analysis is used all the time in biology, but often only implicitly, and therefore not open to oversight.

Darwin and Lamarck share much in their respective views. Comparison of the two explanations of evolution depends on the context, on the level of analysis at which they are evaluated. Logical types are structures that are defined at a given level. The notion of logical type suggests difference, but difference with something significant in common. Sometimes different is only different, but some differences can be in the context of a certain similarity. Similar, as a category, represents the distinction between two exemplars. Human infants appear to build similarity from difference as they come to appreciate the notion of an example.

The notion of similar is in fact triple difference: 1) standing out different from background; 2) different from background in different ways (circle or square); 3) even when not being different (2 circles), there can still be different examples.[24] Examples are similar while different. Darwin and Lamarck, despite differences, also use similar logical types. The concept of a book is different from a particular book, which is again different from a library of books, so we are talking of differences within a logical type. Darwinian and Lamarckian evolution belong to the same logical type: both are heritable—evolution via heritable traits. Even so they can and do belong to different logical types (see below): genetic versus epigenetic are differently inclusive types of characters, the former being the presence of the gene itself, the latter being its activity status. Both types may be heritable. Whether Darwinian and Lamarckian evolution are the same or different depends on the logical types used in the comparison. If one side is mistaken as the unequivocal winner in conflicts, errors in consistent use of level of analysis characteristically blunt scientific progress. Did we really have to wait almost a century to understand how Lamarck and Darwin can indeed be commensurate? In retrospect some critical point was not just missed but was studiously rejected. Yes, we did have to wait that long for Kammerer's rehabilitation because of ubiquitous imprecise use of levels of analysis. A related logical type issue is phenotype and genotype, and confusion there was cleared up only by epigenetics being the nub of it all. The same genotype may have some genes turned on and others turned off in a heritable way. There are genes and then there are active genes. How else would there be different cell types in a single organism? The tension and accommodation between Darwin and Lamarck can be sorted out with precision in levels of analysis.

Denis Noble is a leading physiologist, and as the 2013 president of the International Union of Physiological Sciences he addressed the society, touching on the confusion surrounding the Darwinian and Lamarckian evolution. He said:

> The "Modern Synthesis" (Neo-Darwinism) is a mid-20th century gene-centric view of evolution, based on random mutations accumulating to produce gradual change through natural selection. Any role of physiological function in influencing genetic inheritance was excluded. The organism became a mere carrier of the real objects of selection, its genes. We now know that genetic change is far from random and often not gradual. Molecular

> genetics and genome sequencing have deconstructed this unnecessarily restrictive view of evolution in a way that reintroduces physiological function and interactions with the environment as factors influencing the speed and nature of inherited change. Acquired characteristics can be inherited, and in a few but growing number of cases that inheritance has now been shown to be robust for many generations. The 21st century can look forward to a new synthesis that will reintegrate physiology with evolutionary biology.

Noble[25] cited all sorts of inheritance of acquired characters. In Weaver (2009) and Weaver et al. (2004) there are reports of epigenetic programming by maternal behavior. Noble reports:

> Weaver and co-workers showed this phenomenon in rat colonies, where stroking and licking behaviour by adults towards their young results in epigenetic marking of the relevant genes in the hippocampus that predispose the young to showing the same behaviour when they become adults.

The issue of level of analysis is especially pressing in biology because levels there are exquisitely scaled structures, and the scale defined levels are densely packed. The level structure in physical systems is much less dense; scale differences defining levels in the physical sciences involve orders of magnitude in differences of scale. This means that in physical and chemical systems emergent structures from different levels never work in close enough proximity so as to interfere with each other. For instance, different isotopes of a given element do not interfere with chemistry materially or intellectually. The chemistry is the same because the isotopes are defined and understood to be the same element. It depends on the number of electrons in the outer shell of the atom, and that defines the chemistry. The scaling of different levels in biology is much closer than in physics. This proximity does permit interference between levels, when for instance sex is involved in individual selection. Individual selection should suppress sexual reproduction in favor of asexual reproduction, so as to avoid the gene dilution that comes with sex. Your sexual offspring have only half your genome, when asexual reproduction would transfer your genome in its entirety to the new generation. And yet individual selection may be facilitated by the group selection that is invoked in sex. More out-breeding through sex may increase survival of an individual's offspring if the environment is generally more variable. Individual and group selection clearly are different and stably so, even if they nudge each other through collaboration as well as through interference. By the standards of physics there is no clean shot between individual and group selection, or indeed anywhere else in biology. This difference between biology and physics requires a special explanation.

The density of biological levels comes from the remarkable stability of biological structures once they have emerged, even if other disparate structures that are similarly scaled are present. The identity of individual and group selection as phenomena can be teased apart through experimental precision. Even so they still interfere with each other because they are so closely scaled as to manifest disparate characteristics in just one organism, and work through the very same process of inheritance and selection of characters. Emergence occurs without plans from the physicality of positive feedbacks pressing against physical negative feedback to make emergent structure, and it does so dynamically without plans. While unplanned structures do emerge in biology, the persistence of what emerges in biology comes from stabilizing coded constraints absent in physics but important in biology. Coding for emergent structures is detached from the blind process of emergence itself. An organism dies but its codes will leave behind reproductions of itself in offspring. Natural selection quickly codes for the stability of advantageous emergence in some biological language. As a result there are close packed hierarchies all over the place in biology—for example, taxonomic levels; nested hierarchical structure from organelle to multicellular organisms and populations.

We do not present the above as a contribution to the heated discussion of group selection per se. It is, however, a statement of our sympathies that is necessary because our hierarchical view expects selection at several levels simultaneously. Some of that selection could easily be at the "group" level. Holons exhibit self-assertion as wholes, but are also integrative as the holon plays the role of a part of a higher level context. Selection in our model must occur at least at two levels because of the dual nature of the holon. Integrative tendencies will be selected in the structure that arises at the upper level to which the integrative tendency contributes. The self-assertive tendencies logically occur at different levels at the lower level in the survival of the holon itself. As with so much in biology there is interference between the two levels because failure of the upper level as it is selected out will usually interfere with the lower level holon. The integrative tendencies of holons are not selected principally at the level at which these tendencies are manifested, but rather they contribute to the stability of the system of which they are part and so are selected there. The integrative tendencies of the holon contribute significantly to the selection of the next higher stable level and therefore to the evolution of the second higher stable level. Altruistic behavior in primates or the contributions of worker bees would seem to be examples here.

The above reference to altruism does not address its relationship to genes or whether or not it can be translated into terms of individual fitness. The role of genes in all this is complicated by yet another twist in level of analysis. Gould (1981) spoke of selection of DNA rather than genes. Dawkins argued that organisms may be seen as simply vessels that hold DNA, and that selection occurs at the level of genes struggling for a place on the genome. Gould separated Dawkins' (1978) selfish genes from Orgel and Crick's (1980) selfish DNA. Selfish DNA simply copies itself to other places in the genome and so occurs as repeating sequences that comprise most of the genome in the high ninety percentages. At the level of DNA coded and expressed in protein synthesis, the repeating DNA is meaningless. The argument for selfish genes says they would have an effect on the phenotype and so should be selected. Since individual genes affect multiple characters in bodies and join in their action with other genes, in Gould's words, "the idea of individual genes battling for personal survival makes no sense." Dawkins' error is one of extending the model for gene up to the level of phenotype where selection in fact occurs. By contrast selfish DNA validly takes selection into a realm where meaning of the DNA for the organism is moot. This is an example of Darwinian selection being applied to a larger context in which there is indeed room for selection for meaningless nucleotide sequences by simply replicating faster.

The integrative tendencies of genes are selected in the form of the larger complex character to which the genes contribute. For genes, this complex character may be so large scale that it amounts to the entire phenotype. Quoting Koestler (1967) on the relevance of the integrative tendency to the evolutionary process:

> The most general manifestation of the INT [integrative] tendencies is the reversal of the Second Law of Thermodynamics in open systems feeding on negative entropy (Schrödinger, 1967) and the evolutionary trend towards "spontaneously developing states of greater heterogeneity and complexity" (Herrick, 1956). [His parentheses, our brackets]

Koestler is mistaken as to reversing the Second Law; if you ever find you are at odds with the Second Law, your idea is not just wrong, but wrong on its face. The Second Law in physics is expressed as a closed system running down to more probable states of random patterns. It is not so much that the systems in physics are actually closed, it is that you can get away with that assumption in physics because their systems are contrived to be closed enough. You cannot get away with closure in biology because biological systems are clearly and very open indeed. They eat, sweat, respire, transpire, and expel waste all the time. The appearance, but only appearance, of the reversal of Second Law comes from the system being open. There is a larger scheme to which the biology is open. The metasystem becomes more entropic than it would have been were the living system not present. And that extra disorganization more than compensates for the apparent increased organization over time manifested in the living system. The larger system, the environment plus the life within it, not only does not deny the Second Law, it does not even appear to do so superficially. Schneider and Kay's (1994) title is "Life as a manifestation of the Second Law of thermodynamics." Their paper says life does not deny the Second Law, it depends on it. Life depends on dissipating extant gradients, for instance the steep gradient between food and the detritus in its environment.

The self-assertive tendencies that Koestler describes as the holon's "tendency to preserve and assert its individuality" are selected at the level of the holon itself. A failure of a holon to be sufficiently self-assertive results in its loss of identity. Its parts may or may not survive the selection. Plant community succession may be viewed here as an example of an evolutionary process in which successive species are insufficiently self-assertive and lose their identity in eradication. Plant communities mature as species that establish in the vegetation early facilitate the environment for those that establish later. This process of maturation is called succession. Margalef (1968) compares evolution and succession in Koestler's terms.

A cancerous growth is a case of a tissue holon gaining self-assertion and seizing selective advantage. Since organisms are obligately nested right down to proteins, the self-assertive success is often short-lived. Quoting Koestler (1967) on the vigorous self-assertions of parts:

> The over-excited holon may tend to get out of control, and assert itself to the detriment of the whole, or monopolize its functions—whether the holon be an

> organ, a cognitive structure (idée fixe), an individual or a social group. The same may happen if the coordinative powers of the whole are so weakened that it is no longer able to control its parts (Child, 1924).

The coordinative powers are the whole's self-assertion viewed from below. The gain in self-assertion of the holon below costs the constraining whole its own self-assertion. Selection increasing the holon's self-assertion may evolve the whole to extinction; this is, of course, selection against the whole at a higher level by virtue of loss of its own self-assertion. As individuals assert themselves in breeding and capture of resources, they gain selective advantage; selection of self-assertion here produces classical Darwinian evolution through competition.

Macroevolutionary Events

In a sequence of interdependent responses, a change early in the chain of events has a much more drastic effect on the end result than does an alteration in the succession toward its end. An early alteration affects not only that one early step, but also the execution of most subsequent instructions. The very next operation will say something like "Do this to this thing"; but what if, after the previous instruction has changed, "this thing" is no longer recognizable? Perhaps "this thing" responds differently now. The effect of the prior changed instruction is functionally to change all subsequent operations. A mistake early in a throwing motion will send the missile flying away from the mark, whereas an error at the very end of the action is likely to have only a small effect.

A mutation that affects the action of a gene important early in embryonic life is likely to produce a monster if it produces anything at all beyond a mass of disorganized cells. If the highly evolved growth phase is fundamentally disrupted early, then the many subsequent instructions stand as much chance of producing an effective organism as the proverbial monkey stands of typing at random an entire Shakespeare play. Mackay (1973) has modeled the introduction of a set of inspectors who monitor and preserve the correct but incomplete achievements of a monkey typing pool. With inspectors, the achievements of the random typing grow rapidly. Mackay's monkeys and inspectors working together can produce a Shakespeare sonnet in a manageable period of time, in fact, quite comparable to that taken by the Bard himself. Of course the inspectorate has a plan, but the process of evolution exactly does not plan. Sometimes there is memory embodied in evolved genomes that makes it appear to plan, but evolution does not aim de novo. An example of long-term deep memory manifests itself in Vavilov's[26] parallel series of variation, where different genera of crops that do not interbreed nevertheless show the same patterns of variation. Wheat and barley do not interbreed but they do have similar swaths of variation.

In evolution, natural selection operates as the monkey inspectorate and is rather conservative. If evolution produces change, it generally does so by adding further embellishments and flourishes to the end product, the adult, rather than by fundamentally changing the pattern of development by altering one of the first steps. The evolutionary options of a population are substantially limited by the origins of that group and the general adult form left to them. Evolution by natural selection works through the growth processes of the creature in the breeding population. Change is effected in part by gene mutation as it affects maturation. The change must occur continuously, for it is not possible to evolve from one viable being to another without passing through a series of other intermediate forms, which must also survive long enough to reproduce.

While tweaking the end product is the norm, not all the variation which is presented for selection is produced by changes in the final steps of development. At a level of organization higher than the individual instructions are governors that regulate the rate of the development rather than the course of development. It is characteristic of higher levels in a hierarchy of control that upper levels control the broad outline of the behavior of lower levels without directly influencing the workings of the lower-level processes themselves. For the development of the hierarchy that is an organism, the governors are often manifested as growth hormones. A changed instruction from a governor is likely to produce dramatic departures in form but without intrusion into the orderly process of local step-by-step development. Thus change in a large number of characters is achieved without the disorder produced by a change within the developmental chain.

This change in the behavior of the high-level controllers of development produces juvenile forms that are sexually mature. This type of change, neoteny, has played an important role in the evolution of many groups of organisms. Not by changing developmental pathways, but by stopping the development short, neoteny produces functional forms that are very different from their ancestors. All the stages, from egg to adult, must be viable, and in neoteny the old developmental blueprint is used but in an abbreviated form. Arresting development at any juvenile stage will result in an organism that has many characters altered from its adult ancestors, but the characters will be coordinated instead of disorganized. All animals with backbones can be ultimately derived from the larval form of starfish relatives. Some salamanders never develop beyond the tadpole stage. Dwarf willow trees that never grow taller than a few inches are a botanical example of neoteny. Sexual maturity as a seedling makes for adaptation to environments where any other type of willow would perish. Herbs are sexually mature tree seedlings.

Pattee (1972) argues a similar case, but more generally. He points out that evolution is likely to produce increasing elaboration in developmental instructions, but that organized modification, which is on balance beneficial, becomes more difficult as its complications accumulate. He uses the compelling image of a computer programmer modifying code in a long algorithm. At first, modifications are simply made and their utility is clear. There is a natural selection of lines of code. But soon, problems begin to arise through secondary effects of the modifications, and improvement is made at a price. Eventually deleterious effects balance benefits, and further changes become the machine logic equivalent of genetic drift. Once this point is reached, minor changes are neither adaptive in and of themselves nor do they lead to adaptation. Only a macrolevel change can simplify the systems so further adaptation can occur. The same argument is made by Tainter (1988) where he identifies diminishing returns on problem solving in societies.[27] An example is the greater cost of higher levels of education over primary and elementary school. Tainter's worrying message is that the later stages of societal change lead inexorably to collapse. We see in the evolution of sabertooth animals that the sabers get larger over time, but lead inexorably to extinction. The optimism of contemporary society that there will be a scientific fix to problems in the long run ignores the increasing numbers of people included in the scientific enterprise. At this rate, in a hundred years everyone will have to be a scientist. And there is no way out with increasing sophistication somehow behaving differently. Tainter presents a diminishing return line in his 1988 book looking at the number of scientists and engineers in research and development and number of patents as a measure of productivity. A more recent presentation suggests that high tech does not behave differently.

Pattee identifies a strange reversal in evolution toward complexity.

There is a common requirement for both Von Neumann's self-reproducing automaton and any process which would be acceptable as a measurement, and that is a threshold level of complexity which functions as a simple constraint or rule. It then slowly dawned on me that the original program of searching for complex behavior from simple initial conditions [emergence through natural selection] had been turned around, whether I liked it or not, into the opposite program of trying to find simple behavior in initially complex systems. . . . It is quite reasonable to regard the origin of genetic records as a simplification of the underlying physical and chemical events they represent. And if the basic necessary condition for the origin of life was the separation of genotype from phenotype—i.e., the separation of the description of the events from the events themselves—then it is fair to say that life originated by self-simplification of natural events. . . . Once natural selection has taken over the course of evolution, is there any need for further simplification?

This question brings me back to the two strongest criticisms of mutation and natural selection as the only mechanisms in evolution. They are very old criticisms, but they can be found in a modern context in *Mathematical Challenges to the NeoDarwinian -Interpretation of Evolution* [Moorhead and Kaplan 1967].

The first criticism is that random search cannot be expected to "find" successful organizations because the search space is so immense. Thus a search for just one sequence of 100 amino acids representing one functional enzyme would require of the order of 20^{100} trials which is not within any reasonable probability within the age or size of the universe. . . .

The second criticism is that functional optima or fitness peaks in the adaptive landscape appear to be local, separate optima, so that no evolutionary pathway can be imagined that does not pass through nonfunctional or lethal valleys. . . .

A second, more mathematical view of self-simplifying processes is the topological dynamics of Thom [catastrophe theory]. Here the simplification results from the assumption that it is the discontinuities in the vector field of trajectories which determine the simplified structure at a higher level. Furthermore, these discontinuities may exhibit a kind of stability which is entirely deterministic in the evolutionary sense, but not predictable in the dynamical sense. [Pattee 1972]

Evolution and Purpose

In an earlier section of this chapter the principle of complementarity was introduced so that it could be used to address problems of emergence of characters in the higher levels of hierarchies. The complementarity principle returns here, this time to facilitate a discussion of the evolution of purpose. It will allow the process of natural selection as a mechanism in organic evolution to be factored away from the adaptive significance of the evolved organic forms that have emerged. Adaptive significance is sometimes dismissed as hindsight, structure merely imposed on the world by human observers.

John Harper made the point that adaptation is misnamed. The Latin *ad* means "to" while *ab* means "from." He would prefer *ab*aptation,[28] so as to capture the way that what is conventionally called "adaptation" comes from the past. While accepting Harper's point, we prefer to remember that biology would not exist but for biologists, and biologists recognize purpose. Purpose has a component of anticipation and so there is something of "to," that is *ad* in adaptation. Purpose implies that something is going to be useful in a future circumstance. We do not suggest that the process of evolution aims at the future. The material world in itself does not read future causality because, what to do with the recursion when the future actually arrives? The anticipation in the vicinity of evolution comes not from externalities, but from observers trying to understand their experience. Biology must include an observer, and to say "mere" observer denies the discipline; in a sense, without the observer's interest there is nothing to explain.

Antecedent to every construction of a blind dynamical model trundling into the future is the act of recognition that there is something functional and purposeful that might yield to an evolutionary explanation. An assertion of the sufficiency of blind dynamical selection as an explanation of evolutionary phenomena is an operational pretense to be the all-knowing observer. It is wont to give unreasonable predictions. It leads to contradiction. Levins and Lewontin when they champion dialectical materialism as a philosophy for biology do not appear to mind contradiction. "Its [dialectical materialism's] central theses are that nature is contradictory, that there is unity and interpenetration of the seemingly mutually exclusive, and that therefore the main issue for science is the study of that unity and contradiction." That view is compatible with ours but displays a preference for contradiction over complementarity. The two positions of contradiction and complementarity stand together in holding that there cannot be unity without contradiction in any models of evolution. The only way around that uncomfortable issue is to studiously ignore adaptive significance, which denies the reason most biologists consider evolution important.

At this point it is timely to reemphasize the central role of the observation window in determining the message received. The duality of dialectical materialism can be seen as arising from the fact that both the transmitted signal and the observer's window have a part to play in forming the message, but they do it independently. For any phenomenon, there is one description that relates primarily to the transmitter and another that corresponds to the role of the observer in making the observation. These two descriptions are not compatible, since the one is concerned with the rate-dependence of the dynamics of the observed and the other pertains to the meaningful rate-independent aspects of the phenomenon. Any attempt at a unified description leads to contradiction of the sort that arises in the wave particle dilemma. Adapt as a verb refers to process. Adaptation as a noun does not. Trivially the process of adaptation is associated with a rate-dependence and therefore with the dynamical complement. The mechanism of evolution is addressed by the dynamical mode of description. Adaptive significance, on the other hand, has no rate and is therefore associated with the rate-independent complement. Both complements are necessary if all facets of the phenomenon of biological evolution are to be considered. However, together those meanings of adaptation are contradictory in that evolution does not aim to go anywhere, but still manifests adaptations when it gets there.

No matter how extreme the apparent differences between the more usual models and our position, we do not mean to deny the validity of the standard Darwinian model driven by chance occurrences (although change in the genome is not random); that model is the rate-dependent complement. It reflects the naked dynamics of observed structure. We see the need for explicit attention to the observer-dependent complement. We agree with Ho and Saunders (1979), "We hasten to add that this does not necessarily mean there is an 'inner perfecting principle' or 'vital principle' that must be outside the laws of physics and chemistry." We are not working outside the laws of the physical universe; we are rather discussing an aspect of evolution that, being concerned with the rules of phenomenology, is complementary to the rate-dependent, stochastic model. The aspect of evolution we discuss is not reducible to the same terms as, and cannot be made compatible with, the standard model; our model is a complement. The significant word from Ho and Saunders above is "laws." Pattee (1978) again comes to our aid.

> The functional reason for the irreducibility of either laws to rules or rules to laws is that their descriptive modes are incompatible with respect to rates of events. All of the physical laws of motion are expressible as functions of rates, that is, as derivatives of some variables with respect to time. Time in physics is an irreducible concept, but by no means arbitrary. It is one member of the coherent set of concepts, including space, momentum and energy, that forms the language of physical laws and it should not be confused with timing intervals, the ordering of events or psychological time. The concept of rate, however, may be extended to social dynamics or psychological events.
>
> Rules, on the other hand, depend on order but have no rate dependency and cannot be expressed as functions of a derivative with respect to physical or other time scales.

We are interested here in the rule-related portion of the evolutionary phenomenon.

Purpose and function are peculiar to biological systems or their cultural artifacts. This gives living material a rather strange relationship to time. Waddington remarks that at birth, feet have never been walked upon, but the skin of the sole is thick. Referring to a question raised by Waddington in the context of thickening of the skin on the soles of human feet, Ho and Saunders say:

> The fact that the thickening is already present in the embryo, before the foot has ever touched the ground, indicates that it is not the result of a direct response to external pressure but must be produced by the hereditary mechanism independently of the external influence to which it is an adaptation. Now, it is conceivable that an ancestral population of hominids or pre-hominids was first exposed to the environmental stimulus that caused the soles to thicken (as when the bipedal gait was being evolved). As the response was adaptive, Waddington argued, there would be selection for its canalization; in other words, the response would "deepen" and become regulated so that a uniform response would be produced in the presence of a range of intensity of the original stimulus.
>
> Meanwhile, it would be quite possible for some random mutation to oc-

> cur in the genome of the organisms that would take over the role of the environmental stimulus, so that the specific response would appear in the absence of the stimulus. This latter process is termed genetic assimilation. The combined effects of adaptation during development, canalization and genetic assimilation are to mimic Lamarckian inheritance.

This in no way takes away from their function and role in walking or their reason for being on the ends of natal legs. At least at some time in their existence, parts that are useful anticipate in their very structure an environment in which they will be of future service to the whole. Convention says that after Darwin, Watson, and Crick, all can be explained by evolved mechanism. We say not so, and prefer to agree with Polanyi (1968): "Any coherent part of the organism is indeed puzzling to physiology—and also meaningless to pathology—until the way it benefits the organism (its purpose and function) is discovered." Burgers (1975) says, "Purpose is a part of biological phenomena and deserves to be viewed in its own right. The idea of purpose should not be condemned. It does not mean being directed to a definite goal for some billions of years." Stacey, Griffin, and Shaw (2000) not only accept teleology, they develop the idea of different sorts of teleology used across science and management. There is teleology in all discussion of purpose and design.

But how can future orientation of a purposeful sort be selected out of the physical world wherein even the germs of purpose and anticipation are absent? There is no more purposive behavior in the weak organic feedback loops discussed above than in the cold condensing interstellar gas. So where does the purpose of useful biological parts such as organs come from? We feel that natural selection can only intensify what is already in existence, so how is it possible to select from abiotic systems, even very gradually, structures with purpose? Where in the abiotic soup is the hint of anticipation which leads to the rich functionality of organs and sociobiological behavior?

We find anticipation built into the general model of evolution itself. We have already referenced Simon's example of evolution in physicochemical hierarchies, and so if there is anticipation in evolution then there is anticipation in evolved physical systems. As Simon indicated, evolution depends upon a system's stability rather than upon its being alive. If evolution, as we indicate below, depends on selection of systems that are adapted to future events (preadaptation), then the evolved physical world trades in futures.

The answer to the quandary is that the anticipation and purpose of biosocial behavior arises at a higher level. Change the scale of a dynamical process, and things turn out differently. Change the scale in an informational scheme, and things can remain the same. We have already looked at the essence as being the place where experience information from the past is coded. The scale-independence of coding means that coded experience of a species may be transferred to the concrete biosocial entity that is realized from the essence. The realization does not have the experience of the essence, and so in some sense had no business knowing what to do. It has not seen the past wherein lay the experience that might indicate the same thing might happen again. There is nothing magical in knowing the future from past experience. The point is that it is the experience of the essence transferred unchanged to the realization that tells the future. That phenomenon depends on the way coded information can pass between scalar levels, without being in contradiction to the scale-fixed workings of thermodynamic processes.

Essence in the Other

Zellmer, Allen, and Kesseboehmer (2006) developed a scheme for dealing with the difficulties of creating models. Creating models has the problems of duality discussed above, and so their scheme has two explicit sides. They made reference to Pattee's (1978) laws and rules, and erected a scheme with a meta-observer. The meta-observer is found at level N + 3 in Figure 5.2. The "meta" in meta-observer comes from the observer who decides what is to be studied, less in specific terms, and more in choosing the general discourse. If that observer dies, or if the system dies, or if interest is lost, perhaps because the grant has run out, then everything stops. Plenty-Coups lived in the time of the great decline of the Plains Indians. As Chief of the Crows he said at the end of his autobiography, "When the buffalo went away, the hearts of my people fell to the ground, and they could not lift them up again. After this nothing happened."[29] Cronon comments on Plenty-Coups' words, "All that happened afterwards was part of some other narrative, of which there was neither point nor joy in telling it." After the grant is over, nothing happens.

At level N + 2 in Figure 5.2 are the laws. They are local to the discourse chosen by the meta-observer. The laws of physics would be laws

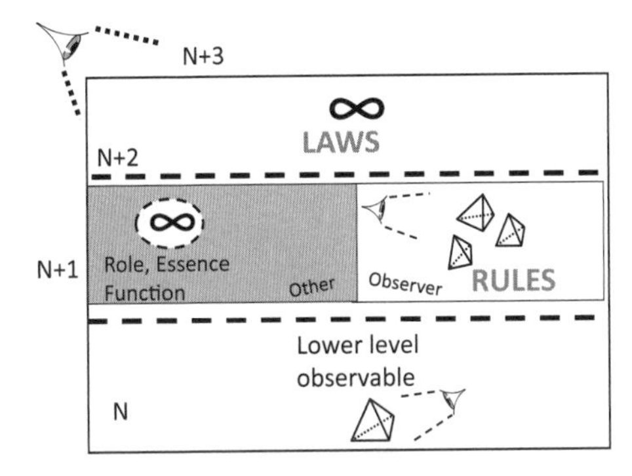

FIGURE 5.2. There are three levels in the scheme of Zellmer et al. At level n is the observation of the object of the discouse, the thing you observe. At level n + 3 resides the meta-oberver who decides what to study in the first place. At level n + 2 are the local universals, the laws that come from the meta-observer's decision about the universe of discourse. At level n + 1 there resides the model and the essence. The model is an equivalence class for the observed entity. The model derives from the observer's rules. It is on the side of the observer and derives from observer decisions. That observer at N and N+1 may or may not be also the meta-observer. Still at level n + 1, but on the side of the other is the role or essence of the realized structure residing at level n.

in these terms, but the laws in biology would be more local and insist on the involvement of carbon. We do not mean laws of biology in terms of Dodds, who was looking for the laws of biology in general. The laws we mean are local to the discourse. We avoid Dodds' realism.

At level N + 1 there are two regions. There is that duality again. The side of model creation is chosen by the observer, who may or may not also be the meta-observer. On the side of the model are situated the things in observation chosen by the observer. The other side is the side of what is technically called in postmodern terms *the other.* We mean *the other* to be about observation, so it is not metaphysical. It is that part of observation in particular not chosen by the observer. The observer chooses to study lions and their spatial position on the side of the observer. But the lion moves; the observer did not choose that, so "lions moving" is on the side of the other.

We find at N + 2 the rules of Pattee. They create an equivalence class that embodies what is in the model, as well as what is related to what and how. On the side of the other is the explanation for the patterns and regularities seen in the model. This we would call the essence or the role of the incumbent found at level N. The observed structure at level N is in a

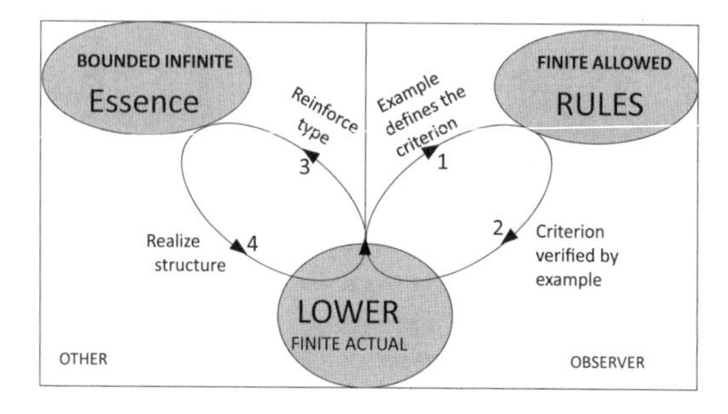

FIGURE 5.3. Expressing the Zellmer scheme using loops for the critical processes. It starts with an experience that is converted to an observation by connectors 1 and 2. The 1 and 2 loop is the modeling exercise where the observed entity is put into a an equivalence class. On the side of the other we recognize pattern in the observed that does not come from the observer. The observed entity is a realization (an example, or model in nature) that is the specific of the essence's realization. The realization of the essence occurs, perhaps through development, in arrow 4. The loop in the other is closed by 3, perhaps through natural selection.

sense what you are looking at. It straddles the side of the observer, and the side of the other. It is the link between the model and the essence that explains the patterns seen in the model.

The parts of the scheme in Figure 5.3 are linked through connections that make two cycles, one in the realm of observer control, and another on the side of the other. The side of the observer has a loop of model construction. The observed entity appears as part of experience. At this point it does not have a name or an identity. It acquires these in a move up to the rules at N + 1. The rules set up an equivalence class to which the entity at level N belongs. That modeling loop is closed by a return back down to the level N. That transition is a check to see that the observed entity fits the class that names it. Cycling around on the side of the observer improves the model and tightens the relationship of the observed entity. Notice it is first part of experience, but once it has been named and classified it becomes consciously observed.

The equivalent cycle on the part of the other is independent of the observer decisions. On the side of the other the observed entity is a realization at level N coming from the essence or role at level N + 1. The essence of an organism is not its DNA, that is just the code for realization. The organism is a realization of the essence. The essence might be a species, or some social abstraction, like the US presidency. The US presidency is realized into an incumbent usually through election.

The identity of the biological entity at level N is in material flux. But its more persistent identity is not material. The essence has experience of the response of a long set of realizations, births and death, elections and endings of terms. The realizations change the essence as realizations of a species might change the species through natural selection. But the essence is still responsible for the successive realizations. The experience of the essence is coded in some way. It might be coded with DNA, or it can be coded in the historical record of citizens' attitudes to the US presidency. For instance, the US presidency was not the same after Nixon's Watergate crisis. That change allowed for an unlikely next elected president, a liberal, evangelical Christian, Jimmy Carter. Ironically, the deeply conservative George W. Bush would have been too bizarre to get elected without the precedent of the fundamentalist Carter (Figure 5.4). The coded information in the essence is transferred to the realization at level N. Organisms have information about a past that is not their own. Organisms and US presidents are not so much material as they are a set of expectations. Your identity as far as your credit card company is not the material you, because the company cares not if it is you or your spouse or a friend helping by gassing up your car. It cares only that you meet the expectation that you will pay your bill. Almost all the atoms in the human body are replaced every seven years or so, even the bones. To what extent are you still you? It cannot be in the material

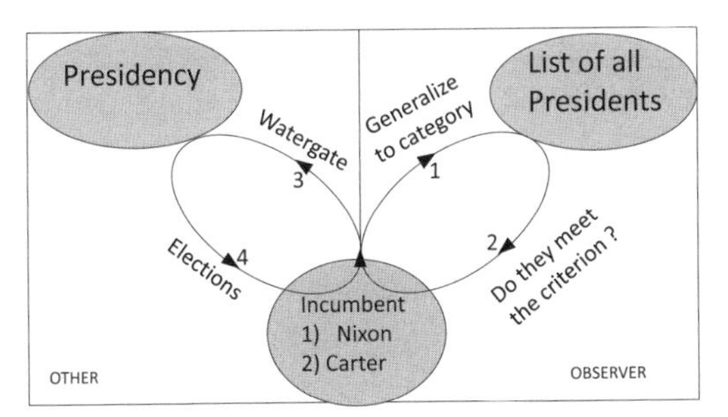

FIGURE 5.4. The Zellmer loops applied to the US presidency. The general role is the presidency. Nixon, an incumbent, is the realization. He has his Watergate scandal and this changes the presidency. As a result a new election creates new sort of president, an evangelical Christian, Jimmy Carter. The model is the list of presidents.

you. It is a set of expectations that you have, and others have for you. There are expectations for presidents and expectations for organisms on the part of other organisms, be they mates, predators, prey, competitors, or whatever.

Rosen turns to the notion of essence.[30] We hasten to add that this is not an idealist position where there are essences in the world external to human endeavors. Rather essences come into play when a human observer creates a set to apply to observed entities. Once the set is created by the observer, the question arises as to what is responsible for, what can offer an explanation for, the equivalence that applies to make a member of the set a member. The notion of the other has been considered earlier. To recap, the other is that which is responsible for an observation that is not chosen by the observer, and it pertains to dynamics. Let us state again that the other is an epistemological device since no position is taken as to its existence outside the observation. There are of course ontological questions of what in an external world is responsible for what is observed, but that is ontology not the epistemology that we seek here.

Consider the essence of dogginess invoked to explain the unity in a collection of dogs. If all the dogs are *Canis familiaris* then dogginess is related to humans. In the section on Primate Knowledge (see chapter 4) we already mentioned how dogs and humans read human faces. We apparently scan from the nose to the left side of the visual field up to the eye on that side. Dogs do the same scan of human faces. Curiously they do not do that with other dogs.[31] Also significant is that wolves do not scan even human faces in that manner. Domesticated dogs are clearly a product of human selection while wolves are not.

If we include a wolf in our original set of dogs, the set that is still to be explained by dogginess, but clearly a different dogginess is involved. But a lot of things about the set of only *Canis familiaris* dogs still apply to this new set with wolves included. The patterns of teeth in dentation are still the same. The paws are the same. The proclivity to form social groups that pack hunt is still there. Clearly dogginess still applies but it has explanations that do not involve humans. Dogs, with wolves included, still form a group that is closely related in evolutionary terms. Dogs separated from wolves only very recently, only some 15,000 years. But we can go even further. There is a certain dog-like set of features in hyenas. They too pack hunt, and the mouth and the paws of hyenas are dog-like in general terms. But in the evolution of the Carnivora, there is the order to which dogs, cats, bears, raccoons, ferrets, and seals belong. That dog-related line separated from the cat side of carnivore evolution some fifty million years ago. The dogs evolve on the bear and seal side of things. Cats emerge on the other half of carnivore evolution along with a very old cat-like group, the extinct Nimravidae. The cat side of things includes badgers, civet, and mongooses. But the startling thing is that the cat and hyena line separate only 30 million years ago, making hyenas much closer to cats than dogs. With hyenas in the set of doggy things, the dogginess they share with domesticated dogs is again something else. Dogginess with hyenas involved is about how a carnivore hunts: doggy packs that run down prey on paws that are blunt and non-retractable, as opposed to ambush and clawing like cats. And hyena dogginess is not a matter of being related to dogs in evolutionary terms that put dogs and wolves together. But there is yet another twist. Cheetahs are perfectly good cats but are distinctive among cats in the way they run prey down. There is a distinctly dog-like chase in cheetahs. And the paws of cheetahs are doggy. At that point one might as well include marsupial "dogs" like the extinct Tasmanian Tiger, which had a very wolf-like snout, even though it is not even placental. Thus the essence of dogginess is a labile thing that applies only in the context of human investigators creating a set. Essences do not exist independent of human modeling. They are tagged to human models, but do seem to allow connection to the undefined other, wherein lies the explanation for our models.

Rosen's essences are thus distinct from Plato's idealistic essences. There are shadows on the wall of Plato's cave, and he cannot see what makes them directly, but in his philosophy there are things as things out there making the shadows. "Thing" and "out there" pressed together make for an oxymoron. Things as things come from us; they come from what the observer decides is discrete enough to be identified as a thing. We choose the boundaries and what is discrete. Boundaries are about perception and modeling, not external reality. Rosen does not intend any sort of idealism in essences, and neither do we. Essences for us are not material but are abstractions. They are not definable, because we cannot experience them directly. Also, the dynamic aspect of essences means they are always changing. This works well in a discourse about evolution, because what is evolving is always open to change.

Essences have some properties like those of chaotic systems, which we consider briefly now. We can create equations that change the state of a system as it is defined. Those outputs of equations give changed states that can used as inputs to the same equation in a process of iteration. Subjected to such repeated calculation some equations settle down to a point where behavior is constant. It is commonly agreed that solved equations are general in that different situations can modeled by equations. Rosen[32] makes the point that well-behaved equations that can be solved are very rare, and so equations used to solve biological issues are not general at all. The general case is equations that cannot be solved. The patterns that repeat in the output of a given equation are called attractors, because the end state attracts behavior to it. A pendulum is attracted to a stationary state where it hangs straight down, motionless. Many other equations do not settle down to a point, but still visit several particular states in a sequence. The particular repeated states appear as points in a mapping over time. There may be two, four, eight, or sixteen points that the system visits. There are even higher even numbers of such points, but as that number goes up, the equations are held in an ever narrower specification. This repeating behavior is called a stable limit cycle. But then there is chaos. Chaotic equations never settle down to any fixed pattern. Rather they move around in general patterns, never repeating a state—that is, never visiting any point more than once. Systems moving to equilibrium, like a pendulum with friction, can be described by a track that ends up at a predetermined point or set of points. Equilibria do not have to be static, and the infinite track of a chaotic system is clearly dynamic but is still in equilibrium; chaos is a strange form of equilibrium. Pendula with friction move to a static point, an equilibrium on an attractor; but chaotic systems also settle down not to an equilibrium point but to their chaotic track, which is also an equilibrium. Chaotic equations are said to exist on strange attractors. Strange attractors are limited to certain regions of behavior, and so are recognizable, but there is an infinity of tracks in those regions on the attractor. Thus chaotic systems are bounded to a local region, but because they never repeat they are infinitely dense. Their tracks around the attractor are infinitely close if we were to iterate the equation over infinite time. This dense packing is related mathematically to the infinitely dense patterns that repeat at ever smaller sizes in fractal systems.

As we have mentioned before, fractals have the peculiar scale-crossing patterns that we see in clouds. One cannot tell how close is a cloud, because local outlines of parts of clouds form the same patterns as whole clouds. Meanwhile in calculus the set has a limit. As we put more and more dummy iterations in calculus the set reaches a limit. In fractal systems no such limit exists. Outline a fractal structure, estimating its perimeter with even shorter segments, and no limit is reached. Fractals were one of the earliest formalizations of the difficulty we have in situations that appear repeatable but cannot be defined. Essences have those nasty properties because they are infinitely rich, but still bounded.

Essences are like roles, they are the bounded infinities like we find in chaos theory. Roles limit what applies, but there is an infinite number of ways to fulfill a role. Many categories in biology can be seen as being bounded infinities. Consider for instance elm trees. No two elm trees have ever been exactly the same, so there is an infinity of elms past, present, and future. But this infinity does not mean anything can happen. Despite the infinity of elms, elms are never oak trees and never will be. The notion of bounded infinity is very useful, because it means we do not have to quit just because of infinitely fine variation. Some chaotic equations are very short, but they still have infinity in the patterns associated with them. This means that some enormously complicated things still have simple explanations.

Returning to the issue of anticipation and purpose, we are now armed with the notion of essence. We have already raised the example of a foal that is a realization of an essence, for instance its species. We raised all this in the first part of this book. When a foal is born it appears to know exactly what to do with its mother, even though it has not even seen any mare before. And first-time mares know what to do with foals. That information, as with all information, is independent of rates. Policy, codes, and plans are not dynamic. The physical happenings in biological and social systems are manifest as rate-dependent processes. Material, energy, and information all flow according to some version of thermodynamic equations. Rate-dependence introduces a certain scaling effect. As a result, it is hard to translate happenings in biology across levels. In physics there is statistical mechanics that does its best to translate micro to macro effects, but there are serious limits there too.

Coded information by contrast is rate-independent. The code need not change when it is applied between levels, and that is the trick that allows foals and mares to know what to do even if it is outside their experience. They appear to know what to do before it happens. Even so there is nothing magic, nothing that actually bends the physicality of time. All that the foal and mare anticipate has been experienced real time by previous foals and mares. Experience becomes coded by natural selection and can be used to expect the future to continue as did the past. Because information from the past is frozen as coded structure, changes of scale do not necessarily change information, so it can be passed between levels. The essence of mares and foals as a group over time has access to information that can be passed unchanged down a level to individual foals and mares.

Evolution is of expectations. You do not know you are an item of prey until another organism treats you that way. And evolution takes note in terms of your fitness. All of this is of course set in the realm of possibilities. The laws in Figures 5.2 and 5.3 at level N + 2 set what is possible. That is why evolution is such a bold notion, and one that has difficult facets to it. It passes through the rule-defined model. Evolution is set in material possibilities, and is manifested in a set of expectations. Scalar thermodynamic aspects to life are related to laws. But the laws apply in subtle ways, as things change with rescaling. The laws do not mean one thing; their implications change between the levels N, N + 1, N + 2, and N + 3. On the other hand, coded information cannot be seen through laws. This is what happens when reduction moves to too deep a level. The causes of a phenomenon to be explained slip between your fingers below the level of meaning. And yet coded information and meaning can pass stably between level N and level N + 1, a crucial part of evolution. The meta-observer at level N + 3 works things out by trying to connect the two sides at level N + 1.[33]

Codes and Meaning

By moving to codes the system gets away from having an infinite regress to faster and faster dynamical models. Every update, if it were to be dynamic, must go faster than the dynamic model that got the system into that situation.[34] The benefit of freezing dynamic situations in information is its predictive capacity at different levels. The cost is, of course, the predictions may be wrong. Predictions are wrong sufficiently often that biological systems build even on mistakes. For instance, predators make mistakes between mimics and models, perceiving the mimic to be noxious when it is in fact good to eat. But it is exactly those mistakes that led to mimicry in the first place. The freezing of patterns in information gives them privilege over simple emergence. Privilege amounts to significance. What is significant is stabilized. The rate independence of coded

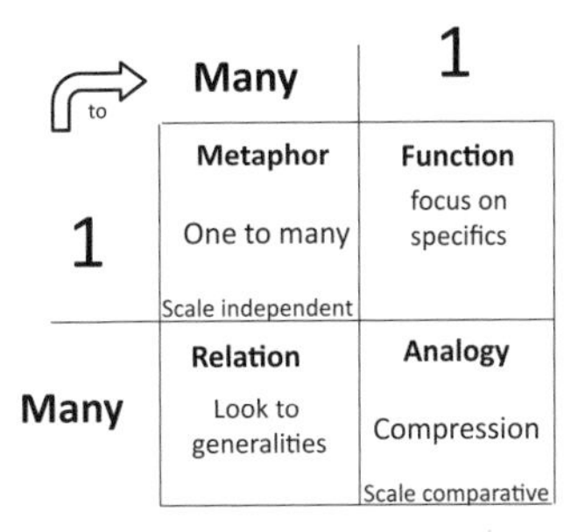

FIGURE 5.5. Lattice based on one to many. The cells are linguistic devices on the one diagonal and mathematical devices on the other.

information allows evolution to work with a ratchet. It encodes history, giving it a larger influence in a longer future.

A useful lattice can be set up to map intellectual devices so that we can shake them down to the base of what they are. The two sides of the lattice refer to inputs and outputs of intellectual activity (Figure 5.5). On the input side the options are one versus many. The output side is the number to which the first device related. Thus there is a one to one segment of the lattice. When one thing relates to just one other, mathematicians call it a function. In a graph, the ordinate relates to the abscissa. If one point on the ordinate applies always to only one value on the abscissa, then the two variables that are mapped onto the axes of the graph create a function. If, however, the graph curls back on itself, there is a more than one to one mapping between the variables. That is called a relation. It is like a function but there is slack in the relationship. In a correlation as a scatter of points, the greater the slack, the weaker is the correlation.

In a one to many mapping we find a metaphor. We have already discussed the modeling relation of Rosen. Figure 1.1 is the particular example of it, while Figure 5.6 is the general expression. There the formal model is a set of scaling equations, which apply to several material systems. A DC10 and a child's paper airplane, both are subject to the equations of the laws of aerodynamics. A metaphor is a single statement that can apply to many situations. As Rosen suggests, if two material systems can be together encoded and decoded into a single formal model, then the two material systems become models of each other, with the critical caveat that the relationship is not coded. The paper airplane and the air-

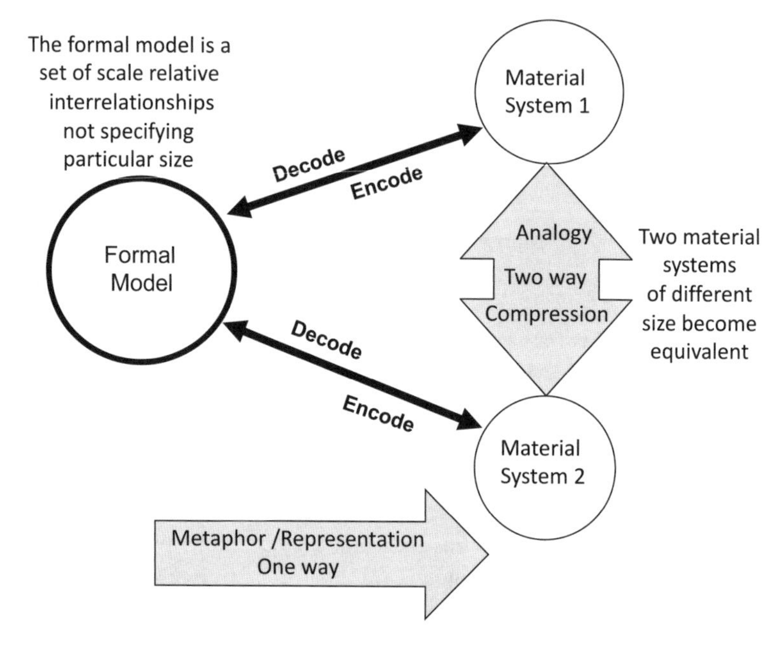

FIGURE 5.6. An austere version of Rosen's modeling relation to show its generality. The formal model is a coded representation and the two analog models are phyical analogs of each other with no coding.

liner are analog models of each other. All experiments depend on analogies between the test system and the one in nature. A metaphor is a representation. Analogy is a compression down to only that which the two analogs have in common: wings, drag, tail, but not paper because the DC10 is not made of paper. Analogy therefore applies to the many to one sector of the lattice.

If we make comparisons across the sectors of Figure 5.5 a new matrix arises. The margins of the lattice are now model and narrative relating to each other. In Rosen's modeling relation, the formal model, that we take as a function, is related to an analogy that sets up an experiment. If we relate a function to an analogy, as in arrow 2 of Figure 5.7, we get an experiment. If we have a general account of a complex situation it takes the form of a narrative. The experiment focuses on a small part of that space, so as to find the answer to some knife edge distinction. Experiments do not show the whole picture, but a good experiment does give an unequivocal, consistent view of some local part of the story. Accordingly the experiment occupies the narrative to model sector in the derived matrix. The experiment extracts a model from the open narrative, so the model can be tested.

Arrow 3 in Figure 5.7 takes the implications of the analogy (an experimental setting) and introduces the slack of a relation (puts it in a narrative). The experimental result closes the possible outcome to some particular. But dealing with the situation does not stop there. Each answer opens up another question. Focus of the experimental result is hurled back into the setting that is the narrative. In arrow 3 the result is set in the narrative of what we think is happening generally. The experimental result improves the narrative, but the narrative still has the slack of a relation. The result of the model goes to the narrative as the lattice says.

In Figure 5.8, the cycle of the modeling process on the side of the observer in Zellmer et al. is contrasted with the cycle of realization. This falls out of the comparison arrows in Figure 5.7. The gray arrows in the two cycles of Figure 5.8 are different. On the side of the observer we are looking at the iterative process of modeling. Experiment leads to a result that is the implication for the narrative. The result applied to the narrative leads to the conception of a new experiment. The cycle is understood as a straight two-way interaction. The cycle in the other is understood in a different way. To understand the cycle in the other we must look not at experiment and result, but in the whole cycle of change between realization and essence. The realization can be understood as

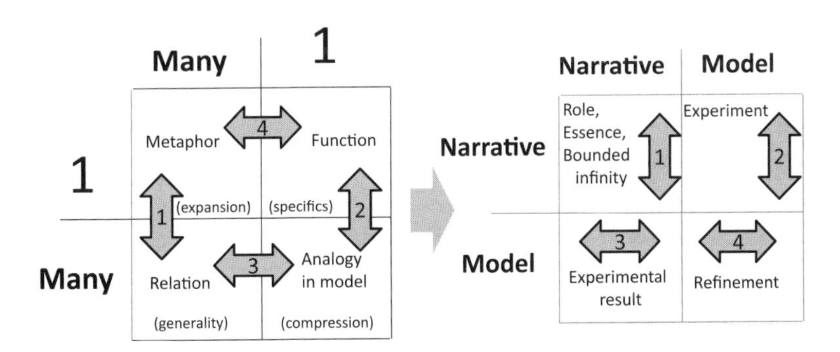

FIGURE 5.7. The lattice of figure 5.4 is tranformed to a comparison. This generates another lattice that deals with narratives versus models. This operation links the two literary devices with the two mathematical operations. Unsuspected at the time we did the calculations, it appears that these comparisons are given account by Zellmer's looping operations and lead to a generalization of what the Zellmer loops do.

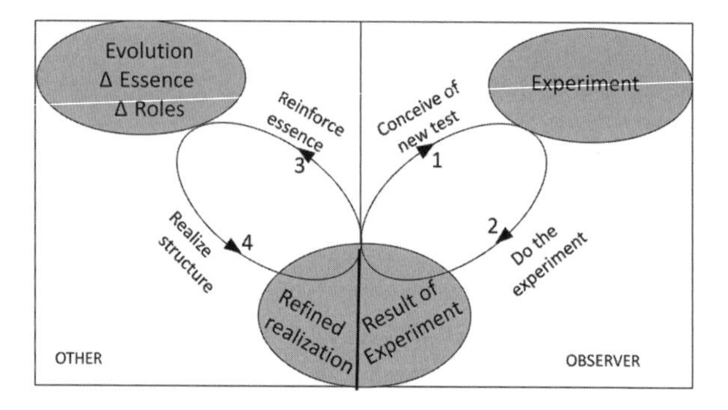

FIGURE 5.8. Zellmer loops applied to performing experiments.

affecting the essence that then effects a new realization. The implication of a realization leads to a new realization. Each update goes around the whole cycle to effect a change. Alternatively, we can see the essence generating a realization, which then updates the essence; another manifestation of a journey around the whole cycle as essence does things that update the essence.

The realization is the model of the essence outside the observer-controlled world. As the realization starts its own update, it amounts to a model replacing a model. The first realization changes the essence that then creates a new realization, a new model. That fits in Figure 5.8 as arrow 4. It is a metaphor in relation to a function. The metaphor is the model, the realized structure that is a model for the essence. It is concretized in the narrow, specific form of the realization (arrow 4). In the derived matrix of Figure 5.7, we see all this situated in a model relating to a model. This is the refinement of the model, as refinement of the process of the realization.

As the essence is updated, it tells a new story to the next realization. That is whence comes the anticipation of the realization. In turn the new realization affects and updates the story in the essence. If all this is in the context of evolution, the narrative of the essence is updated through natural selection. All this occurs in the narrative/narrative sector of Figure 5.7. Evolution updates narratives, so it belongs in the narrative/narrative sector. That is derived from placing metaphor next to a relation, arrow 1 on Figure 5.7. Metaphor is a one-way representation. Evolution works on representations. But in evolution there is slack, which is found in a relation. A relation is not as specific as is a function, but it is still bounded. Evolution is constrained but not determined. Roles too constrain, but there is no singular presciption as to how the behavior of the incumbent is determined, even if it is playing out a role.

Evolution is tricky in that it denies biology a time zero. What we see in biology is set in the context of evolution, which offers significant history ahead of what might be otherwise a time zero. For instance, when we think we see competition, it is in fact, as Connell said, the ghost of competition past.[35] We can show these problems by considering allelopathy. Allelopathy is chemical warfare in plants. We can set this into the diagrammatic scheme we have been using.

The observed entity at level N might be the target plant. A reasonable way to show allelopathy is to take exhudate from the allelopath in water washed off the leaves or roots, and then apply it to the target. The trouble is that the experiment fails (Figure 5.9). The target is unaffected by the presumed poison. As Lovett says, "The complexities of allellopathy are such that the advantage to be gained by a species which produces an allochemical is not always readily apparent."[36] Also bacteria complicate the situation by acting on allelopathic effect.[37] So we might imagine that the poison is not concentrated enough. Concentrate it and it appears to work. But a problem is the concentrate seems to poison the allelopath as much as its target. The way to show the effect is to find a naïve population of targets and compare them to a target population that actually has experienced the allelopathy. Apply the poison at a series of concentrations. If the naïve targets are damaged more than targets that have experienced the allelopathy, then you have shown allelopathy. The argument is that the experienced target population has adapted to the poison. Natural selection will bump up the poison, by concentrating it or whatever, and the target will adapt, also by natural selection. That will continue until the allelopath makes something so lethal that it starts to bother itself as much as its target. The naïve population has not had a chance to respond, and so shows more susceptibility. In other words, you cannot show allelopathy until it stops working (Figure 5.9).[38] And so it is across biology. Everything you see that appears as an adaptation is something that happened in the past.

The intellectual problem in evolution is that it amounts to a shift in levels. Shifting between levels is notoriously difficult, particularly with regard to issues of definition. The individual shows the adaptations, but

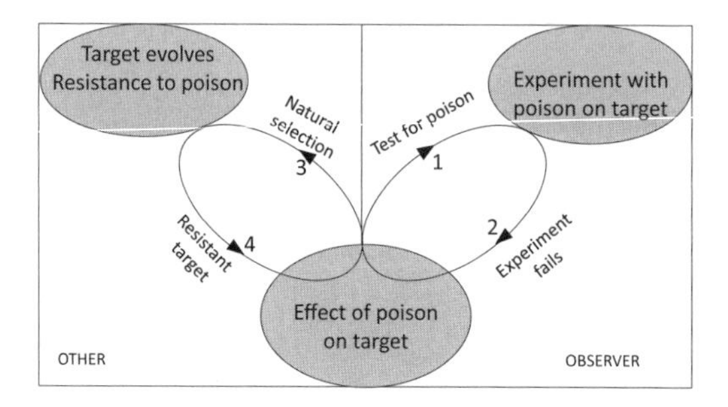

FIGURE 5.9. Zellmer loops showing how change in the essence changes the context of the model, so it ceases to work in isolation. This captures the ghost of competition past and the way evolution undermines allelopathy.

they come from the past. The information that embodies adaptation is developed in the essence, but the essence cannot be defined, because it is always changing. We have a series of devices that facilitate moving between levels, and the mapping of the lattices onto the cycles. Zellmer, Allen, and Kesseboehmer lay out how all this works. The issues raised by the comparisons in Figure 5.7 spell out the way that narratives and models interplay in evolution and how we model it.

As we move between levels we go from specifics to generalizations. The specific might be the organism in general or the thing that is modeled. The generalization might be the model at N + 1 on the side of the observer, or the essence on the side of the other. Table 5.1 looks at the arrows on the Zellmer loops numbered 1–4 in Figures 5.3, 5.4, 5.8, and 5.9. For instance the first arrow (1) in the loop diagrams goes from the observable entity to it being placed in equivalence class that is the type or model. That first relationship may also be interpreted in several ways beyond creating a type or class. It may also be seen as a) applying rules in performing an experiment; b) making a model from a narrative; c) giving meaning to a mathematical function; d) making the general specific to a particular class, or f) moving from tacit attention to a named focal attention. Table 5.1 generalizes Figure 5.8 to indicate the various facets to creating a model or class in the modeling process. It also generalizes what is going on in the other, as the material system evolves or develops. With all the movement up and down a hierarchy of actions in modeling, Table 5.1 spells out many other parallel implications in performing sci-

TABLE 5.1. **Summary of relationships in the Zellmer diagram loops**

Category of activity in investigation	1 Number of arrow on Zellmer loops.	2 Number of arrow on Zellmer loops.	3 Number of arrow on Zellmer loops.	4 Number of arrow on Zellmer loops.
Basic method in Zellmer diagrams. Figure 5.3	Entity indicates class	Check to see if entity is a member	Realization affects essence	Essence is realized in a structure
Experiments in science. Figure 5.8	Rules suggest next experiment	Experiment yields results	Observable behavior in context	Changes in observable
Evolving model evolving narratives juxtaposed	Make model from narrative	Apply model to improve narrative	Model improves narrative in nature	Change model using nature's narrative
Math functions vs relations	Give meaning to a function	Associate analogy with relation	Give model in nature a meaning	Map nature's metaphor to a function
Generality versus specificity. Evolved change	Make the general specific	Check to see if specifics fit special case observable	Relate realization to general situation	Realize general condition into structure
Focal and tacit attention	Tacit to named focal attention	Focal to tacit attention	Tacit to focal attention in nature	Nature's focal to tacit attention

Note: Summary of relationships in the Zellmer diagram loops as they appear in Figures 5.3, 5.4, 5.8, and 5.9. The numbers 1–4 refer to the arrows numbered in those figures.

ence. Table 5.1 gives unity to the role of the observer in working through the science of complex systems.

Models are internally consistent so models may be put in competition one against another to spell out the local implications of the assumptions in the model. But models are affixed to a point in parameter space—that is, they work with exact specifications. But important issues apply to a discourse that is so large as to pull the model off its point so it becomes contradictory or inconsistent. At that point models fail but narratives still work. Narratives are neither true nor untrue, they are merely an announcement of a point of view, as in the statement of a paradigm. Narratives, not models, are the bottom line of science. The relationship between models and narratives is that models improve the quality of science by inserting harder particulars into the narrative. That is what is happening on arrow number 2. "I don't know what exactly is going on, but my story probably has to pass through the unambiguous particulars of my model."

Another facet of science is identifying the generalities that frame the particulars. I see something particular, and as a scientist I try to see the general implications. As I focus explicitly on something, there is always a context that is taken as tacit. Humanities literature shifts to and from tacit and focal attention.[39] In performing an experiment one takes a general understanding and fixes it to a model. The experimental result is particular, but that understanding is then hurled out into the narrative context so as to suggest a new experiment. In science one often flips from a mathematical function that gives particulars to a mathematical relation where one addresses the slack in the context of the function.

On the side of the other in biology we address things like organisms, which have models and narratives of their own. The models might have specific triggers while the essence is an evolving narrative. As humans move between models and narrative, the things biologists study do the same. In evolution the realized entity (perhaps the organism) is the model of the essence (perhaps the narrative of the species). There is a consistency in being an organism, and that makes it a model for its general case, the species or essence. In evolution, the generality of the essence is a narrative that is modified by the realizations through natural selection. Meanwhile the evolved narrative of the essence fixes the form of the new realization. All this is spelled out in Table 5.1 and is diagrammed in Allen et al. 2014. So in Table 5.1 we highlight that evolution in biology is doing something like scientists modeling and telling stories. That "something like" goes on in social settings; however, we wish to emphasize that social change is not a matter of evolution by natural selection. We do not need another round of Social Darwinism or Wilson's Sociobiology.[40]

High and Low Gain in Evolution

Jim Gustafson, a psychiatrist at UW Madison, complained to Allen that the first edition of this book made reference to horizontal hierarchies, but did not develop them. His complaint is that we did not spend much time looking at hierarchies of a given structure moving through and changing over time. In his book, *The Modern Contest*,[41] he used our conception of hierarchies extensively. What he meant in his complaint was that we spend most of our time addressing the vertical changes in hierarchies in structural terms. We look to how things appear different at different levels in the hierarchy. Through the 1990s we developed a notion of high and low gain systems. After those ideas matured, we came to realize that shifts between high and low gain were in fact hierarchies changing over time.

High and low gain comes from looking at economic hierarchies. While biology does address use of resources, as a discourse it tends to stop there, while economics moves on beyond mere consumption. The notion of profit is hardly ever considered in biology. Biologists rarely see their systems in terms of cost to benefit, an idea that comes from profit and marginal returns.

Ecology is significantly a historical discourse, while economics generally is not. Tainter, in his treatment of the Roman Empire,[42] is one of the few social scientists who looks at cost and benefit in a strictly time dependent way (Figure 5.10). At the beginning of a cycle of exploitation of a resource there is insufficient technology to achieve much capture. In the middle period, sufficient technology allows maximal exploitation. In the end, according to ecologists, society or biology runs out of stuff. Ecologists are big on running out of stuff as an explanation and a predictor, which is whence comes their gloom-and-doom posturing. What we have just described is called in economics average return. It looks at increasing effort at first yielding more return, but later giving less return as resources are exploited.

Economists know that you do not ever actually run out of stuff, it just gets more expensive. And they are right on that. Their insight comes from their using a higher derivative of average return, called marginal return (Figure 5.10). It is not a return on how much stuff for extra technology and effort, it is how much extra is there in how much extra effort. That too follows over time a convex curve: a little followed by a lot, followed by a little with limited resource available. Marginal return over time is a concave curve, a hump, as is average return. However, plotted on the same horizontal time axis, the marginal return is about half a phase earlier than its respective marginal return. As the average return is still increasing at 45 degrees, the marginal return is flat on top of its hump. As the average return curve is half way up, the marginal return is only giving you as much extra as did the previous increase in effort. Systems appear to respond and plan based more on the marginal return.

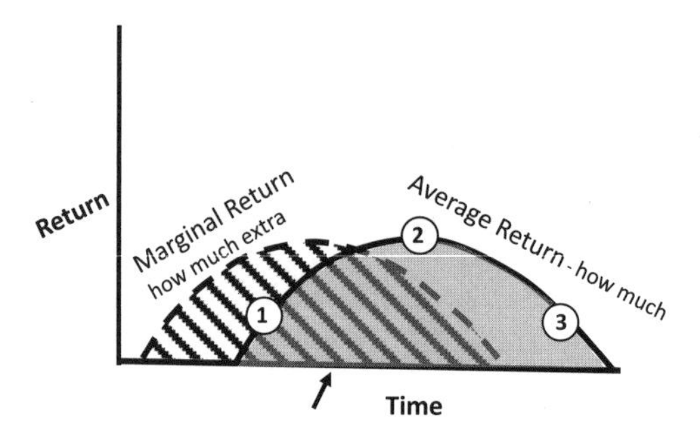

FIGURE 5.10. Showing the economic concepts of average return changing over time to effect diminishing returns.

They quit exploitation as not worthwhile well before the maximum average return.

Tainter developed the notion of diminishing returns on effort in his *Collapse of Complex Societies.* Out of that idea Allen et al. 2001 proposed the ideas of high and low gain. Gain is return on effort and it has an acronym in economics, EROI—Energy Return On Investment. If there is abundant high quality resource in the environment there is high gain on effort. But quality resource runs out, and only lower quality resource remains in the environment. The system gets less gain for a given effort. Superficially it appears that systems are either high or low gain, but in the end, as it so often does, it can come down to different levels of analysis of the same system. High gain accounts of systems invoke rate-dependent flux as descriptors and predictors. Systems so cast are seen to take in ready-made high quality resources as fuel for their processes. Low gain systems take in low quality resources and then process them into fuel. Thus a nuclear power plant itself is high gain, in that it takes in high quality energy in the form of fuel rods, and then produces prodigious amounts of electricity. But nuclear power as a general resource does not start with fuel rods, they need to be made. Low quality uranium ore is mined. It is transported (a form of processing) to sites where it is processed into the right isotope of uranium. That is so hard to do because chemistry will not discriminate between the relatively stable but strongly radioactive isotopes of uranium. Uranium processing has to be done using the small differences of mass between different forms of uranium using filters or powerful magnets. That processing makes the whole business of nuclear power generation very low gain.

High gain systems are predicted on the flux of the inputs and their easy degradation. Low gain systems are concerned with efficiency that gives privilege to certain outcomes that are identified as desirable. Raw flux is constrained to give particular patterns. Patterns and constraints are rate-independent. Thermodynamics is not efficient or otherwise, it is simply what happens. The high quality resources input to high gain systems are local. They are local concentrations that are in themselves rare. When the peaks of quality have been used, only background resource is left. The background is ubiquitous, but the concentration is low. The greater quantity of low grade material is so much larger than that in the hotspots, in terms of production low quality resource simply makes up for its poor delivery per unit input with greater total input of raw material. This appears a fairly universal principle, but it might be that any resource that does not compensate in quantity over quality does not qualify as what would be called a resource. Once the effort to find and concentrate poor quality inputs is not compensated sufficiently by the quantity of low quality material that can be gathered, low quality resource is not a worthwhile business. The end of low gain systems occurs when the quality of available inputs is so low as to offer insufficient compensation. If it is never, even at the outset, concentrated enough to reward collection and processing, whatever the material may be, it is not a resource. Gold mines worked out with an old technology are abandoned. The gold still there is not a resource. But with higher gold prices, and with better technology, the very low quality ore may again become a resource and the mine is reopened. Most mines are a disaster for local economies, because much of the time the good jobs are potential not actual. Enough time is spent waiting for jobs so that the principal effect of a mine is to discourage economic diversification. Comparison of counties with and without mines, with all else equal, shows that counties without mines are consistently better off because of economic diversity.[43]

Allen et al. 2010 worked through the example of termite evolution to drive home this point of switching between high and low gain. They created a response surface on a plane of declining resource quality against effort in planning and anticipation (Figure 5.11). The peaks on the surface are of two sorts, high and low gain. Prudent exploitation of an abundant resource will always lose in competition. If the resource is high quality then profligate exploitation of high quality is the best plan. But the quality of the resource declines over time. There are high gain termites that live in high quality wood and eat it. They literally eat them-

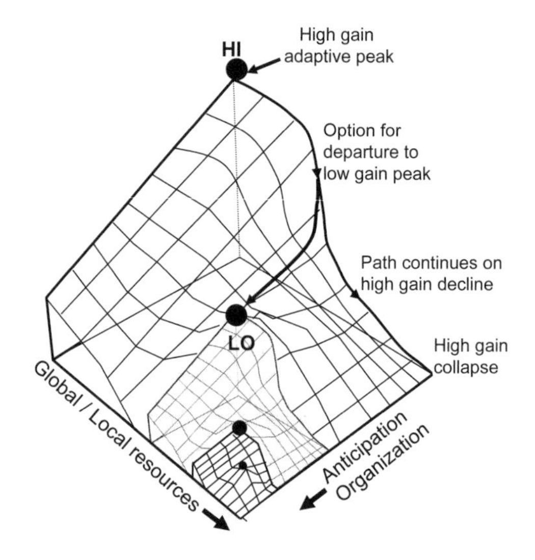

FIGURE 5.11. A three-dimensional space for resource use and adaptation over time. When a system is the first one to move to low gain, it has the option to high gain its newfound abilities. This means that peaks of adaptation can be both high and low gain. Changed from low to high gain, each peak of adaptation can be a high gain start to a new fractal landscape of adaptation.

selves out of house and home. At that point high gain termites have to move to a new chunk of good wood. In contrast to eating the good wood in which they live there is an alternative: increasing efficiency of processing allows for a move to low quality wood resources. This move leads to selective advantage in gathering dead wood fragments and bringing them to the termite nest. These are classic low gainers; the twigs and fragments are only 40% as good as a chunk of good wood in which the high gain termites live. But critically there is so much more total resource in dead twigs and fragments of wood. As a result these are the gigantic termite mounds seen pictured popularly in *National Geographic.* On the response surface, the low gain peak is somewhere in the middle of the surface, half way down the resource quality gradient and half way down the axis of organization and anticipation.

But it does not stop there. There are even lower gain termites that go beyond bad wood, and instead exploit the soil into which the wood has rotted. There is still 5% goodness in soil, but it is clearly much harder to process, even if it is easier to find. These termites live as small colonies of 20–30 individuals, with only the royals and the brood actually in the nest. The increase in size of low gain over high gain termites is because of the larger quantity of dead fragments on the landscape. Soil is beyond dead wood fragments in the poverty of resource it offers. Even though soil is ubiquitous, soil-eating termites invest rate-limited time and effort in digesting it. So poor is the resource that soil-eating termites are energy limited. Stuck with a termite template, soil eaters have limited options. Bacteria do not have the body mass payload of termites and so soil works well for them. The bulk of life on this planet is soil-eating bacteria. But evolution appears almost always to be pressing toward greater efficiency with regard to resource. The mass of soil-eating termites is not great, but over 60% of termite species have moved on to specialize in eating soil. Give termites a common yet better resource and we get the dead-wood-eating termites. They contribute mightily to the mass of termites. Astonishingly it amounts to 10% of terrestrial animal biomass.

But the story is still not over. Close relatives to termites are wood-eating cockroaches. On the response surface we see them in a high quality resource of good wood, but they are still not committed to wood; they are not that good at eating wood. If the good-wood-eating termites are high gain, then in a sense the cockroaches are even higher gain, in that they do not even invest prudent effort into being good at eating wood. So if we move cockroaches onto the high gain peak, then the high gain termites are relatively low gain by way of being wood specialists. This moves them onto the low gain peak in the middle of the surface. This in turn displaces the original low gain termites, but to where? The abundant low gainers are moved to a lower gain peak further down the gradient. All this shifting is a matter of conceptual changes not species replacing each other in nature. Thus high and low gain are relative not absolute statements, linked across a fractal universe. So it appears that peaks can switch from being low to high gain depending on the context in which the assessment is made. This shift from high to low gain occurs because the nature of gain is relative. It is possible for a given resource situation to be either high or low gain because of the nature of the transition to low gain. The first onto a low gain peak can high gain that peak. The first onto a low gain peak is well equipped to use the resource because of its new efficiency. Critically it has no competitors. It high gains the beginning of the low gain cycle. The soil eaters make the same shift onto the very low quality end of the gradient.

As hierarchies change over time, they tend to switch between high and low gain organization. This meets Gustafson's complaint about horizontal hierarchies. Because evolution works by moving across levels, it is well positioned to change hierarchical constraints over time. Evolution accordingly may have a model for its function in natural selection, but it needs the slack permitted in a narrative to capture the whole thing. There has to be the slack of a relationship (not a function) that offers variants as the environment changes. Perfect adaptation gets stuck and is a killer. It takes narrative to account for disparate models. Here we have shifted to interpret evolution as having the same difficulties as human model builders. Unlike physics, biology must deal with things that themselves have models. Human investigators model the organisms' models. Evolution does roughly the same thing as an investigator as it builds the models we investigate.

Preadaptation. To choose the familiar Darwinian level of evolution and to reiterate: if the organism does not adapt in natural selection that generates population evolution, what is the relationship of the organism, in adaptive terms, to the evolution of the population? That is what we explain in the cycle of realization and reinforcement of the essence. The selected individuals that contribute to population evolution through their reproductive success are not adapted, but rather they are preadapted. We wish to separate selection through the differential success of adaptive characters from genetic drift. Adaptation and genetic drift both involve change. In the classic account, those organisms that happen to be well suited to the new environment reproduce and so fix their good fortune in the population. But clearly more than good luck is intensified in evolution. The emphasis on chance is generally valid and often useful. However, for some questions, it is of less general utility than the following stronger statement. Those individuals whose form and function best anticipate the new environment are those who are stable enough to preempt the material otherwise to be used for construction of alternative forms; in reproductive forms in biology, it is those who reach a breeding state. As they breed, the mature (the stable) fix and amplify their anticipatory makeup in the population. This depends on the critical difference between thermodynamic flux and coded particularity. The kicker is that only coded information can pass unchanged between levels.

This emphasis on anticipation does not deny the mechanism of natural selection viewed as a stochastic process building richer function with the passage of time. It is just that such a view is only that—a view—and other perspectives are more powerful and economical for some purposes. The view of evolution that emphasizes chance now appears at odds with the facts (Noble 2013a,b,c). To the extent it is valid, chance focuses on the dynamics of the evolutionary process; our anticipation-oriented view draws attention to the rate-independent significance of the evolutionary end product. A limitation of the perspective that emphasizes chance rather than anticipation is that it does not take the end products of the evolutionary intensification at face value. Organisms of a given form can anticipate, even without experience. The experience is borrowed from the collective, the species, deme, or population. With the stochastic emphasis, purpose and function in the selected individual can be seen only in a historical and indirect fashion; this is useful for some models but is a long way round for many others. The circuitous view of a functional part, such as an organ, amounts to a large change in scale between the evolved biological holon and its observer. In our scheme we are speaking of the gap between level N and level N + 3 (Figure 5.2.). Instead of the observer trying to look at the organ holon with a filter similar to its own functional filter, the classical evolutionary model expands the observer scale so as to take into account a long period of selected stochastic signal. Neo-Darwinian mechanistic emphasis on chance is not absolutely incorrect; but it is wrong if it is set up as somehow necessarily superior to the complementary model. We have quoted others and have argued ourselves that mechanistic models are sometimes powerful in the right circumstances, but we have suggested on earlier pages in the chapter that there is no need to reject models that do not use mechanisms. We have already shown how mechanism is way of looking at things, not a way that the world is (remember mechanism is only a linear approximation). Many structures may be seen as serial at some level of resolution, but there is no necessary utility to seriality except that we can handle it conceptually; we are free to take advantage of levels of resolution that do not emphasize serial relationships. That is the power of neural net analysis, it shows order in very nonlinear situations that are not at all serial. There is order in non-serial schemes. Life lives open and away from equilibrium all the time, and yet it is remarkably ordered and in a predictive way.

Biologists are not used to dealing with the principle of complementarity, a principle adopted by physics only when nothing else would work.

Biologists still embrace unity, and when faced with paradox, they retreat behind apologies that their material is too complicated for one to know what is happening with sufficient precision.

Lazlo (1972) does not see us alone in our position.

> Thanks to Wiener's pioneering work on the concept of "purpose" it is now possible to describe systems in terms of ends and goals. Such descriptions can be, and in fact are, offered by investigators of systems on many different levels. (For example, the human organism is a goal-guided system at one level; its cells and organs are that on another, and a university is such a system on still another level.)

CHAPTER SIX

Functional and Structural Boundaries

The previous chapter has identified an intensified capacity for anticipation as an important characteristic of life that sets it apart from a dead matrix. But the evolutionary process that has brought anticipation to a critical level in living systems also occurs in the abiotic world. How then did living systems manage to gain the freedom to so intensify purpose and function? Pattee (1972) argues that it was the establishment of a language (the genetic code) that simplified the chemistry of incipient life by a system of description. This simpler description itself was then free to evolve the greater complexity of biochemistry proper. We agree, but also suspect that there is a dynamical explanation that can serve as a comparison to Pattee' s linguistic solution. The first section of this chapter follows that line of investigation.

It was indicated in chapter 2 that there are two means whereby a holon can escape constraint. One is to lengthen its time constants of behavior so that they are on a par with those of its environment. At that point the constraining environment becomes a sister holon at about the same level in the hierarchy as the holon it formerly constrained. Bosserman and Harary (1981) developed the idea of holon by first setting it as a sister to its environment. From there they moved to a conception where the environment gradually becomes contextual. The other means of escaping constraint is for the holon to shorten its time constants. With very short time constants the holon comes to live functionally in such a local time/space that the constraints of the larger environment are never encountered (Figure 2.7). A holon scaled very locally reaches such a small scale that the constraining environment passes very little information of any sort, constraining or otherwise. It seems to be this second strategy that allowed incipient life to evolve the distinctive characters that distinguish the contemporary biosphere from its physical environment. By developing biochemistry, life started to move so fast that the organic chemistry environment is outmaneuvered and falls away as only a part of the environment.

The particular characters that allowed this change in scale were enzymes. Enzymes do not change the chemical reaction they address; they merely increase the reaction rate. Reactions have a significant inverse that goes in the opposite direction. The two reactions end up in mass balance, where a greater quantity feeding into the slower reaction makes up for the reaction being the less active of the two. Enzymes bring about that mass balance more quickly. The emergent properties of life come not from new organic reactions but rather from the consequences of new rates for chemical reactions. Thus the qualitative differences in the chemistry of life are attributable not to new reactions per se but to a new scale for these reactions. We have in living systems an example of a Hegelian change of quantity into quality.[1]

While discussing the instability of systems with many variables and high connectivity, Levins (1974) comments on some general systems implications of enzymes:

> A system involves variables the reactions of which proceed at commensurate rates. Those which are much slower are held as constant parameters, whereas the variables of interest equilibrate; those which are much faster are treated as already moving at their equilibria. They are therefore replaced by functions of the main variables. This sheds a new light on the significance of enzymes. It is merely that they increase rates of reaction by many orders of magnitude and make new reactions possible. By doing so, they simplify the biochemical network of slow interactions among thousands of molecular types. The relative simplicity results in either stable or rapidly oscillating systems upon which natural selection may act. But once the enzymes are killed slow reactions creep back into the system; the very high connectivity results in nonperiodic instability; dust returns to dust not by assuming the same chemical composition as its surroundings but by having its own kinetics merge in a sea of commensurate reaction rates.

By increasing the reactivity of living systems, enzymes functionally isolate the system from its environment. Faced with an enzyme system, the primeval organic soup could do little except provide the most general constraints. Certainly the abiotic environment provides constraints as to the quantity of fixed carbon in the emerging biosphere, but apart from limiting global biomass, the soup must have had little to do with the rapidly evolving biochemical system.

Levins' insight that enzymes functionally isolate organic systems frees students of biogenesis from one of their major problems. If enzymes functionally isolate living systems by the rate of their reactions, the requirement for a cellular construction at the origins of life is relaxed. Since organized cellular construction is expensive, it would be appealing if energy-generating systems, such as fermentation, were to be evolved in an acellular mode. Acellularity is not demanding of much energy for its maintenance and would be able to evolve comfortably on a small budget. Processes as complicated as protein metabolism and photosynthesis have been achieved in acellular conditions in test tubes, but this may not be relevant to the discussion here. At life origins, once a metabolic system is evolved, it is then available as an energy source for the calorie-consumptive, cellular mode.

Structural Isolation

If cellularity is not necessary, one might ask what is the special role of a structural boundary such as a membrane. Indeed, there are grounds to question whether structural boundaries are different in any way from the functional boundaries discussed above, apart from the structural requirement that they possess only by definition.

When a light sensor fires in the eye, it tends to suppress its neighbors. If its neighbors are vigorously stimulated, they surmount the suppression and themselves fire. If, however, the neighbor receives marginal stimulus, it cannot overcome the effect of its firing neighbor. All this accentuates and sharpens the perception of boundaries; we tend to see discrete things and this surely contributes to our predilection for structural boundaries. While we would prefer not to cast it in realist terms, even if "reality" is put in quotation marks, who are we to object to von Bertalanffy (1975), particularly when he agrees with us in principle:

> Perception, however, is not a reliable guide. Following it we "see" the sun revolving around the earth, and certainly do not see that a solid piece of matter like a stone "really" is mostly empty space with minute centers of energy

> dispersed in relatively vast distances. The spatial boundaries of even what appears to be an obvious object or "thing" actually are [in a reductionist sense] indistinct; from a crystal consisting of molecules, valencies stick out, as it were, into surrounding space; the spatial boundaries of a cell or an organism are equally vague because it maintains itself in a flow of molecules entering and leaving and it is difficult to tell just what belongs to the "living system" and what does not. Ultimately all boundaries are dynamic rather than spatial.
>
> Hence an object and in particular a system is definable only by its cohesion in a broad sense, that is, interactions of the component elements. In this sense an ecosystem or social system is just as "real" as an individual plant, animal, or human; and problems like pollution and disturbances of the ecosystem, or social problems, strikingly demonstrate their "reality" . . . the distinction between "real" objects and systems as given in observation, and "conceptual" constructs and systems cannot be drawn in any common-sense way.

While "ultimately all boundaries are dynamic rather than spatial," some dynamical gradients (structural boundaries) do map easily into human experiential space. As the passage quoted above indicates, this does not make them especially real in an observer-independent sense, but it does facilitate their investigation and make them more direct in human experiential terms. The meaning and consequence of structural boundaries are more readily observed and so understood than functional boundaries. Dynamic boundaries often cannot be assigned to a continuous part of physical space.

The cell membrane powerfully integrates the enzymatic reactions it contains. In a sense it is the inverse of an enzyme, for enzymes isolate reactions by increasing their rates while cell membranes isolate by slowing down intrusion. Thus isolation involves not absolute separation but rather relative rates of exchange. The self-assertiveness of the cell protects the biochemical holons within. Furthermore, by concentrating end products within, the cell is able to use metabolic pathways that involve complex organic molecules that are exceedingly rare in the environment, but not the cell.

More than many holons, the cell has a surface that is readily recognizable in human perceptual terms through a microscope. The plasmalemma is the surface of a holon. Since the cell membrane is tangible, it makes a useful starting point for a consideration of the general properties of the surfaces of holons. Holon surfaces amount to scale differences and so directly affect rates of exchange. Scale differences in surfaces work to slow down both intrusion from outside and escape from inside. Only integrated signals escape; only modified and ameliorated signals intrude. Especially helpful examples here are cyanobacteria, which are thought to be free-living precursors of chloroplasts in eucaryotic cells. We do see the beginning of that process. There are species of algae that have lost their chloroplasts but survive using an endogenous cyanobacterium (*Richelia intracellularis*). Free-living cyanobacteria in the genus *Richelia* are well known. Signals from the chloroplasts' photosynthesis only escape the cell after they have been integrated with the rest of the eucaryotic cell's metabolism. That integration by the cell's surface works in the other direction too. Conversely, cyanobacteria do not generally occur free-living in low pH water, although the eucaryotic cell membrane ameliorates and modifies hydrogen ion concentration so that chloroplasts exist healthily, even when their host cells occupy very acidic habitats. Filtering and modifying incoming signals is one way that the system is self-assertive as a quasi-autonomous whole. As it integrates the signals from within, the cellular holon exerts constraint over its parts; products from chloroplasts after integration into the cellular metabolism become an important part of the chloroplasts' own environments.

Neardecomposability-. Simon (1962) indicates that hierarchical systems are, as he terms it, "nearly-decomposable systems." The notion is readily introduced by example, one which Simon himself uses. A system that Simon calls near-decomposable can be illustrated by a set of rooms well insulated from the outside; some of these rooms are separated from each other by thin partitions that are not insulated, although some blocks of rooms within the full set contained in the building are moderately insulated from each other, as blocks (see Figure 6.1). Without prior knowledge, this situation could be deduced from reading thermometers, several in each room. Rooms with outside walls would lose heat on cold days. The integrity of those, and all other rooms as individuals, would be identifiable by the strong correspondence of all the thermometers within any single room. Furthermore, the blocks of rooms poorly insulated from each other but moderately insulated between blocks would also be identifiable because there would be some correspondence between thermometers anywhere within a single block. The correspondence would not be as strong as within rooms, but even so blocks could be distinguished. The various levels of organization could be readily distinguished as room, block, and building by comparing the strength of interaction between the parts of each level as implied by the corre-

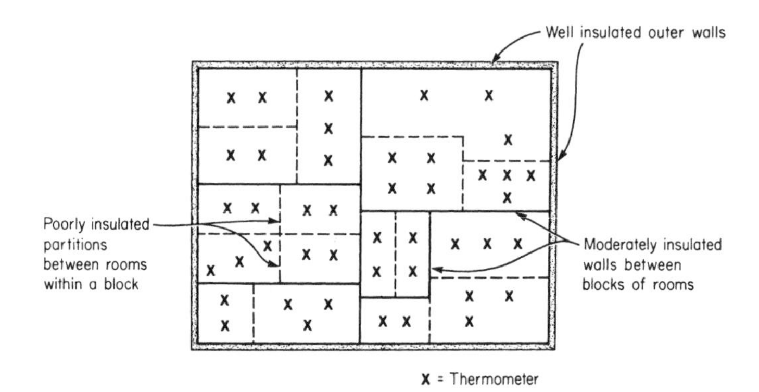

FIGURE 6.1. A depiction of Simon's analogy to rooms with different degrees of insulation. While the outer walls are well insulated, outer rooms of enclosures within will be cooler from heat loss. Outer rooms are more immediately responsive to the enviroment's ambient temperature. Inner chambers will respond more slowly. Each room occupies a position in a hierarchy of rooms. Two thermometers (indicated with Xs) in a chamber will behave in concert. Thermometers on either side of an insulated wall will be lagged relative to the behavior of the other. The lag will depend on the degree of insulation.

spondence of the points in question. In near-decomposable systems, low levels (rooms) are characterized by strong interaction of parts giving rapid equilibration within. Higher levels (blocks of rooms) are characterized by some cohesion within entities at that level but through weaker interactions than occur within entities belonging to lower levels. There are two patterns of quasi-discontinuity involved here: one is between entities at given levels, room from rooms and blocks from blocks; the other is between different degrees of interaction, rooms from blocks but not half-blocks. Rooms from rooms is, hierarchically speaking, a horizontal distinction, while rooms from blocks is a vertical distinction.

Platt (1969) gives a clear account of the term "near-decomposable system" and introduces a helpful example that we use and to which we return:

> These are systems which can be broken up (in thought or analysis) into subsystems such that the interactions within the subsystems are relatively strong and numerous compared to the interactions between subsystems. Deutsch (1953, 1956, 1966) has shown quantitatively how the high level of interaction rates within a nation [an entity at a level in a near-decomposable system], as measured by the flow of mail or communication between the cities, changes to a lower interaction rate between nations.

Compared with levels of interaction within national boundaries, the mail, automobile travel, financial transactions, and other types of interaction have lower rates of interaction across international frontiers. The mechanism of the filter across such borders involves many factors, such as customs control and cultural differences. The small quantity (slower rates) of mail exchange between nations has a spatial counterpart in that within-nation mail flow into frontier post offices will be from a higher proportion of geographically local senders compared to international mail. Detroit and Windsor, Ontario, are less than five miles apart, but on either side of the United States and Canadian border. There will be less mail from US addresses going to Windsor than to an equivalent post office in Detroit, Michigan. But the proportions of local senders to distant senders will also be different. The Windsor post office will receive US mail that comes proportionally from more distant points in the United States. US mail into Windsor is integrated more evenly over the entire forty-eight continental states. Meanwhile Windsor-bound mail of Canadian origin will be less spread over all Canadian provinces compared to the spread of US mail across the United States coming into Windsor (Figure 6.2). The international border integrates mail over a larger area than does the boundary between Detroit and the rest of the United States, because the city limits pertain only to small-scale local events while the international border relates to entire nations.

The idea of near-decomposability implies a special region across which only relatively coarse-grained information can pass (e.g., slow temperature change between blocks of rooms). Near-decomposability addresses the particular levels in a hierarchy that are seen as particularly real. We discussed these levels earlier when we identified the exact human being as being relevant in more dimensions than the human less its skin or plus an extra air jacket. We emphasize again that special levels such as the exact human are arbitrarily chosen as are all other levels in the continuum of levels. They are, however, of more general utility, applying to many models. Further, entities at these special levels with general usefulness appear more stable in more dimensions.

Stability. A definition of stability expressed in hierarchical terms is needed at this juncture so as to allay any suspicions that we are smuggling ontology into the discussion. Margalef (1968) brings into his discussion notions of stability similar to ours when he entertains the substitution of the word *stability* for what Leigh (1965) calls *frequency of fluctuations.* Stability involves the interaction between three temporal

FIGURE 6.2. Detroit Post Office seen from Windsor, Ontario, across the Detroit River, which is the international US-Canadian border. The Windsor, Ontario, Post Office receives predominantly Canadian mail while the Detroit Post Office receives mostly US mail. A comparison of the US mail processed in Windsor Post Office with that processed in Detroit shows not only that there are differences in quantity, but also that Windsor's US mail tends to come from sources more evenly spread over the whole United States. Much of Detroit's US mail comes locally from neighboring Michigan. (Photo: "U.S. Post Office, Downtown Detroit, Michigan from Windsor, Ontario" by Ken Lund is licensed under CC BY-SA 2.0.)

considerations. First are the time constants of characteristic behavior of the holon whose stability is to be assessed. That is to say, characteristic behavior of a holon involves repeated behavior: the return time is the time constant here. The second temporal consideration is the length of time a pattern of characteristic behavior persists. In a sense this temporal issue is how long the holon is defined as existing. The third temporal consideration that we require in a definition of stability pertains to the time periods associated with observation. There are two aspects to the time periods of the observation: first, how often observations are made: the grain. Second, over what period is the whole set of observations made: the extent. The observer determines what are the criteria for stability, while the dynamics of the behavior determines whether the observed meets those criteria. All this harkens back to the distinction between the side of the other, as opposed to the side of the observer, all in the scheme of Zellmer et al.

There are two arbitrary decisions to be made when considering a boundary. First, there is the decision as to where the boundary shall be. This is arbitrary because boundaries can be drawn at will by an observer to divide any two sets of things or places. Margalef (1968) expands on this point. Second, there is a decision to be made about the significance of the difference found on either side of the two sides of the boundary. Significance is not a material issue, as Tainter and Lucas note. They point to changing significance of archeological sites if scientists develop new technologies for measurement, such as carbon dating or microscopic interpretation of charcoal fragments. Even if the material site remains the same, there can be a change in significance. Thus stability or otherwise turns on significance, which is not a material issue. Thus stable or unstable is a relative matter.

Scale and Type with Boundaries

A distinction with regard to boundaries assigns the boundary in question to one of two classes: natural boundaries (which seem real) or unnatural boundaries (which exist only through a priori definition). A natural boundary meets what David Bohm says is of interest to science: that which is robust to transformation. Something is robust if it can be seen as the same with different measurement schemes. For example, a forest can be seen with light, or detected by measured increase of biomass. A forest can be identified with a thermometer that is by day cooler inside the forest than outside, while warmer inside than out at night. Also humidity increases as one enters a forest. The surface of the forest is therefore a natural surface because it coincides with many measured changes as the surface is crossed. It is possible to make any division with an arbitrary line. Unnatural boundaries are arbitrary in this way. They exist only by definition in a particular, even idiosyncratic, observation protocol and do not have reinforcing processes that coincide in the same place in time and space.

If the differences found between one side of the boundary and the other are judged sufficient, then the boundary could be called natural. Sometimes comparison between several alternative boundaries is needed so that the one showing the greatest differences between sides may be chosen as meeting the criterion for appearing natural (but note natural here does not mean ontologically real). Such an approach to a near-decomposable system would probably be able to identify the apparent disjunctions that make the system near-decomposable (i.e., find the partitions between rooms and distinguish them from the moderately insulated walls between the blocks of rooms in Simon's 1962 example).

Margalef (1972) noted,

> To state meaningful relations between two subsystems or half-spaces, the simplest comparison between them amounts to fixing a limit between both, and probably it is also required that the limit be nonsymmetric, and that there be something irreversible about it. Otherwise, what does a frontier mean?

The asymmetry in the international frontier is identified in our thought experiment. While Canada plus Detroit is different from Canada alone, the United States minus Detroit is altered less by such a shift in the frontier. In our experiment Canada was the entity and the United States was its environment; moving the border the other direction would reverse their respective roles (Figure 6.3).

Span. We must now bring the notion of span into our model of hierarchies. In the nested case the span of a given holon is the sum of the parts of which it is made. In the non-nested hierarchy the span of a holon consists of all holons over which it exerts constraint. Simon[2] pioneered the notion of near-decomposability, so he is given to using only those levels that exhibit near-decomposability, those with holons that appear discrete and enumerable. His definition of span is the discrete case, and his example is nested.

> The number of subordinates that report directly to a single boss is called his span of control. . . . Thus a hierarchic system is flat at given level if it has a wide span at that level. A diamond has a wide span at the crystal level but not at the next level down, the molecular level. [See Figure 6.4.]

Thus span is defined by what the observer chooses as natural boundaries—that is, what is determined to be the next boundary inward or the next level down. Span is a matter of appearances, not of ontological truth, although appearances involving natural boundaries are often striking enough to seduce a fainthearted epistemologist.

FIGURE 6.3. The Detroit River is part of the US-Canadian border. While it is a human decision, it has become a natural border. The land-water delineation has caused various characteristics to line up with the border. While the border is a human agreement, it has become a natural boundary as other characteristics have been reinforced by it. A natural boundary is coincident with many criteria: movement of cars, money, origins of incoming mail, and culture all line up with the river. Russia persists in playing the Great Game using Russian ethnics who moved into other countries (e.g., Ukraine) when national borders ceased to function in the USSR. An aerial view of Detroit and Windsor. (Photograph provided by NASA.)

Formation and Consequences of Surfaces

A point that becomes important later (chapter 8) is that the surface properties of holons in near-decomposable systems allow for the isolation of disaster. While the destruction of an arbitrary portion of a diamond would at least damage other portions, the destruction of one

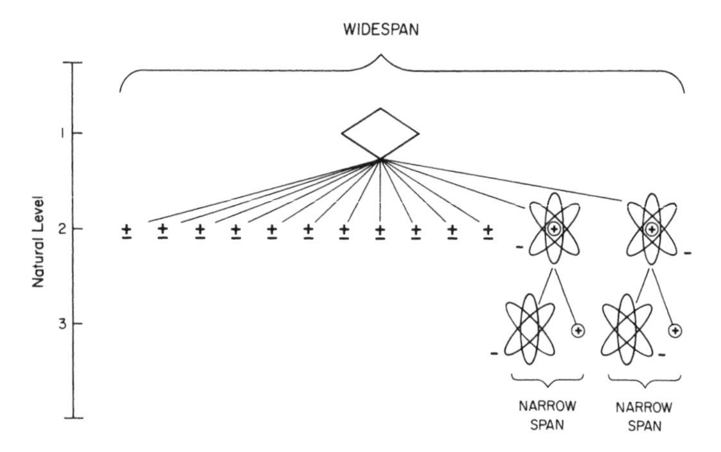

FIGURE 6.4. The span of a hierarchy in a diamond is wide at the level that notes many atoms at the level below the crystal. However, the span is narrow at the next level down, where a small number of subatomic particles make up each carbon atom.

isolated diamond would be contained by its air jacket. This allows the destruction of diamonds one at a time, while also permitting the chain reaction within a single diamond that allows diamond cutters to pursue their trade. If marked scale interfaces are not present, then damaging information can pass between holons. Note here how some conjoined twins are susceptible to each other's diseases and, before modern surgery, death in one usually led to death in the other. Evolution deals with functionally discrete entities, so that one can be selected when an adjacent another is not.

The separation of similar but not identical biological holons appears to involve very particular filtering of signal. Burnet notes that in primitive colonial chordates (*Botryllus*) elements of a single colony may grow apart, dividing the single colony. If those parts of the original colony bump into each other again they will fuse on contact. By contrast if "an unrelated colony II of the same species is brought in contact with colony I, there is a positive rejection and a barrier of necrotic material develops between the two colonies." Cells are triggered to commit suicide because dead cells delay signals longer than living cells! Dead cells place a steep scale gradient between the colonies, isolating each from the other's problems. The situation is different if the other colony is of a different but related genus, say *Botrylloides*. "If another compound ascidian of the same general type, *Botrylloides*, comes into contact with *Botryllus*

I or II, nothing happens. One will grow as if it were just an inert attachment surface." At the range of the genus, intrinsic scale differences are large enough for functional separation. Parent and daughter *Botryllus* colonies will fuse although higher vertebrate familial grafts will not fuse. Burnet speculates that this vertebrate rejection has its origins in the parasitic lamprey type of sucker fish, which represents a stage in the phylogeny of the group. Parasitic sucker fish need to control intraspecific parasitism. Vivipary, a common vertebrate asexual reproductive trait, also tempts parasitism of the parent by the young. As a result rejection must be all the more general (Burnet 1971). Separation to the right degree is crucial for stability. Obligate parasite fungi are sometimes thwarted by the plant cells of their hosts committing suicide as soon as the fungus invades. Living cells would allow transmission to the next cell, facilitating invasion. But the cells that die on immediate contact change reaction rates of infection and offer resistance to invasion.

Even more than is the case for sister holons, low- and high-level holons may be sealed from each other by scale differences. Simon's[3] example of himself and the Internal Revenue Service is a case in point. The IRS cannot discern how he calculated his taxes, whether by using pencil and paper, hand calculator, or a computer. Simon was in a computer science department and so gives further examples from that field. Different computer programs may have the same emergent properties (like response to the user commands) even though different versions are written in different languages. A single program might be written almost identically even though one machine might have in the early days of computing extensively used vacuum tubes while another computer using the program might have been solid state. With the coming of microcomputers, clones of IBM desktops were required by law to be different, but the programmer does not distinguish which is which from the standpoint of writing code because the workings of the computer itself are hidden from the programmer. The programmer works at a much higher level in the hierarchy. The identification of holon surfaces by decomposition is important if the world is to be manageable. Simon points out that "the art of subroutining, in writing complex computer programs, consists of identifying the points of cleavage at which the least information needs to be passed from one subroutine to another." Good subroutines fit neatly inside a natural skin. Thus steep scale gradients give rise to surfaces that do not allow the whole to know the details inside the parts.

Margalef (1972) comments on the way that skins of entities (scale interfaces) arise spontaneously. He observes, "Small random accidents in turbulence start the formation of a thermocline, and then the discontinuity reinforces itself." The effective establishment of structural boundaries such as thermoclines or cell membranes depends to a great extent upon the prevailing conditions of a continuous but steep gradient of scale at the moment when the fortuitous formation of the first part of the surface occurs. Once the surface is formed, the steep continuous scale gradient is substituted by two flat scale gradients with a very steep gradient amounting to a step (the surface) in between (Figure 6.5). Thermoclines form only where there are marked temperature gradients in the first place. Water temperature affects multivariate time constants, and in this way temperature gradients are scale gradients (Figure 6.6). Surfaces confer stability, and evolution is the survival of the stable. The prevalence of readily identifiable surfaces in the hierarchies of life (organelles, organs, individuals, population mosaics) is a reflection of the

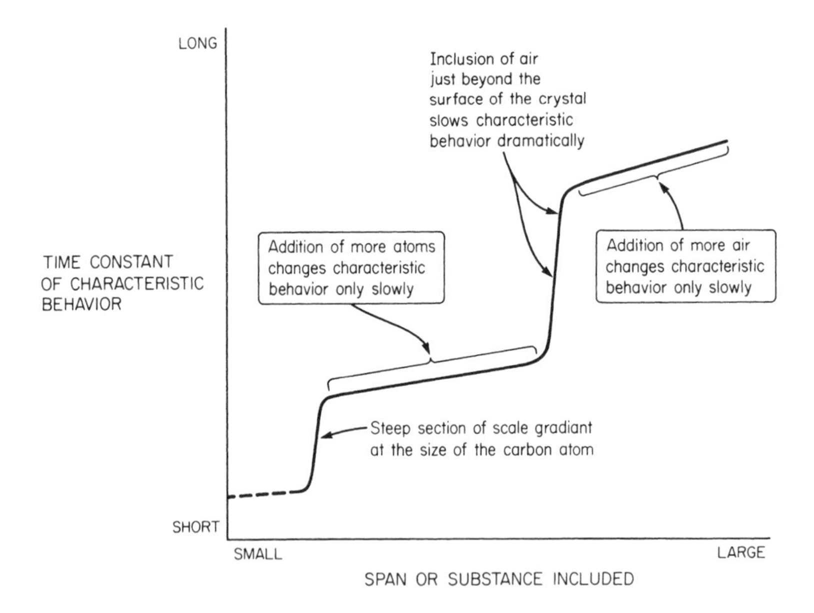

FIGURE 6.5. The space between two observation points is widened as a hierarchy is observed at higher levels. If the expansion is across only one level, the differences in time and space between measurement move up together slowly. But as the expansion crosses levels, the observation points come to be on either side of a new surface. Crossing the surface causes a big difference in the time constants of the observation across the small distance across the surface. Surfaces cause steps in the change of time constants.

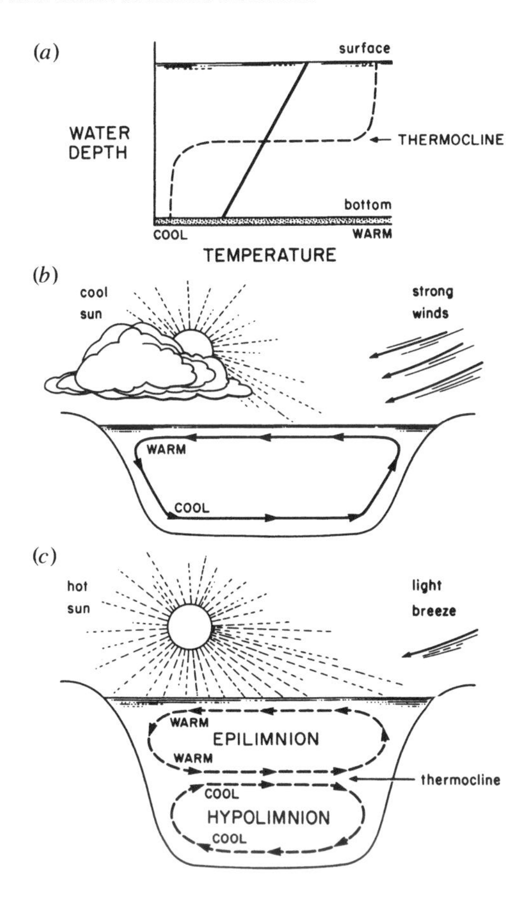

FIGURE 6.6. Thermocline in a lake is the steepest portion of the temperature gradient from warm at the water's surface to cool in the bottom. In cooler windy seasons the whole lake circulates and the temperature gradient is shallow. In hot calm seasons the temperature gradient steepens from the input of heat to the lake. The thermocline forms spontaneously in those times so water circulates separately in the epilimnion toward the top and in the hypolimnion toward the bottom of the lake. Separated from the surface by the thermocline, the hypolimnion becomes anoxic and nutrient rich. Steep gradients generally cause surfaces to emerge, and the thermocline is one such.

importance of localizing fatalities (by means of surfaces) in selection processes. As biological entities ourselves, we are very much in tune with structural surfaces and see them more easily than the processes of selection of which they are a manifestation.

Platt attempts to determine some general laws about the surfaces of hierarchical structures. Holons are n-dimensional, and Platt generalizes to the n-dimensional case:

> A subsystem in a nearly decomposable system in n-dimensions will have boundary surfaces of n−1 dimensions between a high interaction region and a low interaction region. [An intuitive example is the mouth, which is a one-dimensional hole in the two dimensional skin.] The surface may be taken as passing through a family of points where some parameter such as "interaction-density" has a maximum gradient. . . . The boundary-surface for one property (such as heat-flow) will tend to coincide with the boundary surfaces for many other properties (such as blood flow, sensory endings, physical density, and so on) because the surfaces are mutually-reinforcing. . . . This is what makes a collection of properties a "thing" rather than a smear of overlapping images. [A thing will often be identified by a steep gradient, but it is the observer who makes the thing a thing by making a decision.]
>
> All gradients and flows in the region very near to the boundary will tend to be either parallel or perpendicular to the boundary. . . . Evidently in the region close to a boundary the strong-coupling interactions within the systems are parallel to the boundary while the weaker-coupling interaction between the system and the larger super-system flow in and out perpendicular to the boundary. . . . Certain flow transactions are organizationally "cheaper" when spread over the whole surface uniformly; for example, thermal radiation. Others are cheaper when localized; for example, thermal convection which, when the temperature gradient passes a certain critical level, gets concentrated into localized air columns from a biological body as well as from the earth (Bénard cells, thunderstorm updraft).

Platt's example of localized transfer is a higher-order surface where the high-rate of energy exchange across the surface causes a scale gradient within the transfer system. The hot air itself forms a separate smaller holon in the transfer across the larger holon surface. Platt summarizes: "In general the internal language of a system is never the same as its external language to the environment or to other systems."

In his characterization of *Living Systems* Miller (1978) has 19 types of subsystems. There are several subsystems that work information across surfaces, changing the signal from one to another so that the internal language can be compatible with the external language. An input transducer like an eye converts a language of light into electrical discharges in the optic nerve. The internal language of the cell is one of substrates pulsing through an enzymatic cycle (Prigogine et al.), while the external language is one of simple diffusion. Enzymes outside fungal or bacterial cells simply break down substrates that diffuse to the cell membrane. By contrast, enzymes within cells form cohesive hypercycles of activity, quite a different matter.[4] At the origins of life, the cell membrane would have significance only when the language (rules) inside the cell pertained to a differently scaled set of reactions than the reactions occurring outside. Enzymes formed a steep scale gradient away from the slow reactions of the primeval organic soup. When a protein-lipid boundary chances to form in those conditions, it is extended and cemented by the steep reaction-rate gradient. This issue has become a hot topic recently under the rubric of emergence. The title of Steven Johnson's book suggests the breadth of the examples: *Emergence: The Connected Lives of Ants, Brains, Cities and Software.*

The problem of boundaries raises the question as to whether our best model should assume a continuous or discontinuous world. Until recently, physics had wanted a continuous world, and the mathematics of the continuous, calculus, served very well. At Wisconsin, however, Greenspan has developed a model for a discontinuous universe. He uses it to replace the continuous model, and his results are no more contorted or complicated than the heretofore ubiquitous continuous model.

Surfaces and Antisurfaces

In order to give our discussion of structural surfaces as much generality as possible, we contrast them with their inverses, homogeneous space and communication channels. We have shown structural surfaces to occur at points where there are steep gradients of behavioral time constants in the immediately surrounding space. At right angles to surfaces, the ratio between changes in rate constants and changes in spatial position is very large. At surfaces a narrow part of space manifests big changes in time constants. The character of most other places in space—that is, places between boundaries, gives a medium-range ratio where size and behavior rates increase gradually together.

This summary leads naturally to a consideration of places in space where the behavior rates change very little over large distances. These are places where the ratio of change in rate constants to change in position is very small, essentially the inverse of a surface. It is reasonable to suppose that such places exist, for if these are places where rate gradients are especially steep (natural surfaces and boundaries), there should be compensatory places where gradients are especially shallow. If surfaces have special properties, as they most certainly do, then their inverses should have special properties too; but what are the flat places like and what are their singular attributes? We might try seeking places that behave in somehow an opposite way to surfaces. Certainly inverse functions are often useful in solving equations in many engineering fields. In electrical engineering involving computer chips there exist mirror image circuits where a boundary between components in one is converted to a bridge in the other and vice versa. That is one way for so much to be put on a single chip or circuit-board.

Surfaces delineate heterogeneity while the feature of antisurfaces would be homogeneity. If rate constants are essentially the same over great distances, then homogeneity seems to be a requirement. Homogeneity in three dimensions is probably an unstable condition if the rate constants of behavior in the space are significantly quite large. The only vast tracts of homogeneity that come to mind are places like outer space where nothing much happens at all. Fairly soon after behavior rates begin to pick up speed in homogeneous spaces, surfaces arise (star formation). At the level of the events in the Calvin cycle, the cell protoplasm can be considered a large homogeneous space; Prigogine et al. suggest that as light and carbon dioxide (the forcing functions) reach critical levels, waves of sugar phosphates arise in the protoplasm, breaking the homogeneity. At the level of the entire biochemical cycle in the whole cell, homogeneity may exist, but at the level of the organic molecules in the cycle, heterogeneity with its attendant surfaces arises as the reaction rate crosses a threshold. Nonlinearity appears associated with the emergence of heterogeneity. Long enough linear continuity gives way to nonlinear curves that break, making for discrete surfaces.

Surfaces are places that do not facilitate information flow. As we cited earlier, Platt indicates that strong reactions occur parallel, not perpendicular, to surfaces (there is stronger flow parallel to the Detroit River in Detroit, compared to the flow across the river to Canada). Messages transmitted from spaces enclosed inside the surface are smoothed until they cannot be detected (that is how surfaces obscure). Messages encounter the lower-frequency medium on the outer edge of the surface, and their behavior becomes essentially lost. With very shallow scale gradients, however, if a message can exist in any part of the space, then it can travel great distances with minimal modification. Between us and the center of the moon there are steep gradients in density and in transmission of all sorts of energy; we cannot see the core of the moon and know little about it. We learn with the aid of special instruments placed on the moon to help transmission across the moon's surface. Between us and the stars the time constants of the medium generally differ very slightly; messages from the surfaces of stars, if they pass at all, do so easily over great distances. Flat scale gradients allow the passage of signal without modification.

Flat scale gradients are not always associated with 3D radial spaces. There may be flat gradients in one direction but steep gradients in others. Communication channels represent a class of such spaces; navigational canals are a case in point, and nerves are a biological example. The time constants of behavior of a nerve are very similar over long distances, so information entering the space of a nerve travels its length unmodified with ease. Note that the signal travels along (not across) the surface of a nerve. As Platt pointed out, surfaces pass strong signals ("strong-coupling interactions") parallel to the surface. The reason for this becomes clear as we note that steep scale gradients in one direction forming a surface represent an infinity of densely packed thin layers adpressed onto another, each layer at its own particular but single scale.

Transmission of information through and along surfaces is commonplace in biological systems, so much so that some sort of economical expression of layered structure is likely to be helpful. A first impression of the scheme presented in Blum (1973) in this arena appears most peculiar. This is hardly surprising, for it is a bold departure from the descriptive inherited from the physical sciences. Blum's geometry is a distinct alternative device that applies well to biology. A detour through this new geometry that was specially designed for biological spaces is worthwhile for ideas of scale and surface are neatly expressed and amplified by it.

Blum suggests an alternative geometry where the primitives are the point and the disc. A disc is the consequence of growth about the point. A line is not a primitive and must be derived by growth about two points. As the discs touch, they form the midpoint of the line; as the discs overlap, the two points of intersection move away from the midpoint to trace out the growing line. Lines are not asserted, they are grown. If the discs grow at the same rate and synchronously, then the line formed between

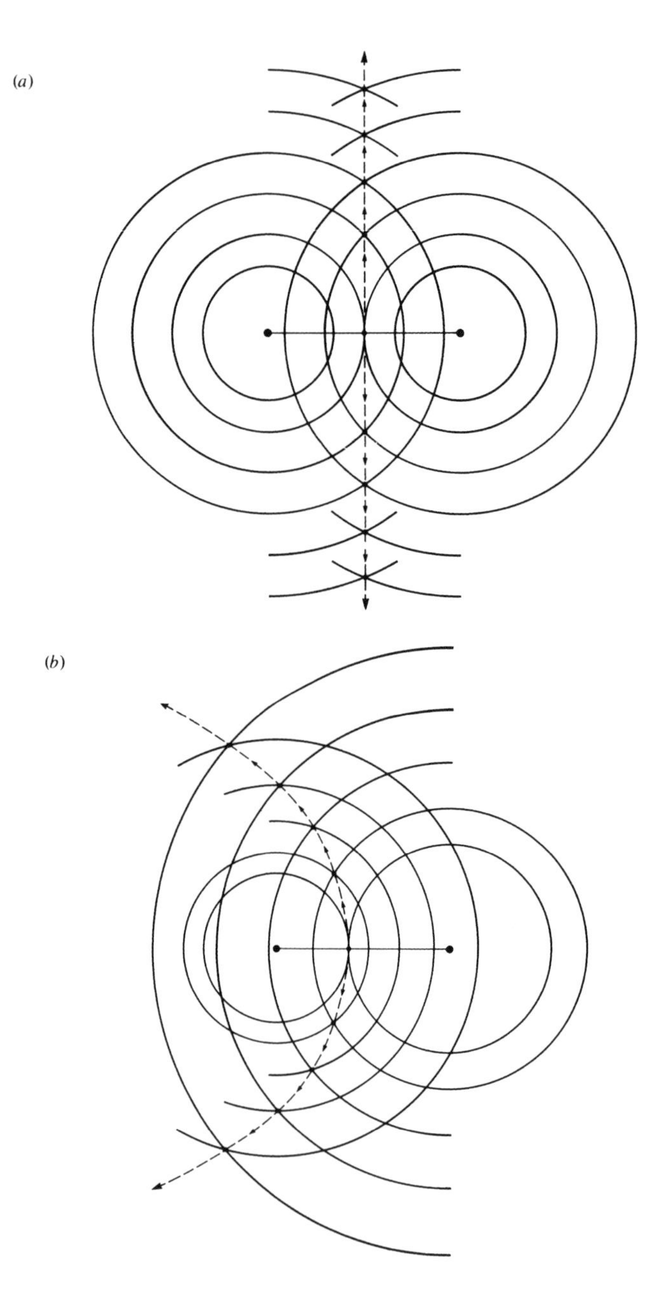

FIGURE 6.7. The fundamentals of Blum's (1973) geometry. (a) Ripples from discs intersect to make straight lines. Curved lines derive from growth at different rates. (b) The curved line arises because the right-hand disc grows faster and curves the line of intersect to the left.

them is straight and perpendicular to a straight line connecting the points at the center of the discs (see Figure 6.7). Asynchronous growth or different growth rates for the discs would produce a curved line. By placing points at various distances apart and by using different growth rates, a wealth of curved and angled shapes may be made easily. These shapes would be called complex using the line-based geometry, the geometry of physics. The power of Blum's approach is that it addresses biological structure directly.

The relevance of Blum's geometry to problems of scale and surfaces is that the two points that make a line can be taken as signal sources from either side of a surface. The surface is the line, and it forms spontaneously as a reflection of the progress of signal against change. Being the collision of different growing discs, the surface or line is the position of a steep scale gradient. A wide range of differences is found in the immediate vicinity of the surface. The line grows perpendicular to the steep scale gradient in a plane of the flat scale gradient. The line grows in the direction that allows easiest passage to signals, the place where layers of expanding discs collide. Remember how Platt says flow parallel to surfaces is common, and now we know why.

When we see a line on the ground there will be two processes centered in different places, both cascading upscale. Commonly fractals are seen as yielding closer and closer scrutiny with repeating patterns occurring at even finer scales. However, fractal patterns form by a seed of the pattern reproducing itself in ever larger terms. So a homogeneous region on a landscape will have a particular underlying pattern that gives a certain degree of complication, a certain fractal dimension. When two different patterns with different seeds (discs) have different fractal dimensions, they move upscale until they abut each other, then we have a line created exactly in Blum's terms.

Perhaps Blum's geometry facilitates the formal description of biological structures because it works in a manner akin to the construction processes in biological material. Three-dimensional spaces are divided into regions of hetereogeneity and homogeneity with fewer dimensions. Reduction of dimensionality is one of the stratagems of living systems. An example would be male moths whose problem is detecting pheromones from females in a massive 3D space. Antennae shaped as combs sweep the signal onto the plane of the antenna. Pheromones pass down to the stem of the comb and so become ordered in one dimension. The molecules pass to the base of the stem and are counted as zero-dimensional points ticked off as they pass. Male moths solve their problems by changing dimensionality.

Space to Evolve

Enzyme systems place life's chemistry at such a small scale that there is plenty of room for the evolution of higher self-assertive structures between the size of the enzymes and the size of the primeval soup. So weak were the effective constraints by the primeval environment on the evolving enzyme systems that much larger-scale structures could evolve before the environment provided powerful constraint. Enzymes are a transition point between two fundamentally different systems or universes. There is chemistry, which is largely a matter of electrical charges, and there is shape where surface and volume are connected by surfaces. Enzymes sometimes facilitate a reaction as they encounter a different concentration of hydrogen ions. This changes the electrical charges within the molecule, and results in the protein changing shape. In its shape-shifting the protein might put two substrates together, or it might tear a substrate molecule apart. The other universe, that of shape, is where biological systems gain leverage by using different spatial configurations. For instance, large flat structures for photosynthesis have been evolved separately as fronds in seaweeds and leaves in land plants to overcome surface to volume limitations. So enzymes link electro-chemical considerations with macro-spatial order.

After billions of years of evolution, life has a structural hierarchy between enzyme systems and the abiotic terrestrial environment. This creates new biotic holons defined by natural structural surfaces, such as cell walls and ecotones. This is hardly surprising since those boundaries that readily meet the observer-defined criteria for natural surfaces will confer the great stability upon those holons. It is these stable holons (those relatively more persistent) that play the principal role in biological evolution. When aerobic organisms first teemed in the waters the evolution of oxygen-producing photosynthetic forms changed the earth's atmosphere from a reducing to an oxidizing state. Life achieved a scale that was large enough to constrain the global environment. Over a longer time scale, James Lovelock's Gaia hypothesis shows how life constrains the temperature of the atmosphere. Without Gaia it is hard to imagine how life survived a thirty percent increase in the sun's output. "Among possible complex forms hierarchies are the ones that have time to evolve."[5] The importance of enzymes and cells in the origins of life lies in their contribution to what we see as hierarchical structure. There is, however, another aspect of life that also relates to hierarchical form, and that is self-replication. It is the subject of our next discussion.

CHAPTER SEVEN

The Self-Replicating Hierarchy

Cellular construction facilitates a highly organized replication. In a certain sense self-replication may be viewed as instability in the single cell.

> Instability must not be confused with lack of persistence. In cellular metabolism . . . it is rather interpretable as spontaneous activity. In cells, where the connectivity of biochemical networks is relatively low, it [instability] results in periodic behavior, cell cycles. (Levins 1974)

Self-replication gives remarkable homogeneity within the population holon. Large populations are sufficiently homogeneous with regard to parts that population biology uses equations that depend on that uniformity. Smaller, less homogeneous populations are more profitably described by individual-based modeling. There must be sufficient equivalence inside a population so that the parts are additive as they sum to N. The self-assertive population in a sense defends the individual from some of the consequences of death by misadventure. Note that a relatively small population of individuals is a less self-assertive population. The individual survives in relatives. With the advent of sexual reproduction, further highly integrated holons with their particular characteristics may be seen as part of phenotypic hierarchy. These holons are the reproductive pair, the recombinant individual, and the breeding population. They are part of the reason for the fast pace and great creations of biological evolution.

The Genetic System

Parallel to the phenotypic hierarchy, and interacting with it, is a genotypic hierarchy. Here we find messengers and masters in protein synthesis, the Mendelian genes and polygenes, gene complexes, duplicated segments, and crossover segments. The hierarchy of genetic components has level structure that is well considered as temporal. The order in different structures persists for a certain time giving genetic holons at particular levels. Each of the levels in the genotypic hierarchy holds genetic signal constant for certain periods of time, and is associated with certain rates of genetic change.

There are parallel genotypic, physiological, and phenotypic hierarchies. The conservative new synthesis in evolutionary thought largely ignored the physiological hierarchy that works between the genes and the phenotype.[1] Noble[2] has blasted conventional wisdom of the gene control by introducing the parallel physiological hierarchy into the scheme of things. Those parallel hierarchies are reminiscent of James Kay's linking of ecosystems thermodynamic hierarchies with social hierarchies in addressing dysfunctional ecologies in Katmandu.[3]

Genetic signal is manifested in strings of a molecule, DNA. Those strings are arranged in cells as chromosomes. The number of chromosomes is generally specific to the species. In some species there are only two chromosomes, but there can be hundreds of them, as in mosses. Often there are a score or so. The linear arrangement of chromosomes allows orderly replication but with the capacity for rearrangement, a hard combination to achieve. Life is about reliable copying—that is, memory, and trying on variations. Evolution is characterized as descent with modification. Genetic material arranged as chromosomes allows for reliability and creativity. Physically parallel arrangement of genetic variants allows chromosomes to switch pieces of DNA to give new combinations. The trick is to get the balance just right. Variation is metered through cell division and creation of new generations. Short life spans may allow for too much variation so there are other devices for slowing down change in chromosomes. All this is not simple in that sex may speed up evolution by offering new combinations, or it may slow it down by scrambling winning combinations. In general life is as various as it is because it is very conservative and has a stable memory. It is not so much that DNA can mutate, it is that is it dragged kicking and screaming into change. It has backup copies and devices for guessing at what has been lost. In *The Hitch Hikers Guide to the Galaxy* Douglas Adams tells of a species that, when it wants a cup of tea that is out of reach, it does not move closer, but rather evolves a longer arm. Not enough stability and memory there. Too many new variants and life descends into a shambles of bright ideas. The hard drive on a computer is more important than the speed of the processor. Memory is more important than innovation.

While caution should be observed in applying ideas of evolution to social settings, at a higher level than the process of natural selection (social evolution does not work via any sort of natural selection), something like evolution applies to social circumstances. As in biological evolution, change in social systems is conservative. Much of how people do things is guided by the way things have been done. Biological evolution works on similar conservative principles most of the time. But sometimes there is a radical change, as when willow species switch from insect pollination to wind pollination and back again. The move to wind selects against petals and nectaries, which disappear. But then out of wind pollination selection occurs pussy willows that are insect pollinated. There a different part of the flower comes to function like a nectary. Absent petals, the yellow attractive parts of the flower are the stamens. There must be some steep gradient of advantage, akin to sexual selection, to cause such a peculiar thing as a willow mimicking a petaloid flower. In social settings such radical change can occur through what is called *jugaard*.[4]

The strong gradient in jugaard is social need in places that cannot afford the technology. Often in India there is simply not the financial resource to perform heart surgery, given the massive need for heart surgery technology. The output of ECG machines in India is with an adapted bus ticket machine. Rather than use an expensive heart-lung bypass machine, Indian surgeons did not stop the heart for the time of the operation, they simply left it working and paralyzed the part of the wall where the surgery was being performed. Heart-lung bypass machines are not only expensive, but they also damage red blood cells, which cause many mini-strokes. The phenomenon has a name amongst heart surgeons, "pump head," to describe the changes in personality that often follow the use of the bypass machine. Soon enough, First World surgeons started copying the Indian surgeons.

Another example of jugaard occurred in Kenya where land line telephones were never fully functional. But cell phones are common and work well. Mobile phones revolutionized Kenyan society by groups of farmers buying one together so they could all escape the monopoly of the local buyer, and so reach the market. Women became functional phone booths. Most revolutionary were cell phone minutes that were usable and electronically transferable.[5] That allowed people with small assets to access global banking. With regard to jugaard, banking in Kenya started to use mobile phones for transfer of funds. In the First World it was always done by land line, but changed to mobile phones under the influence of Kenyan practices. So most innovation is a matter of fidgeting with standard practice, but sometimes, with a steep enough gradient, radical organization can occur in life and in society.

In biological systems using genetics, the issue is how to create new structures while transferring information. Cells divide through two means, mitosis and meiosis. Regular cell division is called mitosis, where identical daughter cells arise from evenly dividing genetic material. Somatic growth of lumps of cells is by mitosis whereby organisms grow bigger. There is accumulation of identical cells. Mitosis is conservative and robust.

The other sort of cell division is associated directly or indirectly with sexual reproduction; it is called meiosis. Mitosis represents the balance of replication and division. Meiosis solves a problem that is somehow the opposite of mitosis. Meiosis is the antidote to increasing the number of chromosomes as a result of sexual fusion. The product of sexual fusion has two sets of chromosomes, one from sperm and the other from the egg. Sex usually involves haploid cells with only one set of genetic material, one set of chromosomes. The fusion of haploid cells makes a diploid cell with two sets of chromosomes. Most vegetative cells are diploid. Meiosis halves the number of chromosomes so a diploid coming into meiosis becomes haploid. Meiosis involves diploid cells that contain genetic material fused in the previous sexual union. That union gave rise to the cell or line of cells now undergoing meiosis. Meiosis involves equivalent chromosomes from the two respective lines of the parents from the previous generation. In meiosis the equivalent chromosomes from the previous respective parents must find each other in the cell and come to lie beside one another. This coupling facilitates the chromosomes exchanging segments. The process of exchange is called crossing over. In crossing over, segments of one chromosome in the pair become inserted into the other, and vice versa. Thus chromosomal material that arose separately in the previous generation of parents can get onto the same chromosome. That search for the other chromosome in the pair is the reason meiosis is more challenging than simple cell division in mitosis. It can easily fail, causing sterility.

Crossover segments discourage crossing over within the segment, and so hold constant genetic configuration in the segment, sex and meiosis notwithstanding. Even so, more than one crossover in a pair of chromosomes is common. The chromosome represents a vital holon between the gene and the genome. Without the chromosome as a filter, either the gene would be permanently locked together in just one genome or there would be messy duplication and deletion of genes in a shambles. The linear construction of chromosomes allows for organized segregation at the same time as recombination in crossing over. Reorganization occurs with crossing over, and the results depend on where a pair of homologous (equivalent) chromosomes crossover. Thus a whole series of functionally important holons between the gene and genome can be formed. It is the chromosome that allows the near-decomposition of the genome into many apparently discrete levels. Each level varies its genetic signal according to its own distinctive time schedule, each presenting a functional surface in the context of which natural selection may readily operate.

Chromosomes are of various number and size; this allows different species their own characteristic genetic time scales. There are various devices for changing the degree to which flowering plant fertilizes itself (inbreeding). The pollen grows down the style to deliver the sperm to the egg; a longer style increases outbreeding and so increases the frequency of new combinations. In division of cells there is an assortment of genetic material between the daughter cells. When meiosis separates one pair of equivalent chromosomes, other pairs separate independently. The independent assortment of non-equivalent chromosomes is a powerful filter that scrambles the particular signal of the antecedent cells. There is a hierarchy of genetic holons, each holon with its own periodicity.

Sometimes chromosomes lose a section and then insert it back into the chromosome, but backwards. When it comes to regular cell division in mitosis, the inverted segment makes no difference, normal sequence versus inverted. The same is not true for meiosis because the inversion can make a mess of segregation of genetic material. Crossing over in parts of inverted chromosomes that have not been inverted works just fine. But crossing over in the inverted segment fails. Directionality is physically reversed in the pairing of inverted segments. Crossing over in the inverted segment causes deletion and duplication because of the reversed directions. Evolution selects for not crossing over in the inverted segment. Thus when an inversion occurs the inverted segment has a slower recombination rate. It preserves its gene sequences longer, lowering the segment's frequency, and so moving it up the genetic hierarchy. Remember, we said that short generation time can disrupt genetics with too much variation. It is telling that chromosome inversions occur to separate fly species.

If the piece of chromosome broken out is reinserted not backwards in the same chromosome, but into another chromosome, the phenomenon is called transposition. Then at meiosis the two chromosomes must pair also with regard to that moved segment. That means that four chromosomes, not two, are involved in pairing. Segregation of the four chromosomes is not problematic so long as there is no crossing over. Crossing over in thus anomalously connected chromosomes means they must segregate in just one particular way. Segregation is thus not the normal independent assortment of chromosomes. The donor and the recipient transposed segment must separate in the same direction, thus slowing down variability. If they segregate in different directions the recipient of the transposed piece goes to one daughter cell and the transposed segment goes to the other daughter cell, even though they are still attached. This denies physical separation at the end of the cell division. *Rheo discolor* is a common border plant with purple spear-shaped leaves; it is a relative of *Tradescantia*. It has all its chromosomes translocated, so at meiosis a ring of connected chromosomes forms. This is viable only because the species has a fixed pattern of segregation. The translocated segments join all the chromosomes into one genetic unit. Thus translocation works at a scale above the whole chromosome, for it determines the behavior of more than one chromosome pair at segregation. In this way translocation creates a holon above the level of the chromosome pair, but below the level of the whole genome.

Above the genome occur holons associated with breeding systems and populations. There even appear to be genetic holons above the level of breeding barriers. The great Russian geneticist Vavilov[6] formulated a law of homologous series of variation, where parallel patterns of gene expression are predicted to occur on either side of ancient breeding barriers. For instance, there are patterns of variation that are similar in wheat and barley, even though those genera do not normally interbreed. The parallel patterns come from a genetic inertia manifested as to what variations can arise most easily. That way there is no need for interbreeding for parallel patterns to appear. Some genetic changes arise more easily than others, and that would explain parallel patterns of variation without direct interbreeding.

Each of the above genetic holons represents a safe stage in evolution akin to the stable configurations of watch parts in Simon's watchmaker story. The genetic stable configurations then interact with the phenotypic safe holons of enzyme, biochemical pathway, organelle, and cell, on through to levels as large as those in the higher levels of taxonomy. The biosphere is near-decomposable at many phenotypic and genotypic levels. The stable holons are so densely packed that evolution with its intensification of purpose and anticipatory potential can move very rapidly. When the genetic system suffers defeats in extinction it loses very little ground, just like Simon's hierarchical watches. By interaction between levels, living systems become variable to a second or third order or higher. It is variability that can be selected that drives evolution, and the capacity to vary variability builds into life a new higher order capacity for anticipation of the environment. Evolution depends on variation, which is the basis of selection in natural selection, but amount of variation can be engineered.

Voeller (1971) has shown that ferns have a mechanism for outbreeding and thus suppression of recessive genes, which are usually deleterious. This outbreeding system works well except at the margins of the species' range, geographically well away from most of the individuals in the species. At its range a fern is faced with an environment that is likely to be very different from the norm for which its spore-producing parent has been selected. At the range of a species a genome in desperate straits might profit from a long-shot gamble. It is probably worth exposing recessives as a long odds bet. Inbreeding tends to expose recessives to selection, because half the heterozygosity is lost in diploid organisms.

Inbreeding in a fern is a dramatic case of variability in variability. In inbreeding the diploid genetic variation is cut in half; half the heterozygous genes become homozygous. But in ferns inbreeding is between haploids. The egg and sperm are identical, having the same haploid parent. As a result, in the next generation the normally heterozygous diploid becomes homozygous at every locus. Not half, but all heterozygosity disappears, exposing all recessives to selection. The homozygous inbred genome is likely to give a most distinctive phenotypic variant in the diploid generation. Thus genetic homogeneity increases the range of variation presented to the environment for selection. Note that domesticated corn produces more distinctive phenotypes than wild corn precisely because the domesticated genome has less genetic variability. Less genetic diversity in an individual exposes remarkable phenotypic variants. Built into the reproductive mechanism of the fern is the capacity to predict otherwise unpredictable conditions for its offspring. Inbreeding in ferns results in individuals where genes normally suppressed by outbreeding are exposed as double recessives that appear in the phenotype. While variation within the homozygous population is zero, the particular form of the homozygote is an extreme phenotype, a remarkable variant. This is a manifestation of great phenotypic distinction only in environments that are likely to be challenging. It is the response to an expected unpredictable future. Ferns can thus dramatically vary the variability of their variation in variable environments, just when it counts.

The same thing happens in *Daphnia*. Water fleas reproduce asexually within one year so as to be well suited to prevailing conditions. At the end of that year sexual individuals arise. Their presence anticipates that there will be a change in the environment between years that needs to be met with the variation that meiosis and sex engenders. The anticipatory capacity built into the realized structure of the phenotype of life forms is effective at many levels.

Complex characters are holons that are the product of intricate interaction between many genes. In such characters the genes may be fairly simply additive as in the height of humans. In characters like camouflage the many genes may each fulfill special roles. Genetic recombination is a mechanism whereby new variation in complex characters may be achieved. It seems paradoxical, then, that a capacity for greater recombination is inversely related to variability in complex characters. Once a very peculiar complex character is achieved, further recombination will usually create a character which is less distinctive, not more particular. In changing environments, recombination becomes a necessary evil. It is necessary because the new environments require recombined characters; it is hazardous because particularly helpful characters that would remain so in the new environment are likely to be recombined to an unexceptional, less helpful condition.

Most eucaryotes do not go further up the hierarchy in their control of recombination, but fungi have a mechanism for preserving the integrity of the whole haploid genome while retaining options for new combinations. In fungi there are at least three genetic conditions. One is the normal haploid state; another is the typical diploid condition resulting from fusion of haploid nuclei (which immediately undergoes meiosis); while the third is a cell that has two separate haploid nuclei from different parental lines. It is this last condition, called the *dikaryon*, that allows the fungi to make new combinations without disrupting the haploid genomes. The two nuclei in a dikaryon may eventually fuse to produce a normal diploid. In a return to haploidy by meiosis, the disruption of complex characters occurs but new haploid building material is formed. Sometimes the two nuclei in the dikaryon may not fuse at all, and on encountering cells from other lines they may exchange nuclei with those cells to make a new dikaryon. It is thus possible for the original haploid condition to be restored without meiosis and the disruption meiosis causes. Meanwhile new combinations will be formed in the process of exchange. Evidence in *Saprolegnia* genetics indicates that there can be exchange between cohabiting haploid nuclei even to a level as fine as a single gene.[7] Genetic engineering that inserts new genes would not be possible unless that sort of thing happened in nature. Fungi not only control genetic variation with gene and chromosome holons, but also recombine at the whole haploid genome level. In fungi the haploid genome is unusually self-assertive but without interfering with recombination. The surfaces of the twin nuclei allow assertion without intrusion upon the option for recombination.

CHAPTER EIGHT

Scaling Strategies

The full significance of self-replication in hierarchies becomes clear in the context of environmental perturbation. In the face of environmental perturbation self-replicating hierarchies display complex scaling strategies. It is these strategies that give contemporary life forms their impressive resilience.

The Paradox of Perturbation

A given fatal perturbation often occurs importantly at a single scalar level. A disturbance can affect a holon only if the scales of each are similar at least for that local time period. If the perturbation is small and ephemeral, it passes unnoticed; a solitary leaf-eating caterpillar on a tree would be a case in point. If, however, a disturbance is long and large, there are two ways in which holons survive unscathed, but in both cases the perturbation becomes the state of affairs in or around which the holon normally operates. In one case the disturbance is so long-lived that the holon lives out its life knowing nothing but the disturbance. In the other condition repetitive cycles of long disturbances have sufficient periods in between to allow the holon to complete its life or finish its business in the spaces between disturbance. The holon never experiences the cycle of disturbance. Cacti and drought show the first case; desert ephemerals and drought are examples of the second condition. Another example involves all of life; an effective strategy of life in the solar system has been the manner in which it slips, although full blown, between the birth and death of the sun. Gaia[1] suggests that life gains control of the hierarchy at the surface of the Earth and its atmosphere. The output of the sun has increased thirty percent in the three and a half billion years of life on Earth. And yet the ocean has neither frozen nor has boiled away. It appears that life asserts control over its setting, such that life itself is the top holon in the hierarchy.

If, however, the holon and the disturbance use almost identical scales, then the holon will be perturbed by the transition from no disturbance to disturbance. It will enter the disturbed regime with its course not yet run and with its business, of say reproduction, unfinished. But there is a complication here; two holons that interact using similar scales will begin to assume a common identity. Hutchinson in his paper "The Paradox of the Plankton"[2] noted that disturbed and heterogeneous environments tend to have higher diversity in terrestrial systems. Each piece of the diverse environment offers yet another species an existence. However, water is a very homogeneous environment and yet the diversity of plankton that live there is high. Hutchinson says that evolution away from each others' niches may allow coexistence in what J. H. Connell called "the ghost of competition past."[3] That can promote diversity. We already mentioned character displacement,[4] but in plankton something else is happening. Two species that are very similar to each other in their environmental demands cannot afford much competition because to hit the other species the perpetrator will also compromise itself. This is Hutchison's explanation of plankton diversity. Plankton holons become replicates of each other, even if they are different species.

We discussed something similar in chapter 4 when we noted that the scientist using the same scale as the observed holon becomes a large part of the observation. Clearly the "disturbance" perturbs less as it becomes incorporated; in fact it ceases to be a perturbation at all. Incorporation is achieved through evolution. The effect of the perturbation has been to change the formerly perturbed holon to a brand new system in which the disturbance is taken for granted. This is one of the possibilities for biota in the face of perturbation, and it seems to be manifested often. Once the perturbation is incorporated, it is removal of disturbance that would be the perturbation. In human society, elections are a case in point. Note how the different election schedules produce different emergent properties of pragmatism versus high-sounding principle in the elected branches of power in Western democracies; wider-spaced elections are related to longer-term views expressed by elected officials.

The grasslands provide examples of incorporation of perturbation.

The prairie community is not a community that is perturbed by fire; it is fire-dependent. We can only guess at the course of the incorporation of fire into the community, for we cannot watch the process retrospectively. Biotic disturbances have also been ameliorated by incorporation into stable higher-level grassland holons. Grasses have been able to incorporate their grazers because the plants have evolved meristems, growing points, which are low to the ground and are not damaged by the herbivores. A moderate amount of grazing rapidly shifts a community in favor of grasses. Apparently grasses make a profit from being grazed by saving on the costs of competition with forbs, non-grassy herbs. Ecologists often deal with resource use and sometimes resource loss, but they rarely deal with the higher level of analysis at which profit works. Economists understand things often in terms of profit, and so have insights to give to natural scientists. The trick is that forbs have their meristems at nibbling height, and so have difficulty competing successfully when the grasses keep fueling the perturbation. Forbs not only lose capital in grazing, they also lose meristems that are the means of production. Grasses, with their meristems at the base of leaves, below the level of grazing, lose only capital to grazing; they get to keep their means of growth and recovery. Grasses without forb competition or grazing would do better still if they were not grazed, but in the presence of competition from forbs, grazing is a cost of doing business. Having evolved to profit in that regime, lack of grazing can lead to self-shading with accompanying reductions in productivity or loss to shading by taller woody plants that are at no disadvantage in having their meristem high in the absence of grazing.

The horse becomes part of the grass holon. Natural selection is based on relative rather than absolute resource capture, and spoiling the environment for others may be selected, even at the long-term cost of accepting perturbation as a bedfellow. Economists would understand this more easily than do ecologists. The notion of gain introduced in this book allows ecologists to understand like an economist.

If incorporation is so common an evolutionary event, one might ask why it is that Darwin does not mention it explicitly (he mentions a lot of other things) and that modern evolutionary theory gives incorporation such short shrift. We think it is that scientific models in natural history often reflect the social climate in which they are conceived. Von Bertalanffy[5] sees the gene paradigm as a democratic principle of equality before a genetic law. Note here Vavilov's problems as a chromosomal biologist in Stalinist Russia.[6] Soviet values preferred society to be able to change people with environment into willing communists. The notion that there are fundamentals to humans that cannot be changed by environment was anathema to Stalin. In Darwinian evolution, implications of "survival of the fittest" were developed in an environment of Victorian private enterprise. Actually, Darwin did not originate the cliché, the phrase coming from Spencer. Darwin liked its original usage as it referred to fitting into the natural order. The athletic fitness metaphor seems to be more recent. Competition is a central paradigm of modern ecology in an environment full of competitive academics. A principal effort of evolutionary biologists has been to find clean examples.

Few have asked whether competition is important in the first place, and it took Simberloff[7] several years of complaining to put a stop to competition experiments in the early 1980s that had not actually demonstrated competition. At the time revolt against the idea of competition (Wiens 1977) indicated a changing social climate. International corporations and cartels make a suitable backdrop for a timely ecological and evolutionary paradigm: "if you can't beat them, join them." Taking over firms instead of competing with them is now the regular pattern in post-competitive corporate capitalism. This recognition of the importance of social milieu is shared with the postmodernists view of science.

Perturbations that might come irregularly can be given a more regular pattern by the biotic alliance. Chaparral comes in with fire, and volatile compounds keep the fire coming. Grasses encourage grazing and fire so that they both become regular components of the system. In this way biota ride with their disturbances, indeed become one with the disturbance, encouraging a special frequency that is incompatible with surrounding holons whose competitive edge is compromised.

All this discussion leads us into a problem with the definition of the term "disturbance" or its twin with more mathematical overtones, "perturbation." A perturbation must have a scale close to that which it might perturb; yet when this happens, the perturbed entity metamorphoses so as to incorporate the perturbation. This metamorphosis is then translated into structural terms (adaptations to the disturbance) by natural selection that fine-tunes and cements the new relationship in a new holon of organism-disturbance. Biome vegetation may vary considerably in its species, but biomes are physiognomically recognizable.[8] The physiognomy of biome vegetation arises from the species being a physical manifestation of the incorporation of the prevailing disturbance in that biome. All the species have performed an incorporation.

Clearly a perturbation can occur only a few times before it is tuned out of existence or rather into a new state. Perturbations must be rare and will occur only as transient phenomena. They must have a scale that is sufficiently similar to the perturbed holon so that it is felt, but the scale must also be sufficiently different so that incorporation does not occur. That is not to say that the perturbation does much choosing, for it loses nothing by being incorporated. The problem is the ecologists' business, in that what was called a perturbation no longer disturbs. It should be clear from the above discussion that a regular perturbation is a contradiction in terms.

The Power to Disturb

A perturbation is an event that usually does not occur at the scale in question. Reception of a very high or low frequency message throws the system into either excited disintegration or a stall. Some examples follow:

1. Cancer is a particularly troublesome disease for an organism, because the mode of operation of malignancy is cellular and subcellular rather than organismal. The natural organismal adjustments that usually interfere with disease agents pass significantly unnoticed by the cancer cells. By killing old adults, cardiac diseases release resources to the new generation and thus, being of evolutionary advantage, come to bear from a larger scale than the organismal level at which medicine works. Most curable (spontaneously or medicinally) diseases are more or less contained at an organismal level.
2. Invading species are ecological examples of trans-level catastrophe. Should they establish, continental alien species do not fit quietly into their new communities. The prodigious success of invading species introduced into Australia may be viewed as an interaction between systems (continental biota) separated and thus differently scaled in geological time and space. The invader places different weights on signals it receives from its new environment and new neighbors than do its indigenous competitors. Usually this is fatal for the invader, but if the invader does well then the invaded community members have to do a lot of scale changing. Holling's work on lumpiness of sizes of animals in a given region notices that endangered and invasive species appear at the edge of lumps.[9] That is to say, they are of sizes that are marginal to the local norm. Endangered species are close to being disallowed by the periodici-

ties of the environment. Invasive species, if they can survive, have little competition from the bulk of the native species, which are too large or small to present difficulties for the invader. This is all a matter of relative scaling.

A problem arises at this juncture. Forces for destruction abound, and damaged structures affected by these forces are commonplace, all of which seems inconsistent with incorporation as a general principle. Let us return to fire as an example. When a tree burns, it suffers disturbance, true; trees have burned for millions of years, also true. Red oak trees cannot reestablish without the full light of a site recently cleared out by wind, fire, or logging. Observation of a single circumstance at two scales will give two very different experiences that could easily appear contradictory. We have come to expect that contradiction with a hierarchical approach. At one scale, fire is a perturbation, but at another, it most certainly is not. The central argument is that a disturbance at one level may be a stabilizing force at another; this state of affairs demands the use of appropriate scales, only one of which will apply for a unified model of a given level of organization.

Compartments for Survival

The effect of a perturbation can be blocked because the disturbance and the things to be disturbed are too differently scaled. Destructive high-frequency messages are smoothed out of existence to give a mean value if integrated by a larger receiver. Low-frequency messages, on the other hand, may appear as constants because the message is not monitored for a long enough period for change to be apparent (e.g., individual organisms die before continental drift matters).

A biotic hierarchy has resilience in the face of perturbation through self-replication. For a pattern of perturbation to extinguish a reproducing hierarchy, all levels must be extinguished simultaneously. That degree of coordination is relatively rare in abiotic hierarchies. In nonliving hierarchies, unlike their biotic counterparts, there has not been selection for the special communication channels that would be needed. A self-replicating hierarchy can be extinguished piecemeal only if reproduction can be slowed to a critical point or stopped altogether.

It is hard to destroy a self-replicating hierarchy because there is likely to be a level surviving that will simply replace the rest of the levels in the system. An extensive change in environment can put all levels out of action. One self-replicating hierarchy can be destroyed by another self-replicating system, where the destructive hierarchy can attack at all the levels of the victim. We have a name for that: competition. But mostly the victim evolves to avoid competitive encounters in what is called character displacement. We have many examples of extinction on islands because the universe there places an upper limit on avoidance strategies. Parasitism and predation destroy entire hierarchies less frequently because there is a tendency for the process to be self-damping. As parasites and predators impact their food supply, hosts and prey become harder to find.

The wave of extinctions associated with human intrusion into natural ecosystems would seem to be peculiar to modern times. While there is a camp that stubbornly blames humans for megafauna extinctions,[10] there is always a counter argument that marshals evidence from climate change.[11] This back and forth in the literature is symptomatic in that neither argument is singularly correct. Another sign that there is politics over reasoned argument is the enormous number of authors on some of the papers involved.[12] Another recent paper also indicates politics with again a huge number of authors. That paper by Metcalf et al. (2016) is closer to getting it right in that they acknowledge multiple causes and multiple levels. Even so, they do express some surprise that they have to argues both sides at the same time. The mistake is thinking that to argue at only one level of analysis gives the true explanation. When biomes are disrupted, there will be extinctions and this gives a complex system.[13] The disruption has complicated effects at multiple levels so no single explanation applies. Tainter[14] was faced with a host of arguments for societal collapse, one scenario for each society, for instance Romans and barbarians. He worked out the general argument where all societies progress along a path of diminishing returns on effort, gradually leading each into a marginal condition. Which stressor finally takes out a particular society is happenstance. The general argument is excessive infrastructure to deal with diminishing returns. The general argument in extinction of species is specialization in blind allies like the universal extinction of the saber tooth specialization. The history embodied in inheritance gradually becomes so burdensome as to lead to extinction. Species extinction is a relatively rare event in the short and middle term, although universal in the long term. This would lend support to Wiens,[15] who suggests that competition is not commonly a significant factor.

The difficulty for an invading species is that, until it can develop a sizable population (preferably sexual), the invader has a shallow hierarchy. Resilience increases with hierarchical depth. Cultivated species have the colony protected by human husbandry as a safe holon to which the hierarchy can intermittently retreat until tiers of feral holons can be established. Extinction of colonies of escapees may occur many times over until good fortune allows successful invasion of the natural ecosystem.

In replicating hierarchies there is a double standard for message transmission. Disturbing information is thwarted; constructive information has special channels for communication. Evolution by natural selection is facilitated by well-defined near-decomposability; otherwise there could be difficulty in selecting one without all. Even at higher levels such as speciation, some sort of barrier is necessary. If the disturbance is itself biological then it has its own replication to use as a weapon. Disease has its own constructive forces from its own point of view. For instance, the 1918 influenza epidemic was particularly destructive, killing young adults differentially. Natural selection of the disease on troop ships returning home favored making the victim as sick as possible so that escaping bodily fluids could infect nearby yet uninfected individuals of which there were many close by on the ship. Quarantine on arriving in the United States continued the environment that favored a particularly deadly form of the disease. As quarantine was lifted it became more difficult for the virus to move between individual humans who, out of quarantine, were then more diffuse. Selective advantage for the virus then switched from killing quickly by making the host particularly sick to keeping the host alive until the more difficult task of infecting less proximate and less frequently available uninfected individuals. Disease has its own ways of remaining robust by reproduction.

At the death of an adult the population as a whole provides a replacement. Furthermore, constrained juvenile holons move up the hierarchy to replace the destroyed adult's function. The juvenile is at once an individual and a representative of the population, two separate and differently scaled conditions that are not mutually exclusive but complementary.

There follow two examples of scaling strategies as a defense against perturbation:

1. Redwoods provide a convoluted example of scaling strategies with resolution in a complex of tiered strategies. Adult trees are not perturbed by fire as it passes through the lower forest strata; big trees have the fire-resistant mech-

anisms of thick bark and gigantism, and the fire is scaled to insignificance. However, should fire be withheld, the lower strata will begin to gain height, and when the fire does occur, it will cause a fatal crown fire in the adult redwoods themselves. Thus fire at the longer phase becomes a very definite perturbation for the individual. Nevertheless, by reproduction mature trees use large scales of space to continue to provide an unperturbed environment for redwood DNA. Gigantism reverberates through all of redwood ecology: (i) only large fires affect the trees; (ii) such fires occur at very low frequency; (iii) as a result, gigantism is linked to longevity, a low-frequency signal in reproduction. Reproduction can be viewed as a system of averaging that is both fire and fungus resistant, and at that level fire ceases to be a perturbation; it becomes part of the stable environment in which the species lives. Further, the old trees give seed more adapted to conditions millennia ago while young adults produce seed that is well adapted to the contemporary climate. Old redwoods often occur in fairy rings. A tree that was there a thousand years ago has been replaced by stump sprouts, which become new trees with the same genetics of the old tree now rotted away. The stand is thus using asexual reproduction in a still higher level holon to hedge bets against climate change.

2. A single parameter such as reproduction of the individual may play a role at two levels of organization simultaneously. The fate of a litter of baby hamsters depends on the food supply to the parent at the time the litter is born. If food is abundant, then the young are reared in a normal fashion, as a contribution to the population holon. If, however, food is in short supply, the parent kills and eats the young. Lotka[16] suggested that birth does not contribute to population growth directly, it only provides vessels. If there is food to fill those vessels then a new individual survives. Sometimes birth only generates a food resource for adults. In ants there is a loss of totipotency where only some individuals can reproduce (*totipotency* means being able to do everything any individual of that species can do). Loss of totipotency can come from police in the ant colony that steal eggs from workers and use them as food. Thus eggs can be for reproduction or they can be a useful source of protein. Eggs for reproduction is a manifestation of high gain. Eggs for food is a low gain strategy.

   Alewife (*Alosa pseudoharengus*) is an invasive species of herring in the Great Lakes region. It is quite hard to control because it too switches from high to low gain strategies. If there is abundant food, the fish population booms in a high gain takeover. But should alewife find itself with insufficient food, it becomes cannibalistic, a low gain strategy where fish in the population become food for the population. It will switch strategies as it needs, making it a particularly robust species.

For hamsters, food is a constraining holon at a physiological level of organization. This would be well and good, but if the food supply behaves at a higher frequency than the individual life span, then the hamster depends upon a highly variable compartment, the behavior of which is likely to be fatal. Somehow the hamster must alter its relationship to the food compartment either by changing its own relaxation time or by stabilizing the food supply. As is characteristic of selected holons in such circumstances, the hamster runs the food through a buffer compartment, its offspring. Signals from the food supply are integrated over a long enough period so that the adult hamster reads food messages that are essentially constant. Thus the food holon is seen by the hamster to behave in such a way that it protects the organismal holon; environments (food here) must behave more slowly than that for which they are the environment in a stable hierarchy. Eating babies stabilizes the food supply.

Resource Conservation and Capture

When a retention or capture system appears very effective, that efficiency may often be explained by coordination between processes in compartments with different cycling times. Slow compartments deal with conservation and large-scale processing, while faster compartments effect fine-grain capture, sealing the system to avoid loss. If a perturbation may be avoided by existing at the level that remains unaffected, then when the system is engaging in resource capture, living resources may actively avoid capture with their own scaling strategies by existing at a scale where capture is difficult. Some examples of resource capture strategies that are fined-tuned for scale follow:

1. The resource value of seeds for predators is at a molecular level, the chemical energy in seeds, and yet seeds may have hard seed coat defenses at the macroscopic structural level. The predator too must operate at both levels. Mastication prepares the food for the gut where the large surface areas of food and gut allow the low-level enzyme and absorption systems to effect the capture.
2. If the resources are nonliving, then scaling strategies become important when competing biological systems preempt the resource by speeding up capture. By wide seed dispersal and rapid development of roots and leaves,

weeds have scaled themselves into a fast-moving compartment that captures resources before other growth strategies can occupy the open site.

3. Particularly scaled organization exists in biological structures above the organism level. In some forest ecosystems of the Eastern Deciduous Biome, mean nitrogen retention is an impressive eighteen hundred years.[17] This is achieved by a multileveled structure wherein arthropods are a minor part of the energy budget of the system, but they play a vital role in nutrient recycling. By consuming dead leaves and roots, the arthropods change the surface-volume ratios. This encourages fungal growth, which traps the nutrients and prevents them from being leached out of the system.
4. There is evidence that patterns of feeding and growth involving fish, zooplankton, and phytoplankton in some lakes concentrate phosphorus in safe and disjunct compartments. Organisms with middle-range rates of handling phosphorus occur in relatively small numbers, so constituting a step in the reaction-rate gradient for phosphorus dynamics in the whole system. Fish live longer relative to phytoplankton and so lock up nutrients longer. Also, the physiology of fish is logarithmically slower per unit size than that of plankton, dramatically slowing their nutrient cycling activity by plankton standards. Increasing the fish component might be expected to slow phosphorus cycling in the lake. However, the fish take out the large zooplankton component, and this has two effects. First, the larger zooplankton are a slower nutrient cycling compartment relative to the smaller plankton on which they graze. Therefore the removal of larger zooplankton speeds up the system; second, their removal eases grazing pressure on the phytoplankton so expanding the fastest moving compartment. Thus the fish speed up phosphorus cycling by as much as two orders of magnitude.[18]

   In doing this, the fish eat a step into the gradient of physiological and size scales in the lake. In this way they compartmentalize phosphorus in the nutrient-starved epilimnion. This allows a rapid cycling phytoplankton compartment to track a variable phosphorus input and fix it into the biotic segment of the ecosystem. The fish, on the other hand, with their long turnover time can hold phosphorus in the system from one year to the next. This ensures a phosphorus baseline for the system no matter how severe a temporary decimation of plankton may be. The compartments are neatly separated to maximize phosphorus trapping while ensuring phosphorus retention in the system.
5. Compartmentalization offers biotic hierarchies a certain resistance to disease and parasitism. However, this state of affairs can also be viewed as a compartmentalized host resource from the parasites' or disease organism's point of view. The problem for the infective agent is to glean resources from

the individual host holons while managing to slip from that compartment to the next.

The front line of defense for wheat against rust fungus, an obligate parasite, is at the cellular holon. At first contact with the parasite the cell commits suicide, so trapping the parasite inside the dead holon and preventing further infection. It is a matter of stopping the chain reaction at strategic points. Once the rust is able to avoid triggering the cellular allergic reaction, it is then able to use resources from the infected cell to mount an infection on the adjacent cells. Once it can move from cell to cell, the rust is able to occupy levels above in the host hierarchy from whole plants up to continental-harvest holons.

The common cold appears to have solved the compartmentalization problem and is further assisted by agents for symptomatic relief. Nowadays infectious human holons fully consumed by the disease go to work anyway on cold remedies instead of staying at home. Aspirin is thus an ally of the disease. Apparently, higher holons are needed to short-circuit host defenses, for after a few years on an isolated oceanic island the common cold dies out. It succumbs to a defense mechanism at the cellular level that is ubiquitous across all hosts on the island.[19] More usually, given the unnaturally high connectedness resulting from the density of humans on the planet, the cellular defense cannot be erected fast enough to isolate the disease in any given holon. The oscillatory behavior of plague in classical times is the sign that global connectedness was passing through an important transitional stage. Oscillation is a sign that a system is heading for disruption of rules as negative feedback turns into a positive feedback through amplifying oscillation.[20]

Relevant here is the shift in scale of the human population since the Neolithic revolution. The increasing connectedness of the world population has allowed success to infectious prokaryotes and viruses. These may be eliminated in oceanic islands because islands are small enough to allow extinction, but there they are replaced by important eukaryotic, even multicellular, parasites, the type of pest load with which the human creature coevolved in Africa. Host defenses at the cellular level, which work well against microbes isolated on islands, are less effective against multicellular disease agents since cells are too small to constrain and contain such invaders. Eukaryote infections integrate their responses to environment over too long a period for the strategy of defense used against viruses to work. In the presence of global travel that connects the whole human species, global immunity after epidemic becomes a problem for viral diseases. A new strategy has gained ascendancy globally in the last forty years, which is to attack the immune system. AIDS is a case in point.

As we have mentioned, a defense against disease can be a scorched earth policy of local suicide. Some parasites overcome the suicidal defense mode by being only facultative in their parasitism. By living on the dead host saprophytically, they occupy a safe holon from which new infections may spread. Many fungus diseases use this mode: *Armillaria mellia*, the honeydew fungus, infects from dead hosts, making its control in forests exceedingly difficult. Yet other diseases come from reservoirs of infection in different host species. The existence of such a safe holon for malaria in wild game makes it especially difficult to eradicate as a human disease. Malaria seems to be something with which we coevolved; some diseases, however, survive in other compartments and only blunder into the human being. Rabies, tetanus, and botulism are all examples of disease agents whose selection has ignored the human creature. As parasites they are failures; they end up trapped in the human corpse.

6. Prokaryotes, bacteria and the like, have a scaling problem in that they have only one bundle of genetic material. Too big a cell means some functions are too far away from the genetic instructions of how to perform the function. Eukaryotes are a solution to a scaling issue that limits prokaryotes. Nick Lane[21] points out that life is about hydrogen ion gradients associated with membranes to do the work of organizing living systems. Some bacteria use nitrogen and sulfur to drive those gradients, but it is hydrogen ion gradients that are the universals of biological function. The small size of prokaryotes keeps gradients of hydrogen ions at the cell membrane close to the functioning of the mass of the cell and its centralized genetic material. Prokaryotes do not have bounded nuclei, but do have concentrations of genetic material in the middle of their cells. Were prokaryotes to get too large, the genetics of how to use the hydrogen ion gradient with the environment would be too far from the genetic system with instructions as to how to work the gradient. Endosymbiosis, the incorporation of free-living prokaryotes into eukaryotic cells to become organelles, allows eukaryotes to increase in size massively. The prokaryotes have become the organelles in eukaryotes. The critical change is the transfer of most of the genes of the organelles to the nucleus of the eukaryotic cell. Only a remnant of the wild genome is left behind in the organelle. Any genetic information still in the organelle actually codes for respiration and the like, the explicit job of the organelle, the mitochondrion. The executive functions of the trapped bacteria are now in the host cell's nucleus, which can be quite far away from organelles. So mitochondria

> have the instructions for respiration, which must be immediately present for mitochondrial functioning. Less immediate integrative functions are coded away in the nucleus. So eukaryotes have instructions for dealing with membranes and hydrogen ions duplicated throughout the cell, but in their endosymbionts, organelles. The eukaryotic cell only has multiple copies of the mitochondrial part of the genome, and so does not have the payload of having to duplicate its whole genome. It seems that when bacteria do make multiple copies of their whole genome they are still limited in size and function. It is a genetic hierarchy that is the trick that gives eukaryotes the edge. Different codes in the cell exist at separate hierarchical levels.

Endosymbiosis works the same in wholesale takeovers in business. Smaller companies are engulfed, but keep their own instructions for their crucial local functioning. There is the same division of executive function and building site information. Dairy Queen is a franchise that was taken over by Berkshire Hathaway. Which ice cream Dairy Queens offer on a given day is decided by the franchisee; larger integrative decisions are made by the higher level companies. Firms taken over often have almost complete autonomy. However, the larger firm offers protection, rather like the eukaryotic cell. Eukaryotic cells can live in acidic conditions that are not suitable for many bacteria. Endosymbiosis allows bacteria cum organelles to live protected by the host in acid conditions. Berkshire Hathaway protects Dairy Queen. There are peculiar arrangements of multiple membranes around mitochondria in some organisms that indicate a history of multiple endosymbioses.

On the topic of intrusion of environments, in very harsh environments, intrusion such as drought stress can be regularly encountered; in such circumstances biochemical modification becomes relevant. Many plants capture carbon dioxide first as a three-carbon chain; C_4 plants capture the resource in a particularly efficient four-carbon molecule. The point here is that this unusual procedure divides the photosynthetic organ into two tissue holons. The mesophyll holon captures the carbon dioxide while the bundle sheath holon (cells around the transport cells) reduces the captured carbon to sugar. Thus a spatial barrier limits carbon dioxide escape, so minimizing losses from photorespiration and maximizing photosynthetic gains.

Other plants from even drier habitats (Crassulacean stoneworts and others) must so restrict water loss that water is sealed inside to the point that carbon dioxide entry into the leaf is a limiting factor. Carbon dioxide is sealed in like water. These plants present a temporal barrier to water loss by opening stomata during the cool night instead of the day. In order to allow this reversal to be effective, parts of the total photosynthetic reaction are hierarchically split at the lower level into day/night holons. Carbon dioxide can then be absorbed from outside the plant in the cool of the night.

Summary and Broad Implications

From this and preceding chapters it is clear that not only are the time constants in biological systems various, but also relationships between differently paced parts of a hierarchy are what give complex biological behavior. Further, we have tried to show that there is a perceptual component to biology, the time constants of which are the key to what we know and what we can know. These perceptual time constants are worthy of our attention.

Any observation is scaled by temporal patterns. These patterns relate to the level of resolution of datum values and the extent of the entire observation period. Observations made at regular intervals measured in minutes, days, years, and the like might seem to be the simplest pattern to use, but this is often not so in biology. Time passing regularly in minutes might be very irregular for the process under observation, some minutes containing much activity and others containing none. An insistence on measuring time in minutes may have the effect of stretching and contracting different time periods as far as the observed holon is concerned. Behavior of the observed in these circumstances appears very complicated. More than this, depending on when the observation is made, perception may stretch time on one stem of the hierarchy while contracting time on another. For example, with the slowing of physiological processes in winter, the time units of physiology lengthen, so making winter of brief physiological duration. On the other hand there is much death in winter, making it a very long time (i.e., much happens) for populations in a demographic context. A main purpose in biology is to track back from biological time to our experiential time framework. In classical physics this is easy since we are used to mapping cosmic time onto experiential time, and we even have mechanisms (calendars and watches) to help us do just that in our daily lives. In biology we have little idea as to what are the relevant time frames. Identifying the appropriate tempo-

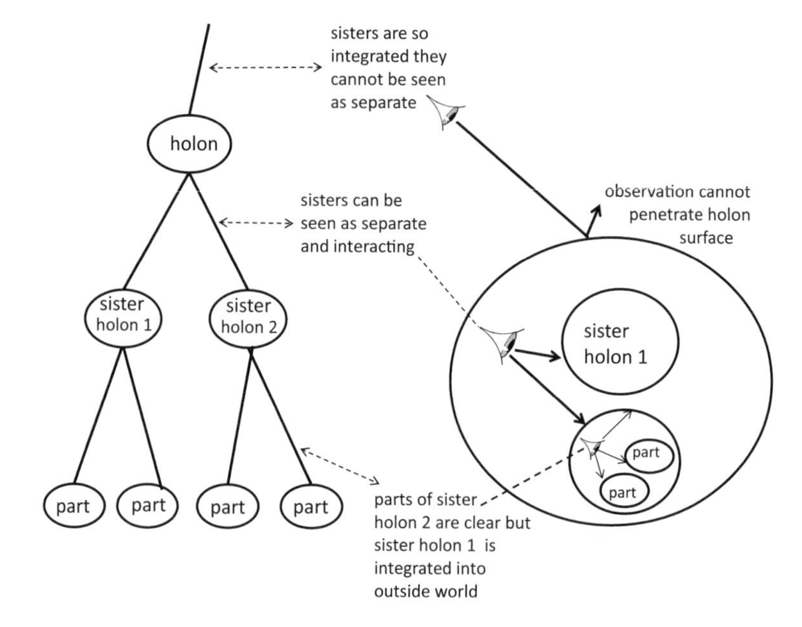

FIGURE 8.1. Showing the translation between a stem depiction of a hierarchy, as opposed to a holonic representation. How observers see what they do is better presented in holonic form. There is a difference between looking in and out. Seeing radiation of holons emphasizes the integration of holon parts. Reflection of holons shows more how surfaces generally obscure inside from outside and vice versa.

ral constructs is one of the most challenging aspects of biology. However, it is only when we know these time frames and reflect them in our biological models that biology can become tidy and workable.

Figure 8.1 shows how different levels of observation allow the observer to experience different hierarchical levels and the holons that occupy those levels. Viewed from outside we often see signals from the environment that cannot enter the holon and so are bounced back. That is what we see when we observe using the special narrow spectrum of light. However, sometimes we pick up a holon from the signals the holon itself emits. There is a distinction in night vision. The older methods from the 1930s used the small amount of light present in a dimly lit place. That light is converted to electrons, whose effect is amplified as in a stereo amplifier. Modern night vision uses a longer spectrum in the infrared range. In that range biological entities emit infrared light in their own right. The image and the identity of the thing observed may be similar in light as opposed to infrared, but there is a fundamental difference in the use and source of the signal. The surface of the holon works differently. Infrared being a longer wavelength, modern night images are fuzzy. Snakes have regions of their skull that pick up infrared, and they strike at night at glowing prey items. They do not have to pick up details, only where is the rodent generally.[22]

Spyware can change the image by penetrating a surface. It may also change the signal in that the biological entity in question will behave differently if it does not perceive it is being observed. The use of a one-way mirror in police work and psychiatry gets around the phenomenon that R. D. Laing noted. In *The Politics of Experience* he notes that he generally does not see the client, he only sees the client being observed by a psychiatrist; and that is different. Laing was a favorite of 1960s postmodernists.

The proximity of the observer makes all the difference. A football match looks different from the sideline than it does from an aerial camera in a dirigible, but it may remain identifiable as the game in question. There comes a point, however, when observation is made so closely and with such fine resolution that the range of windows through which the phenomenon is identifiable is exceeded. It is at this point that the observer has penetrated the skin of the holon associated with the phenomenon in question. Once observers look so closely that they can see only one or two players, the game disappears. The penetrated holon becomes a collection of stable parts, and the integration of the parts is essentially lost. The game can be inferred only with great difficulty. If the observer further reduces the time constants of the observation, then all evidence of the whole is lost because of filtering. The time and space scale of the observation becomes so small that the daughter holons (e.g., football players) may only be seen one at a time, and the other connected daughter holons (all other players) lose their identity. The other daughter holons become integrated into the rest of the environment of the daughter holon that is in view. At this point the observer has penetrated the skin of one of the stable daughter holons and is looking out (Figure 8.1). At one time, we can only directly see connections that are associated with a certain band of frequencies. The depth of focus of our perception (in terms of frequency) is greatly enhanced by a hierarchical approach to perception and model building. Sometimes the observers may stand so far back that they move backward through the skin of the next larger holon. A dirigible above the clouds cannot see the football game. Alternatively, as the observer moves back, the signal from the holon of interest becomes gradually weaker until it is lost in a sea of noise. Move so close to the game as to penetrate the ball, and a pass appears as internal darkness.

Images of a radio receiver may be helpful here, where the crackle of background static gradually overwhelms the station that is being tuned out. Note here that static noise represents the intrusion of a large number of anecdotal accounts of unstable holons. In the football example, the observer in the dirigible on a clear day becomes gradually confused as to the progress of the game as the balloon goes too high.

Biological structures need very particular time frames if subtle biological interconnections are to be found. By doing most of our calculations in biology in sidereal time, we may substantially miss the point of what is going on in biological systems. Because we do not try out biologically sensitive time frames, we might reject hypotheses unnecessarily for lack of data; more seriously, we fail even to formulate hypotheses because phenomena lie unseen.

PART III

Scale and Complex Systems

CHAPTER NINE

Identifying the Scale in Community Ecology

We only see systems under analysis not systems in reality. At this point we need to show what "systems under analysis" means. So this chapter speaks of some particular analyses, often in ecology, but many another arena of discourse would do. The point of the whole discussion is not to discuss ecology per se, so we ask the readers' indulgence. The point is to explain what we mean by analysis, and level of analysis. We just know about ecology, that's all. We could have spoken about business instead of ecology with regard to big data, or analysis in the social sciences. The point is that the difference between a system in reality and a system under analysis should be clear; any technical specialty could suffice to pursue the issue. There are some technical discussions but we try to keep it all as general as possible so the main point remains clear.

Well before complexity was recognized as a distinct issue with its own special properties, there was work in the field, but it was unselfconscious that it was dealing with complexity. All this has been spurred on by the coming of cheap and massive computation. It makes one full of admiration for those who did such computation with a pencil and paper. At first fast computation was simply a means of summarizing data sets that were too complicated for humans to countenance. Humans are very quickly overwhelmed with increasing numbers of variables. The situation is made worse when the pattern in the data is elaborate. Modern computer-based methods of pattern analysis appear often in discussion of complexity. However, it is not easy for a layperson to see what the methods are doing, and how they work. There is a significant body of mathematics in ecology that is more obscure than it needs to be. So let us unpack multivariate pattern analysis so the intelligent uninitiated can understand what is going on.

Allen learned early in his lecturing that it is impossible to teach an elegant solution to a problem the students do not have. So before teaching multivariate ecological analysis he would create an artificial facsimile of a data set that had just sixteen ecological variables, underlain with simple close-to-linear patterns. Students consistently could not see the pattern until an analysis revealed it. That gave them a problem, and promised an analysis to solve it; they were ready to be converts, and so were easy to instruct. For us, multivariate analysis is one of those few fully elegant and exciting things one ever learns, like Mendeleev's table, or the alternation of generations in plants from mosses to flowers. It offers unreasonably powerful summaries.

The artificial data set simulated sixteen species growing in sixteen patches of vegetation (that both entities and attributes were sixteen in number is happenstance). The data set was created before the students came to class by arranging glass beakers in a four by four grid on a lab bench. Allen then scattered an item (perhaps toothpicks) across the beakers in a pattern by placing the largest number of the item in question starting with a corner beaker. Toothpicks will have been one species growing in various densities in stands of vegetation, sixteen beakers. Successively fewer were placed in beakers successively farther from starting corner. The corner beaker opposite the starting point may not contain any of the item. Other items, say paper clips, were other species. Paper clips were arranged (grew) in a similar manner, but starting with another corner of the grid of beakers. The paperclip species apparently had different environmental preferences compared to the toothpick species. Yet other patterns for other items started with concentrations along one side of the grid. The beakers across the opposite edge might have none or only a few of the item in question. The items were intended to mimic sixteen different species of plant occurring together in stands. As with real plants in real vegetation, some items were small and numerous, such as toothpicks representing herbs that occurred many to a given beaker. Other items were like trees: large and less numerous. There was only room in the beakers for three big corks. The students collected the "field" data by counting items in their assigned beaker. Sixteen types of item were arranged in linear gradients running in different directions across the grid. Simple as the pattern was, the students seeing the whole data matrix they had created could never, year after year of teaching, see

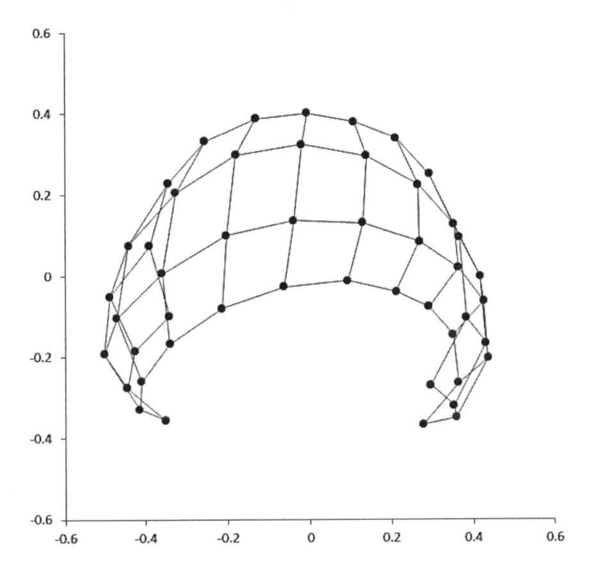

FIGURE 9.1. Peter Minchin answered our request for a demonstration of curvature of a simple grid with distortion from heterogeneity. He describes it technically: "This is a simple model with two underlying environmental gradients, in which all species have symmetric, unimodal response curves. The beta diversity of the first gradient is three times that on the second (in my "R" units it is 1.0 × 0.33 R). The sampling is on a regular 12 × 4 grid. The ordination illustrated is the first two axes of a Principal Coordinates Analysis using the Bray-Curtis index with abundance data standardized by species maximum." The environmental grid shows curvature in two dimensions.

the pattern from simple visual assessment of the data matrix. The summary analysis result was somewhat similar to Figure 9.1 without the curvature, except it was a 4 by 4 grid of the original arrangement of the beakers on the lab bench. Without multivariate analysis, complications overwhelm human capacity for recognition, let alone understanding

The analytical method was a simple device that started by placing each beaker in a sixteen dimension item (species) space. The dimensions are one for each item. The score of a given beaker on a given dimension is fixed by the number of items of that "species" found in the beaker. Do not panic; hyperspaces of many dimensions, sixteen in our case, are easy to understand. One can use Pythagoras' theorem on right triangles to calculate the hypotenuse on a plane, say on a lab bench (Figure 9.2a). The numbers of each item (perhaps toothpicks) is a dimension. The first two beakers will have different numbers of the first and second items to be considered (perhaps arbitrarily toothpicks and paperclips). The relative geometric position of the two beakers on the first species axis is the difference between the scores for that species in the two beakers under consideration. The different scores for the second item gives the difference on the second axis, which is laid down at right angles to the first. The distance between the beakers on the plane of the first two axes will be the hypotenuse of the right triangle formed by the first two species on the toothpick/paperclip plane. The distance between the beakers is the hypotenuse on the plane. Under Pythagoras' theorem the hypotenuse is the square root of the summed squared lengths of the other two sides as offered by the data. If toothpicks are item A and paperclips are item B, the distance (d) between the two beakers is:

$$d = (\Delta A^2 + \Delta B^2)^{-2}$$

the square root of the sum of the squared differences on the two axes.

But one can go further. Take the hypotenuse just calculated on the plane; then use it as the base of a new right triangle with the new axis reporting differences between the beakers for a third item, the third dimension, perhaps big corks. In Figure 9.2a, species C would be big corks. The third axis is vertical, rising from the end of the old hypotenuse. That creates a new right triangle standing on its edge. The new hypotenuse is the square root of the sum of the squares on the other *three* sides, two for the other sides of the first triangle on the plane AB, and one more for the side that gives the new triangle height, C. The new hypotenuse is a line in a three-dimensional ABC space, and it is tangible; you can see it (Figure 9.2a).

The hard part is now done. We have drawn a line across a three-dimensional space by simply using a pair of two-dimensional spaces at right angles to each other. It is not much of a challenge to do the same operation again to create a line across a four-dimensional space. All we have to do to put a hypotenuse in a four-dimensional space is to set up a new right triangle by using the three-dimensional hypotenuse, the one we just calculated, as the base of the new four-dimensional triangle (Figure 9.2b). Simply lay the three-dimensional hypotenuse down on the original plane of the table. Forget the old "other" sides, because they are already captured in the length of the 3D hypotenuse. Do as we did before and the new triangle on its edge exists in a 4D space. You can just keep going into as many dimensions as you like. The hypotenuse is the length of the distance between two points in an n-dimensional space.

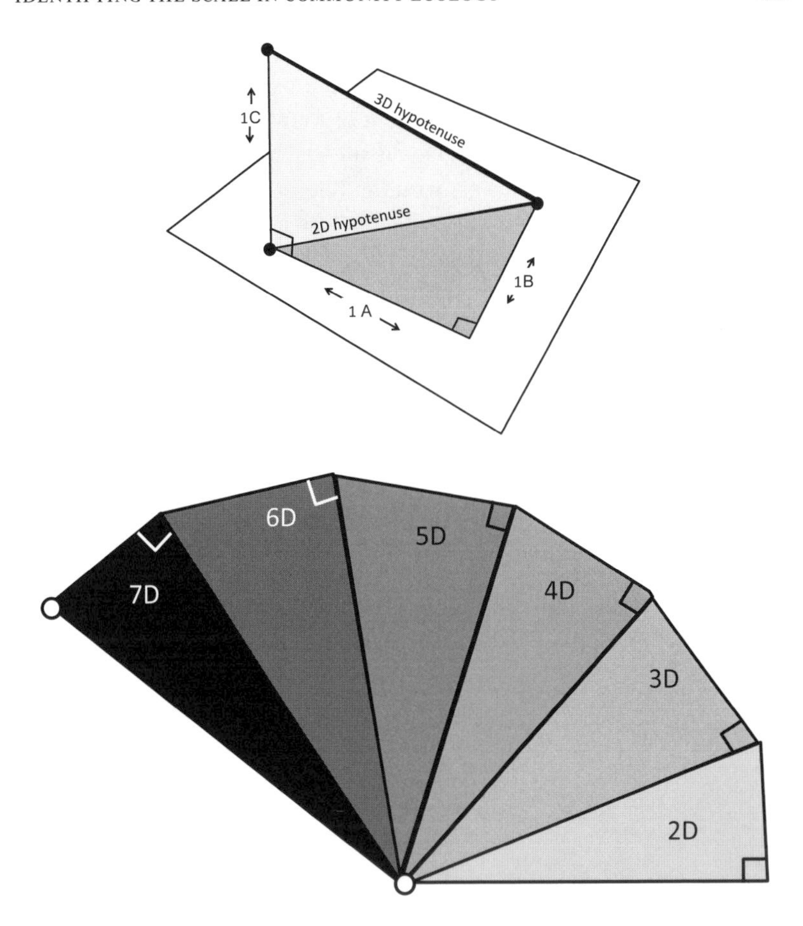

FIGURE 9.2. a) First a right triangle is created from the different values in two stands on a species AB plane. The hypotenuse of the first triangle is used as the base of a second triangle, which shows the distance between stands in an A, B, C volume. The grand hypotenuse is the distance between the stands in a 3D space. b) The 2D triangle is the first triangle on the AB plane in panel a. The triangle on its edge in panel a has been laid flat and becomes the triangle labelled 3D here. Repeating that operation over again, a triangle in a 7D space is ultimately created here.

Each beaker in the exercise is placed in a sixteen dimension space by the numbers of each of the sixteen types of item it contains. The difference between the contents of each beaker and any other of the remaining fifteen beakers is a distance in a 16D space (it so happens the sixteen items, dimensions, positioned sixteen beakers, but sixteen species and sixteen sites are a happenstance coincidence).

A teaching device for the conception of hypervolume spaces is to play in class a game of tic-tac-toe (noughts and crosses in the United Kingdom). Figure 9.3 shows an arrangement of four-by-four grids. There are four rows of four by four grids. This is the playing table for four dimensions. The top left grid shows a winning line in one dimension. Also in the top left grid, on a dominant diagonal, is a win in just two dimensions on a plane (the standard game but 4×4 not 3×3). If the opponent blocks on that line, the player can still win by shifting to a new grid, the one to the right where move 2 on the line is shown. Move 3 is one more to the right. The final move is score 4 on the last grid in the top row. The winning line now goes diagonally through a four by four cube, the cube across the top. The game at that point is a three-dimensional game. A blocked line on that diagonal through the cube still allows a win by moving to a four-dimensional winning line. The fourth dimension is accessed by using the lower rows of grids; it is a matter of playing in four cubes in order instead of one. Figure 9.3 shows a winning line that passes diagonally across a space defined by four cubes, not one. One year a group of too-clever-by-half graduate students tried to play a three handed game

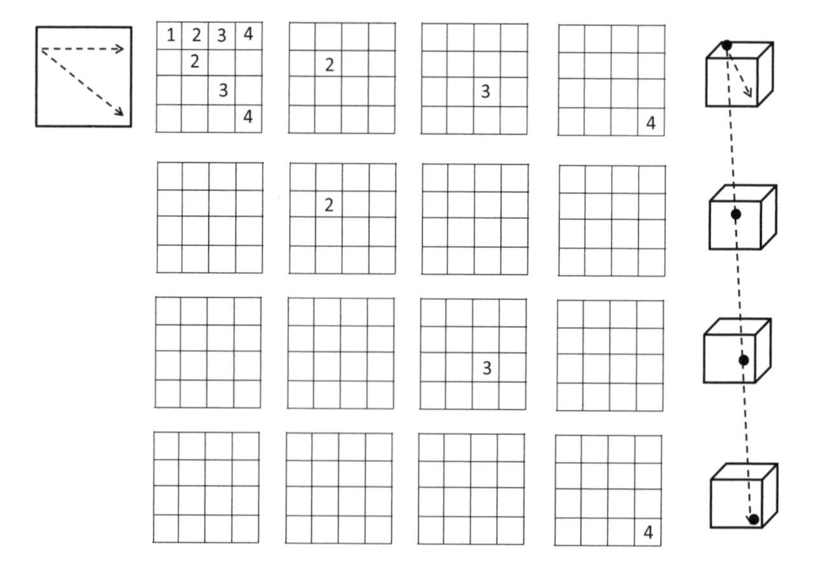

FIGURE 9.3. The tic-tac-toe board used in Allen's class to play in four dimensions. The figure shows lines blocked but then moving into a higher dimension. The marginal plot in the upper left is the summary of the winning lines in 2D. The four marginal cubes on the right show the winning lines using the full 4D of the figure.

in five dimensions. Four of Figure 9.3 was their game board; blocked in the fourth dimension, they moved to another board, another Figure 9.2. It ended up a rather Buddhist exercise in that one of the players found he had won without noticing. High-dimensional tic-tac-toe is also a good device for teaching the use of vectors. The game in vector notation is challenging for humans, but trivial for computers.

The simple analytical technique that yields the pattern that Allen had used to arrange the beakers casts a shadow from the 16D space to a smaller space. Hold your hand in front of a screen and shine a light on it. Your 3D hand has a 2D shadow situated in the smaller space of the screen. Rotate the hand and the shadow changes. From some directions, with the palm facing the screen, the shadow shows it is a hand, with fingers and a thumb. From other directions the shadow is not obviously a hand, as when the edge of the hand faces the screen. Then the shadow is of the edge of the hand. Viewed on its edge, the palm-view is foreshortened so that the identity of the hand as a hand is lost. The analytical technique simply rotates the beakers in the 16D space until a useful shadow is cast in a smaller space.

A first step in the projection to a smaller dimension space is to put all the distances between the beakers in the 16 dimension into a triangular symmetric distance matrix. A distance matrix is familiar from roadmaps where inter-city distances occupy a triangular chart filled with distances between cities on the map. Another lab exercise uses the triangular distance matrix at the back of a US road map to calculate a map of the United States. The longest distance on those matrices is often Boston at one end and Los Angeles at the other. That distance is about 2590 miles, so we use a line 25.90 inches long to make the US map. Cites were projected to the Boston/Los Angeles line using intercity distances in inches in the distance matrix to make the projections. The distance matrix contains all the information on the map, but in a way that is difficult for humans to read, although easy for the computer to manipulate mathematically. To extract the information, points (other cities) are projected to the reference Boston/LA line by using intersecting arcs. Projection is from the intersection of arcs coming from opposite ends of the summary axis. The city to be positioned is a known distance (look in the triangular distance matrix) from the reference cities of Boston and LA at the ends of the 25.90 inch summary axis. Those distances are radii centered on the respective end of the summary axis. We do not know where exactly the city to be positioned is with regard to the reference cities of

Boston and LA, but we do know its position will be on a circle whose radius is the distance to the reference city. As arcs from the circles from opposite ends of the summary axis intersect, we can drop a perpendicular to the summary axis. The perpendicular in the method is like the line, perpendicular from the hand to the shadow on the screen. The reference line summary axis holds the shadow one-dimensional line. The perpendicular positions the city in the one-dimensional space of the summary axis from Boston to LA. This can be done for all the cites to be mapped in turn.

With the first axis completed, a second summary axis is chosen roughly at right angles to the first summary axis. In constructing the US map from the atlas distance matrix, the second axis was usually a line roughly at right angles from New Orleans to somewhere in the northwest, usually a city in Montana. Using the positions of cities on the two summary axes, a map of the United States is constructed. LA to Miami is artificially long because one has to drive on roads around the coast of the Gulf of Mexico. This has the effect of pushing Florida out in the Gulf Stream. Similarly the towns around Appalachia are pushed apart by the length of mountain roads. The straight line interstate roads west of the Mississippi tend to contract the West. The first part of the technique used on the beakers is to calculate the distance between all beakers in the 16D abstract species space of all the items (toothpicks, paper clips, etc. are the species). We make a distance matrix for beakers in the 16 dimension species space. The distances are the hypotenuse of right triangles in 16 dimensions for all pairs of beakers. Casting a shadow in the beaker lab exercise is achieved by putting a line through the 16D space and projecting to it. There are lots of criteria for the direction of the line, depending on the particular method. We use the direction of the longest distance between any two beakers in the whole distance matrix, the largest number in the triangular distance matrix. There are various alternative methods, but each will have some reason why the line chosen will capture lots of information from the original 16D space. In our methods the reference beakers are placed at opposite ends of the line, which is the length of the distance calculated between them. The beakers are projected to the first summary axis much as the cities were projected to the Boston/LA line. A second axis roughly at right angles is used, chosen on the basis of residual distances in the distance matrix, much like the New Orleans/Montana axis. In the beaker analysis, we chose two beakers projected to the place in the middle of the first axis, but still with a long distance between them in the distance matrix. They will have been two beakers whose differences are not well considered on the first axis. Using the summary axes the beakers were plotted in a 2D space, and Allen's method of filling the beakers is revealed. A grid like Figure 9.1 appeared every year the class was taught. This sort of summary is often called an ordination in ecology. In the lab exercise, when the students had performed that geometric maneuver for all the beakers, the grid pattern Allen had used to arrange the beakers emerged in the ordination. The students were amazed that they could not see such a simple pattern in the raw data matrix.

There are many methods for projecting data down to a smaller space, each with its criterion for orienting the line used for projection. Some use a Euclidean space that works to show some version of the greatest variance in the data, like the one we just described. The advantage of using Euclidean distance is that the squaring and square-rooting of Pythagoras' triangle sides preserves distance between points in the rotation to find the lines of summary. The most common methods in the complexity literature today use something called stress reduction. Non-metric multidimensional scaling (NMDS) is an example that uses stress reduction in erecting the lines for projection. Strictly speaking, some methods do not use distances per se, but the effect of all the methods is something like projection to a smaller-dimensional space. The idea in all of these methods is to preserve the original pattern as much as possible. Each method has its own criteria for "pattern" and "as much as possible."

There is a whole literature in ecology that tests different methods and their criteria against known patterns. Those papers reveal the ability of this method or that to reproduce in the smaller space the pattern of the full-dimensional original. Peter Minchin (1987) is the authority on using grids (like Allen's beakers) to test the emergent properties of various methods on their ability to retrieve the original pattern. Figure 9.1 shows his results for a test of principal coordinates analysis using an underlying flat grid of 12 by 4 sites. The thing to notice is curvature of the original flat linear grid. If the data are heterogeneous, principal coordinates display much curvature of the original linear gradients. Minchin prepared Figure 9.1 using a moderately heterogeneous data set. Heterogeneity is described by the rate of species turnover from one end of the gradient to the other.

For Minchin deviation from the original pattern is validly identified as distortion, which usually takes the form of bending an original straight gradient around which the data are organized. Fundamental notions of complexity enter this discourse on curvature in a manner that is not generally understood. The curvature of the gradient coming out of analysis sometimes is in fact distortion, but sometimes it is more adeptly treated as signal coming from two levels of analysis. Consider a gradient where species replace each other in series down an environmental gradient from dry to wet (Figure 9.4a). Species appear successively better adapted to wet conditions as we move down the gradient. Drought-resistant species will be replaced by those better adapted to wet conditions. The wet to dry ends of the gradient share no species with each other. Curvature may arise because the ends of the gradient have something in common; they both share almost no species with the middle of the gradient. This mutual negativity makes the ends similar in some respects, and that pulls them together. But sometimes commonality between the ends of a gradient can have positive ecological meaning. There are species that do not occur in the middle of the gradient but do occur at both ends. Those species experience something similar at both ends. Stress in extreme environments may make different extremes similar in some respects. The curvature in analysis in this case has some real biology behind it, and is not distortion. Opposite extremes on some criterion may be similar on some other criterion. Collapsing that space in an analysis can pick up two levels of organization at once. The paradox of similarity of extremes can be explained by the presence of two levels of analysis at the same time. Methods that pick up both gradients are in fact displaying a greater depth of focus; they are better at showing complexity where they display two levels of analysis as curvature.

An example of this manifestation of complexity occurs when the ecology of black spruce is analyzed along with species that avoid extreme environments where black spruce thrives. When a gradient analysis ordination is performed on forest data in Wisconsin the amount of curvature of a wet to dry gradient depends on how large is the universe of discourse—that is, how wet and how dry are the extreme environments in the analysis. If we investigate only a middle part of the gradient, from moist mesic to dry mesic, there is little curvature. That is because the ends of that short gradient share species with the middle of the gradient. A straight gradient of wet to dry appears under local analysis as species replace each other down the local wet to dry gradient. Moderately wet and moderately dry environments are homogenous with respect to intense shading. In moderate environments there is high competition of

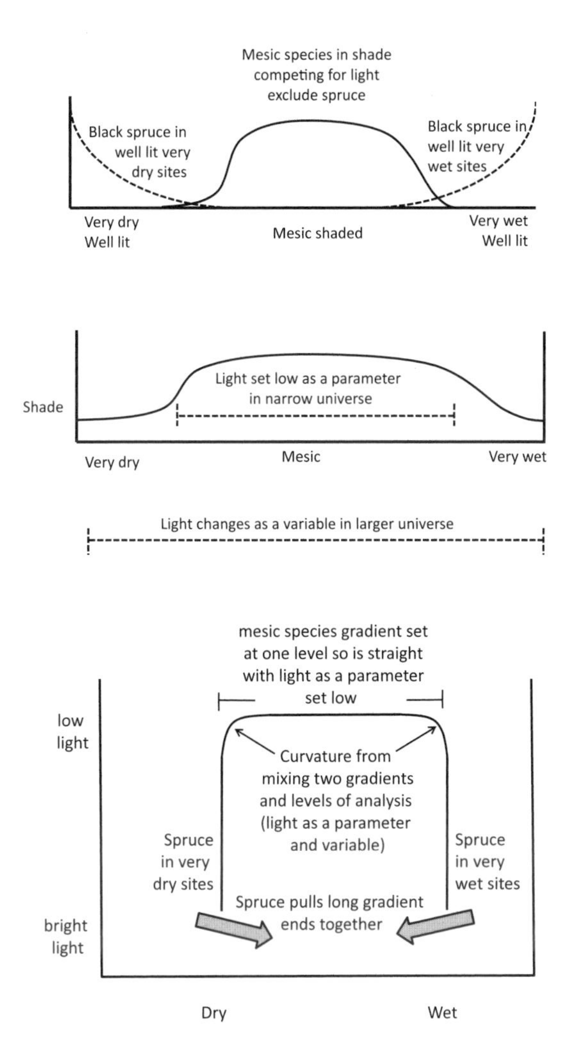

FIGURE 9.4. a) A long gradient showing how black spruce is excluded from the mesic zone. The bell shaped curve in the middle shows how the mesic species compete for light in the middle of the gradient. b) The environment behind panel a. Competition for light in the mesic zone is a parameter set high. Along the full gradient competition for light is a variable that is low at the ends and high in the middle of the gradient. c) There is curvature that arises from the mutual absence of mesic species from both ends. It also arises from the presence of black spruce at both extremes. The figure shows how expressing two levels of analysis together creates curvature. Not all curvature can be dismissed as distortion; some of it is material and meaningful.

light everywhere. Perform an analysis that goes on to include extreme environments from bogs with standing water to very dry sandy ledges on cliffs and the gradient is curved for ecological not mathematical geometric reasons. It is inappropriate to call the curvature distortion because there is a level of analysis issue in the pattern.

The key is that black spruce grows in standing water and on dry sandy ledges, but not in forests of middling moisture.[1] The analysis of just the mesic vegetation in the middle of the gradient is where all stands of vegetation involve high competition for light. Light competition is set high as a parameter across the mesic universe. Black spruce cannot establish in low light. But at extreme wet and extreme dry sites there is little competition for light because of an open canopy in such harsh places. Black spruce can establish in those sites. Including very wet and very dry sites in the study expands the universe beyond the mesic zone (Figure 9.4a). Across the middle of the gradient, competition for light is a parameter set high. From end to end of a gradient that goes from extreme moisture to extreme drought, competition for light ceases to be a parameter and becomes a variable. There is low light in the middle and high light at both ends of the gradient (Figure 9.4b). Loucks[2] shows how black spruce relates to other tree species in a fundamentally different way from the manner in which mesic trees relate to each other. Black spruce works in a more heterogeneous environment. Complexity arises when a change in level of analysis is forced on the investigator. A low level system may have a parameter, a constant. But that same factor may manifest itself as a variable across the universe of a high level of analysis. The answer to the question, "How do tree species relate to each other?" is "It depends." The answer to complexity questions is always, "It depends."

There is no simple way to look at multiple levels of analysis simultaneously. The jump between levels cannot be given a unified account. But the complexity can be decomposed if the two answers can be seen as contradictory complements that must be kept separate to avoid contradiction. The two analyses say that in the narrow case competition is constant and high, but in the broader universe the degree of competition changes. Parameters become variables at a higher level of analysis as a general principle. Nonlinear gradients often mean that there are two gradients involved but they have been pressed into one (Figure 9.4c). More or less water means different things depending on the context. With careless analysis comparisons appear to be to a muddle of things. Nonlinearity is a characteristic of complex systems. It gets a bit mystical in the literature, but it need not be so. As nonlinearity becomes strong, the steep part of the curve enters a new universe with demonstrably distinct phenomena. We are not saying nature is one way or another; we are saying that analysis of a system can make curvature appear. We experience systems only through analysis of some sort, and that is whence comes curvature and conflicting levels of observation and analysis. Linearity or curvature are a matter of analysis not nature. Imagine looking through a fisheye lens and moving through a landscape. The center of the image will be at first flat with no distortion. But as the lens moves into the center of the space, the planar image moved to the side starts to curve. Eventually what was planar in the center disappears out of the sides of the image. Nature cares not about fisheye lenses.

As heterogeneity increases in a material system, at first there is curvature in its mathematical representation. With further heterogeneity the increased curvature demonstrates the complexity of chaotic behavior, which is infinitely rich and elaborate. This is the case even if the underlying equation is short and simple. Chaos arises when there is a lag and a big fast growth term. The heterogeneity is increased by increasing the rate of change in the equation relative to the lag in the equation. But press the heterogeneity even further and the system develops what physicists call phase locking.[3] The elaborate interaction of different processes in the observed universe suddenly linearizes. This is what happens when elaborating patterns at the micro-level disappear into macroscopic phenomena that are more or less linear. At that point there is complete change in level of analysis. Much is made of complexity in chaotic systems, and the points made there are valid. The changes in chaos lead to situations that are undefined, because zero and infinity enter the mathematics of the situation. Both are undefined numbers. Normal reductionist science is likely to throw up its hands, whereas complexity science keeps going when things cannot be defined.

A principle emerges here in systems analysis, which is "Different as opposed to what?" The question queries the context. Scientific statements of far or close, or fast or slow, or more or less all carry with them an "as opposed to what?" All science is in a context of some sort. We can show changes in "as opposed to what?" with a simple two by two contingency table indicating presence and absence of species in a set of sample sites. The statistics here is elementary, so the discussion is not of statistics, but is of changes in level of analysis in a fairly familiar setting. Statistics are commonly understood at a basic level; changes in levels of analysis are not. Sometimes the species co-occur, sometimes they occur alone, and sometimes neither is present. The "as opposed to what?" is relative to the sites where neither species occurs because that condition fixes the size of the grand total of sites investigated, the universe of discourse.

These occurrences can be expressed as contingency tables where the occurrence of species A is compared to the occurrence of species B (Figure 9.5). In a contingency table the proportions in the columns and rows should match the proportions in the marginal totals if there is no correlation. Correlation is expressed as deviation from expectations. Are there more or less hits for mutual presence of a species than we might expect given the proportion of marginal totals? More than expected mutual occurrence gives positive correlation, while less than expected mutual occurrence gives a negative correlation. Working in round numbers for transparency, if species A is present half the time in the total set, we would expect the cell for mutual occurrence to be half the score for presence of species B. If there is no correlation the cells inside the table agree with the relative proportions inside the whole (Figure 9.5). The notion of "as opposed to what?" arises as "occurrence as opposed to the occurrence in the whole universe of sites observed." With all else equal (co-occurrences stay the same), as the universe gets bigger from the inclusion of sites where both species are absent, correlation tends to become more positive. That is because the species are more rare but the mutual occurrence stays the same. Conversely in a smaller universe correlation tends to become more negative. Correlation is always "as opposed to what?"

So let us work with two species of tree, northern hemlock and sugar maple, to illustrate the same point in nature. Both are commonly present in mature old forest in northern Wisconsin. The species come into a stand of trees in waves, so if the over-story is one of the species, the other will be tend to be absent or only seedlings. If we make a table to reflect that ecology, almost all samples will contain one or the other species but not so often both. In our example the species will be present together less often than expected (no trees and only seedlings of the other species) given the small size of the universe containing only northern mature forest.

But now let us change the correlation between the species from negative to positive, not by changing the mutual presence, but by changing the size of contextual universe (Figure 9.5). To reflect that ecology let

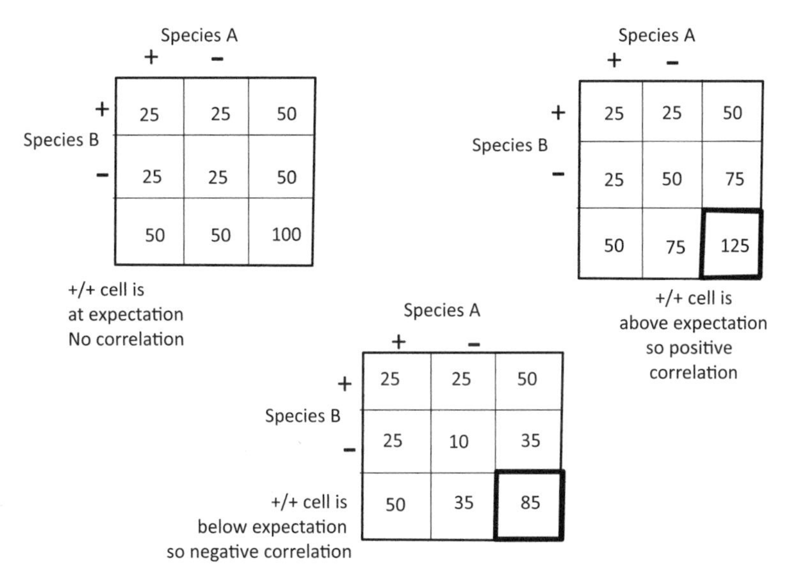

FIGURE 9.5. All three contingency tables show the same value for mutual occurrence, and so they are equivalent expressions of species presence. The marginal grand total of 100 in the upper left contingency table makes the mutual occurrence meet expectations. Therefore there is no correlation in the upper left table. However, in the upper right table all is the same except that the grand marginal total is bigger, making the universe larger, which makes the correlation between species positive. In the lower contingency table, once again there is no difference except that the grand marginal total is smaller, making the universe smaller, in which correlation between species is negative. All correlation is "as opposed to what?"

us sample from all regions of the state. The "as opposed to what?" becomes as opposed to the vegetation across the state, not as opposed to other northern forests. This increases the grand marginal total in the contingency table. Given how uncommon are hemlock and sugar maple in southern Wisconsin, they occur together in the north more than expected. In the north the species still tend to occur together less than expected, but in the larger universe of all of Wisconsin they will occur together more than expected. Thus the two species can be seen as both positively and negatively correlated, depending on how the size of the context brings which aspect of their ecology to the fore. The ecology reflected in the positive correlation is that both species tend to occur in northern mature forest. The "as opposed to what?" is changed to give a different result.

We are describing a scale change here, a move between levels of analysis. Seasoned users of statistics and contingency tables may be scoffing that all this is too elementary. And that would be the case if biologists and ecologists generally understood nonlinearity, scale, and level of analysis, but they generally do not. This has not been a lesson in introductory statistics, it has been a middle level discussion of complexity theory. It is yet another warning about the dangers of reification. All this explains why we have been so cautious about invoking reality in scientific investigation. Both positive and negative relations occur in the material situation, even though those two situations are contradictory. We see only systems analyzed, not systems in reality. When the investigation is for the moment over, and your friends agree with you, to suggest science approaches reality is acceptable and does no harm. But start observing and investigating again and the limits to observation in relation to reality again become troublesome. The fact that we do not observe reality becomes important.

Data Transformation

We have given a summary of the principles involved in multivariate analysis. But we must go further. Data transformation changes the grain of a data set, and that amounts to changing the scale.[4] Multivariate data reduction can be controlled by data transformation as it cuts into the system at several levels of analysis. Another means of rescaling is to change the size of the data matrix, so as to analyze more or less of the total data collected; this changes scale by altering the extent of the data and so the context of the datum values. One reason for data transformation includes the need to bring entities or attributes to some form whereby comparison is reasonable. In Allen's grid of beakers there were transformations applied, such that each "species" would be made comparable to the others. Instead of the actual value of the number of toothpicks in a beaker, the score was expressed as a proportion of the beaker with the most toothpicks. This would bring the scores of high scoring items in some sense in line with the items that were few in all beakers. Such a transformation in real field analysis would bring values for a grass species in line with the numbers recorded for trees, a smaller number. A similar result might come from bringing all the species rows in the data matrix to a common sum of one. A common sum of one hundred converts scores to percentages. Double transformations would go on to bring the columns, the beakers, to some common sum. Thus not only would small numerous items be brought into equivalent terms as large items with smaller sums, but beakers that were full of many items would be brought in line with beakers less full of items.[5]

Sometimes quantitative data are reduced to binary form, where all zeroes are kept and all other values are set to one. The comparison then is of species lists not quantities of species. This transformation has often elicited meaningful results.[6] Relative instead of absolute abundance emphasizes peaks in species performances, thus focusing on the effects of short segments of multivariate gradients.[7] Meanwhile transformation to presence data emphasizes a coarse environmental grain at the tolerance range for the species. Transformations to presence data emphasize the limits of species survivability. The critical organizer of presence data is the distinction, not between different values of the species, but between epsilon and zero. Favorable conditions in the field would allow a population with very low values (epsilon) to multiply and become more common. But a population at value zero (i.e., it is absent from a species list) will have to gain a foothold in some sort of invasion process before numbers can increase at all. By contrast, data of abundance, not just presence, changes the emphasis from persisting in a site to how well the species does once it is present. The biology underlying those two ways of analyzing data is different. The change is in the level of organization projected to the foreground of this versus that factor.

Yet other transformations have been used, usually to take into account the biological meaning of the transformed matrix entries. Substituting absolute species scores with their logarithm may be defended by reference to the logarithmic growth potential of plankton species. Ruth Patrick's use of Preston's lognormal species curve has certainly been useful in relating species diversity to degrees of stress from the environment.[8] Margalef, in a discussion of communities of swimming organisms, considers data collection and data transformation.[9] He emphasizes that large-scale structures can be detected only by choice of suitably coarse-grained parameters.

Allen, Bartell, and Koonce[10] take sequential algal presence data and subtract samples from the next sample in the time sequence. The data then become not what is there but what is different. This gets rid of the momentum of the species. Momentum appears as the resistance to change of status in species over short time periods. Remember Newton's law that bodies at rest tend to stay at rest. In equivalent terms, species in data tend not to change unless some force for change is applied. The absolute value was taken, so a distinction between entering and leaving the community is not made. At the community level of organization, community change is achieved by both species addition and species loss. This transformation allows the expression of a relationship where the state of the environment generally indicates, not the state of species, but rather the change in species composition. Since species composition rarely catches up with state of the environment, the environment state is always being chased in terms of species composition. There is a general condition reflected here: that which is causing is seen as one derivative lower than that which is caused. An environmental state causes a change in species composition. The analysis had to go all the way up to the third derivative of the environment mapped onto the fourth derivative of the species before they could capture that general relationship.

In all data analysis there is a level of analysis. There are various organizing principles underlying a raw data set. The pattern that is revealed under analysis puts one set of causes before others. It is not often apparent a priori what a given level of analysis will reveal. Level of analysis is generally understood after the event. This is because the relationship between the particular data set and the transformation, along with the level of analysis, generates the pattern seen in the outcome. Thus level of analysis is not some ideal in the external world, but becomes pertinent only when it is applied to some extant data. We have suggested that hierarchies are conceived for reason and understanding, and are not a fact of nature independent of observation and understanding. We have raised the notion of essence elsewhere in this volume, and have been at pains to emphasize that essences do not exist independent of a human-contrived model. Essences are explanations of behavior and equivalence in models, and do not arise before a model is posited. This is exactly equivalent to hierarchical levels of analysis not existing in the absence of recorded data.

The whole point of data analysis is to bring to the fore simple patterns in what is observed. Often the simple pattern is linear, something humans can readily recognize. Underlying data are patterns of regularity in different gangs of factors. The gang members can be related easily if they vary together in simple pattern, linearly for example. For instance, species loss has a certain symmetry with species appearing. But loss and gain are not simple opposites, because extinction and invasion are not exact reverses of each other in biological terms. One relates to death and the other to movement of organisms, but even so the relationships between those factors is close to linear. In analysis we seek to lock equivalent process together to see upper level models. While close to linear is easy to see, curvilinear patterns are a bit more complicated and so more difficult to see and explain. Even so moderately curved relationships are well within human ability to capture pattern. Gangs of causal factors can become so differently scaled that their relationship cannot be grasped. For instance, species increasing in abundance is caused by a gang of processes that is set in the context of the species being present in the first place. So the gang of processes causing invasion usually will not have a recognizable relationship to increasing abundance. A propagule arriving in a lake on a duck's foot is remarkably independent of the conditions in the lake that will allow the species to grow or not. Oceans are dominated by invasion factors, whereas lakes are dominated by local extinction factors.[11] At that point, one gang of factors tumbles out first, under analysis, and takes over the results of the analysis, be it a gradient analysis, or a hierarchical cluster analysis.

One of the points of data transformations is to change the order in which the gangs of underlying processes dominate the analysis. In gradient analyses the axes come in an order that puts all later axes in context. Axes emerging first in an analysis are free from interference from other axes. Later axes emerge from an analysis of the residual information left in the data after the earlier axes have been extracted. Data transformation lets various gangs of processes jump the queue. A transformation to binary data coaxes the influence of local extinction or invasion to the fore. Those patterns are then seen in their own terms, and not just in the context of quantitative values in the data. The point of data analysis is not directly to preserve information in the data. It is exactly the opposite. That is to say, analysis is performed to throw away other simple patterns so that just one pattern of interest comes to the fore, and can be understood. It has been bemoaned that binary data throws out valuable information that is then lost to the analysis. No, the very aim is to get rid of those other factors so that binary patterns can come to the fore and their biology can be addressed cleanly. Yes, patterns are revealed, but the whole process of analysis is to get all other factors out of the way. The reasoning behind data analysis is generally not understood, even by those who perform multivariate analyses with canned data analysis computer packages.

Perturbations and Attributes

In chapter 8 we discussed at length the incorporation of perturbation into ecological systems by evolution at various levels. Regular disturbance comes to be expected by the entities at a higher level; fire disturbs as a surprise, but fire regime is expected. Data transformation with multivariate analysis can be used to identify the level at which a perturbation becomes incorporated into the system as a force for stability at a higher level. Data transformations also identify the level of resolution of the important variables in a given system. We now present a specific example of analysis that uses various transformations to achieve elucidation of community hierarchies.

Allen,[12] working Grant Cottam's permanent quadrat data from the Curtis Prairie, used transformation with multivariate analysis to identify the effect of fire at different levels of organization in the community. The prairie was restored from a cornfield in the 1930s. Sampling of five of the permanent quadrats commenced in 1951. The data were collected as cover data that estimated how much ground each species occupied. Several analyses of the data were performed, but two analyses make a particular point. These involve data for all five quadrats through thirteen sample dates between 1951 and 1974 (ten years not sampled) using (i) raw cover data of species abundance and (ii) presence data derived from a binary transformation of the cover matrix.

Scatter diagrams of samples in the smaller shadow space of ordination for both the presence data and cover estimates are given in Figure 9.6. When years that the prairie was burned are superimposed on the presence ordination, the burn years are only weakly concentrated at one end of the first axis. Vegetation pattern coming from individual fires appear to be overshadowed by some other factor. By contrast on the quantitative cover ordinations there is a fairly clear demarcation line between burn years and non-burn years. If the number of fires in the preceding five years is superimposed, the pattern is reversed; the presence ordination gives a clear trend from zero to three fires with distinct lines of demarcation between different burn frequencies, while the cover ordination gives altogether a less clear picture. The differences can be explained by level of analysis.

Fire may change the biomass of species, but it does not eliminate species, thus leaving the species list the same. Therefore on the binary data

Mean Cover for All Five Quadrats in a Year:
0 = no fire, 1 = fire in spring

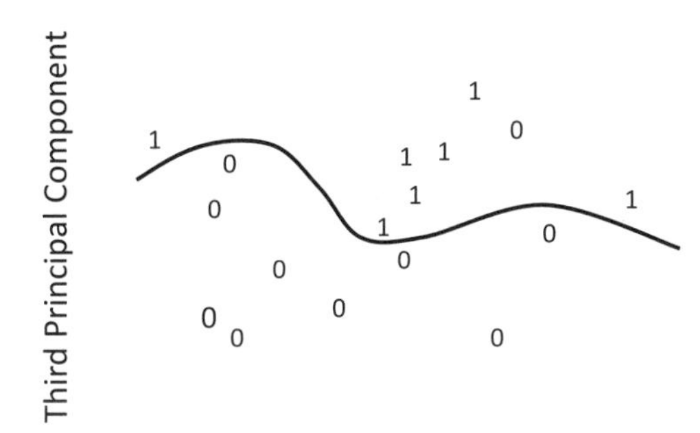

First Principal Component

Binary Data for All Five Quadrats: number of fires in five years

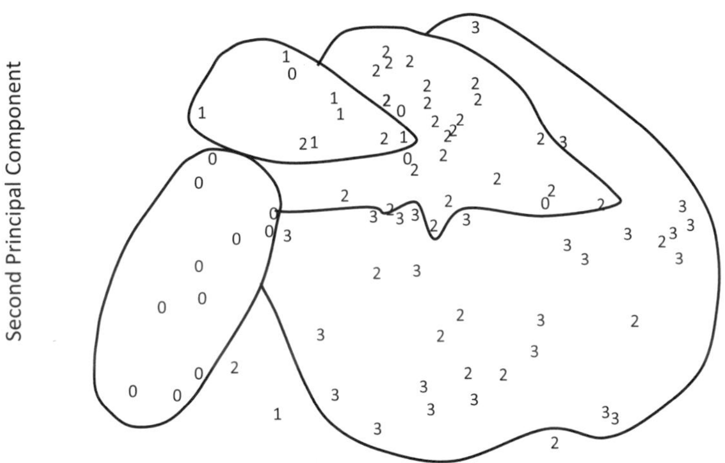

First Principal Component

FIGURE 9.6. a) The analysis of quantitative cover in Cottam's prairie data. The 5 quadrats are averaged over one year. 1 denotes fire that spring. 0 denotes no fire that year. Fire versus no fire gives distinct states of the vegetation. b) The analysis of binary qualitative presence in Cottam's prairie. Here stand quadrats were all kept separate so there were 5 for each year. The numbers 0–3 mean the number of fires in the previous five years. There is a trend. Fire is remembered by the vegetation species list over one or more winter seasons.

an individual fire leaves no mark; the species list is the same before and after an individual fire. Species lists do change, but over a longer time that is influenced by fire frequency. Fires do not remove prairie species that are adapted to fire. But over the longer term, prairies unburned are susceptible to invasion of non-prairie woody species. So it is absence of fire that changes the species list. That is why fire frequency maps onto the binary data with its species lists. With binary data it is not fire that sends a signal in the species list, it is continued absence of fire. That takes a longer time than a year and so invokes a longer memory of fire withheld not fire occurring. Meanwhile prairie species biomass is changed by individual fires, principally because of change in nutrient status. Fire drives off nitrogen and releases other minerals for growth; grasses come on strong after fire. But nutrient status is evened out by winter, so a fire last year before winter is forgotten; a spring fire is remembered in the fall in terms of relative biomass, but is forgotten by the next spring. The memory of individual fires is only in terms of relative biomass of species, and is forgotten each year.

The meaning of these comparisons becomes clear if scale is taken into account. The binary transformation makes a coarser-grained data set than the matrix of cover estimates from which it is derived. Therefore the presence ordination is a message from a larger-scaled holon. If that larger-scaled holon contains fire as a stabilizing force, then individual fires will carry little information and so will not map clearly, but fire frequency will show significant trends. The results concur in the binary analysis showing weak pattern of individual fires, but strong patterns of fire frequency.

The smaller-scaled holon seen in the finer-grained cover ordination will include fewer ecological entities, and one that is missing is fire. Fire being excluded from the system, when it does occur it becomes a signal from the outside (a perturbation). Therefore individual fires show as clear signals in this holon while fire frequency is of little importance. This means that individual fires greatly affect the cover of species in the fire year (burns were in spring) while they do not influence cover in the next year. The signal for an individual fire is largely forgotten over the following winter.

A summary of the results is presented in Figure 9.7, where the top holon is the fire community. The presence ordination identifies this holon. One stem leads to the abiotic fire holon, and the other to the biotic multispecies holon. The cover ordination identifies the multispecies ho-

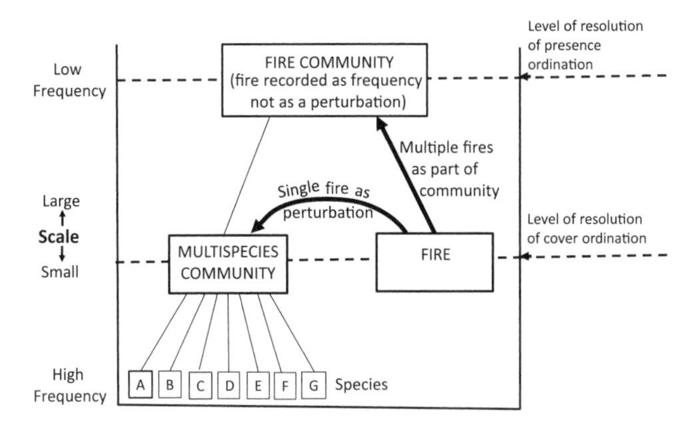

FIGURE 9.7. The summary of the prairie analyses. With cover data, fire is forgotten over each winter at a lower level of analysis. But with presence data fire as an event is not singular, but is remembered over several years. Fire becomes part of the prairie as a fire regime.

lon with stems leading from it down to species guilds or individual species. Thus the important variables in the cover data are species, and they show a clear pattern on the cover ordination. On the presence ordination the important constituent holons are not the species; they belong one level lower down. The significant daughter holons of the fire-community holon are not species but include multispecies parameters such as levels of diversity. Thus diversity maps clearly onto the presence ordination. We found it refreshing that, for a change, the notion of diversity should answer rather than pose questions.

So data transformation under multivariate analysis is able to show the hierarchical structure in field data from ecology to sociology. This has been a fairly complicated process for us to explain. There is hierarchical structure in field data but it takes care and persistence to capture it. All this should reemphasize that we never see systems as they really are, but rather we have access to complexity only under analysis. Often the only device to explain what we see is narrative. Teasing apart hierarchical levels often leads to contradiction between levels. Narrative, unlike models, sits comfortably beside contradiction.

CHAPTER TEN

Hierarchy as a Context for Modeling and Simulation

The multivariate analysis of masses of data in the previous chapter is a statistical approach to complexity. By contrast one can model complicated situations with equations, and even touch on complexity. This is called an analytical approach. The advantage of the analytical approach is that the various terms in the equation can be considered as parts of a complicated set of relationships or processes. As of the 1970s mathematicians and meteorologists found equations that were short and tractable, which nevertheless behaved with infinitely rich behavior. A new science of chaos theory arose from these equations. Much is made of chaos and complexity. Chaotic equations are not generally useful for complex situations, but they did advance theory by establishing minimal conditions for complex behavior.

In biology chaos opens up new considerations that were not available in the first half of the twentieth century. When Allen learned biology there were problems that were simple enough to solve, and the complex insoluble issues were left alone except by biological philosophers, who in a sense never solved actual systems. Before chaos there were simple systems with simple solutions, which we found. By contrast there were impossibly complex systems that never repeat and so were not susceptible to science. We left them alone. But with chaos comes an intermediate consideration. There are infinitely complex systems whose behavior never repeats, some of which do have simple solutions. Some chaotic equations are in fact short and tractable. These were commonplace but not recognized as being soluble. Translating this to a mundane concrete example in biology, we have the oak/elm case we have offered before.

Every elm tree is different from all elm trees there ever were, or are, or will be. But that infinite variability does not mean that an elm is free to become an oak tree. Despite the infinities that defy definition, any good naturalist can consistently distinguish any elm from any oak. Chaos gives us courage to go on anyway, even if the situation is infinite and so undefinable. It is not so much that we use chaos in everyday biology, but chaos does give us the minimal situation of workable but undefinable situations. Chaos implies the notion of bounded infinity. In earlier chapters we referred to Zellmer's scheme for dealing with complexity. There the essence is a bounded infinity and so cannot be defined, although it sets limits (Figure 5.3). As in roles, there are boundaries, but the outcome of the equation cannot be predicted, except by playing it out. Chaos is helpful less as a calculator and more as a new signpost toward what is possible in our understanding.

The geometry of chaotic bounded infinities leads to fractal geometry. That allows investigation of awkward undefinable situations. In calculus equations have limits as to their values, which is what happens when the term *lim* arises. Chaos is related to fractal geometry, the geometry where equations never have a limit, where the same pattern repeats at different scales to infinity. We gave fractals a nod in chapter 3 where we considered the coastline of Britain (Figure 3.2).[1] There are two sorts of fractal pattern, one where the same exact pattern repeats, and another where only the same degree of complicatedness appears at different scales. The perfectly repeating patterns are called self-similar fractals. The other sort is only fractal. An example of a self-similar fractal is the Koch diagram. It has a simple rule. Start with a triangle. Take a side, divide it into three segments, and then add another segment of the same one third length. This causes the edge of the triangle to buckle so a small triangle kinks out of the middle of that side of the triangle. Do the same for the other two sides of the original triangle. In the next iteration we address the four straight segments of the line that was kinked in the first three to four segment substitution. The smaller straight segments are themselves kinked with the same three to four substitution side. The kinks get kinkier at a smaller scale. In the next iteration do the same for now smaller straight line segments, and so on. The straight line segments get smaller as the three to four substitution is made repeatedly downscale. The exact formula for the Koch diagram persists, where exactly three segments are replaced with exactly four (Figure 10.1). The same applies for all fractal equations, but not for fractal measurements.

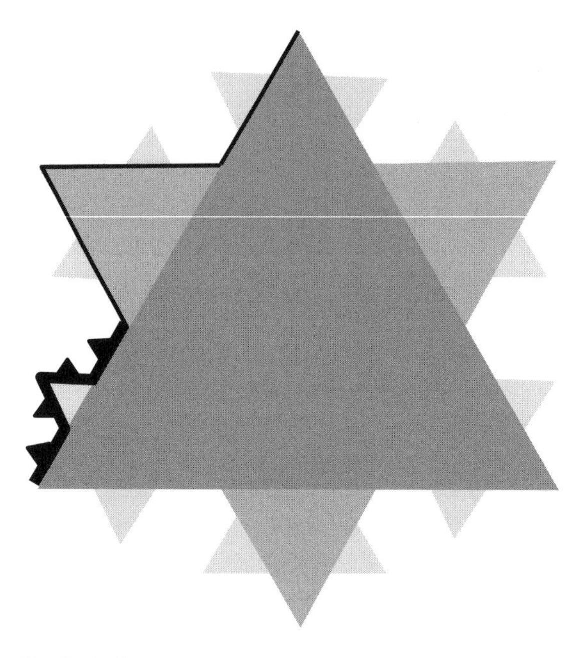

FIGURE 10.1. The Koch diagram showing the process for building it. Start with a triangle, divide the sides into three segments and add another third. This causes a kink on that segment. In the next round, take the four shorter segments and divide each into three; then add a fourth segment to each. Do the same in the next round for all the new segments. The thicker lines in the bottom left would apply all round the whole shape.

An example of a fractal measurement is clouds, which are fractal but not self-similar. One cannot tell how close is a cloud, because you cannot tell how big it is. All clouds of all sizes look the same, but not exactly. In chaotic systems the starting point, the initial conditions, make a difference in the values that come out of the equation in later iterations. Chaos, and so fractals, are infinitely sensitive to initial conditions. Cloud formation cascades upscale. The reason for clouds being fractal but not self-similar is that the process of making clouds all the time has initial conditions, which are very similar, but not exactly the same. As a result, all clouds form from the same processes, which make them all cloud-like, no matter what size. But there is an ongoing set of exact initial conditions for each cloud and cloud part. As a result the slightly different exact initial conditions stop clouds from being self-similar, while still allowing the generalized cloud making processes to show fractal patterns. The regularity of clouds comes from the unified process that all clouds share in the making of clouds. All clouds are different, while all still are cloud-like. The details of particular clouds come from the infinitely small slight differences in initial conditions at each stage of cloud formation. Clouds are not self-similar in the way of equations because clouds are jostled in the process of the infinitely detailed physical creation.

Equations are commonly considered to be general solutions, because a large number of solutions can be calculated, one for each initial condition. Rosen[2] says that solved equations are not a general condition. As equations become longer and more complicated, an actual solution to the equation is usually not possible. The general condition is that a given equation cannot be solved. This is bad news for the strategy of using equations and expecting to understand a situation by tinkering with the terms in the equation. But all is not lost. Some of the time we can retreat to a computational solution.

Monte Carlo Approaches

An equation has certain parts that matter as an expression of some behavior of the system that is to be investigated. For example, the number of components in a system and the number of connections between those parts may be of concern as to the stability of the system. There are, however, other parts of the particular use of the equation that do not matter, and are simply some arbitrary starting point. An example here might be a four-part system where the number of connections matters; four parts is fixed as is fifty percent of the possible number of connections in this example. However, which component is connected to which may not matter. In a specific example, Gardner and Ashby[3] wished to see if there were general effects on stability of increasing the strength of connections between parts of the equation, or the number of components in it. An equation to solve the problem of the relationship to stability would be very long and could not be solved. So they set up examples of systems with a certain number of parts and a certain number of connections, and they looked to see whether or not that particular example was stable. Imposing the limit of a set number of parts, and a set number of connections, Gardner and Ashby randomly created a large number of particular examples that they tested for stability. From that they could work out the average percentage of stable systems with that number of parts and connections. Having tested systems with four parts at fifty percent connectance, they generalized by probing test systems with four, seven, or ten parts, all across a range of percentages of connectance.

Say they were investigating a four-part system; there are 12 possible connections. They made 6 connections of the 12 possible. The measure for connectedness that they used is called connectance, which is the percentage of possible connections that are actually made. After setting up and running a thousand systems of that type, they recorded how many remained stable. With that value they had a robust value for the general stability properties for that percentage connectance (50%). They did these calculations for different degrees of connectance from 5% connections to 100% of possible connections. It would not be possible to calculate the chances of the system remaining stable from direct analysis, but it is possible to work it out by trying it out at random enough times for the outcomes to be statistically stable. This is the Monte Carlo approach (James Bond and spins of the roulette wheel). This is one of a set of solutions that are not analytic but are computer-based and called numerical. Statistics as a discipline these days is divided into two camps, the analytical as opposed to numerical. The former relies on proofs and is accordingly very satisfying. The latter takes otherwise insoluble situations and simply forces an answer by sampling, but it does get an answer.

At higher values of connectance Gardner and Ashby's systems become unstable (Figure 10.2). Also they note that larger systems (they use 4, 7, and 10 parts) are more unstable and have a sharper decline in stability with increased connectance. The general take-home message appears at first to say increasing diversity decreases stability. However, the critical aspect of the experiment is random connections between parts. Since there are many stable systems with high diversity, the final message is that connections in material ecological systems are not random. They are in fact hierarchical, where a nexus of local connections forms a subsystem that stabilizes the whole.

Moving to Higher Levels

Buzz Holling is one of the great ecologists of the twentieth century. The multivariate material considered in the previous chapter moves between levels, but often moving upscale to higher levels. By contrast, analytical approaches generally identify processes below the level of the whole and then work downscale to achieve more precision in a reductionist mode.

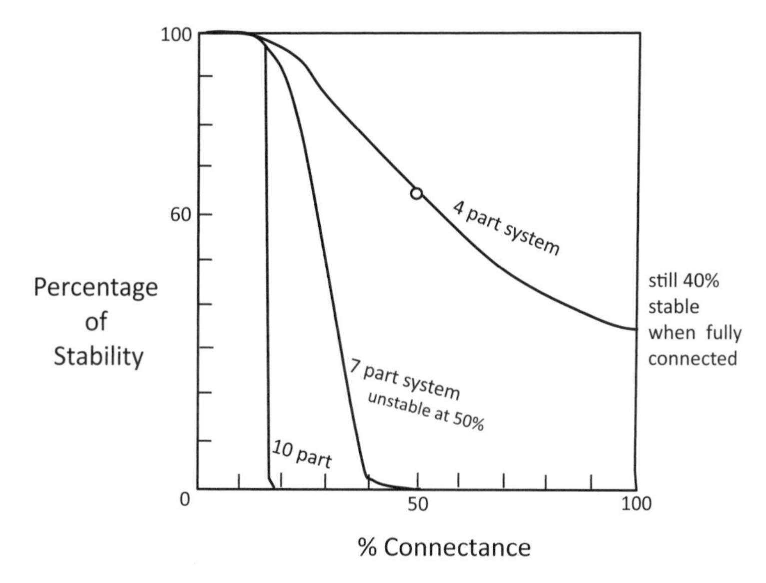

FIGURE 10.2. After the figure in Gardner and Ashby (1970), plotting their Monte Carlo experimental results. A four-part system is made to have 50% connectance. Calculate whether that version of the system is stable. Make many other systems whose actual connections are different, but which have the same level of connectance and number of parts. See if they are stable. There emerges an average stability for the whole set of iterations. The small circle on the four-part line is that average stability for a 50% connectance. Then do the same for various other degrees of connectance for four-part systems and plot average stability across degrees of connectance (shown as a line). Then do the same operation on seven- and ten-part systems. They become unstable sooner. In general randomly connected systems become unstable sooner if they have more parts or higher connectance.

Holling works with equations but moves upscale throughout his career, always looking for context and meaning. He started by solving for ways that predators control prey. Having solved for ways of control he shows how predator controls fail in general. Failure of control moves the equations upscale so as to consider the region around the zone of stability into the zone of explosive growth. Having lost control the system enters an epidemic pest outbreak. But the epidemic lasts only a short time. After the epidemic predators reclaim control in a predator/prey cycle of delayed negative feedback. Holling generalizes that cycle. Ultimately he moves upscale to look at how animal ecology is structured in general, as he addresses size distributions in biomes. Upscale gives understanding in terms of meaning and context, and so addresses generally more important issues.

As an experimental population biologist Holling works first on prey/predator relations in small mammals feeding on pine sawfly insect resting stages underground (Holling 1959a,b). His first concern is to achieve a general understanding of how predators feed relative to prey density. He develops three equations as hypotheses.[4] The experiments are with artificial and real biological systems. The first is simply a linear increase in predation as density of prey increases; more to eat gives more eating (Figure 10. 3). The implication is that predators can restrain prey up to any density (type 1). Of course they cannot but it makes a first model. The second equation (type 2) again shows rapid increase in consumption with increased availability prey at low density. But then with increasing prey density predation begins to saturate; the predators cannot take full advantage of increases in prey. There will be some other limiting factor, like number of nesting sites, which limits predators independent of how much food they have available. The prey numbers are less controlled at high densities. The second type of equation asymptotes, indicating a continuous process of saturation in the ability of predators to respond to prey by eating more. At full saturation, the predators do not respond to more prey. The third type of equation (type 3) at first has low consumption at low prey densities. The assumption is that the predator cannot find the prey so as to focus upon them. Once there are enough prey to excite focus, the type 2 process takes over. Type 2 is the one that fits most empirical situations. Type 3 is a sigmoid curve, at first flat, then

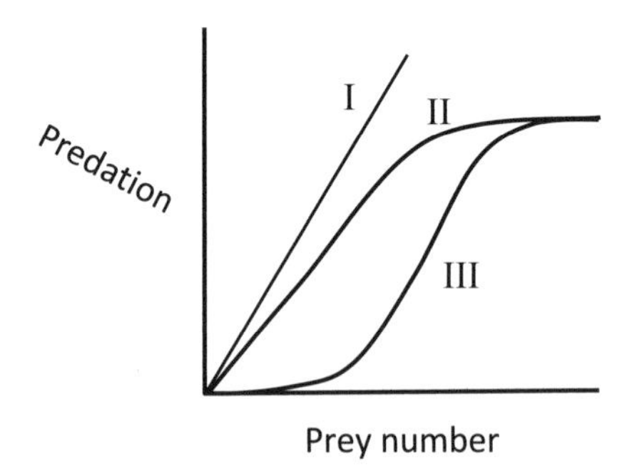

FIGURE 10.3. Increasing prey density gives different intensities of predation: more to eat gives more predation. Holling developed three equations for different types of predation. Type I increases linearly. In Type II predation asymptotes to a saturation point. Type III is like Type II but at lower densities prey cannot be found. Sometimes Type 1 is pictured as unlimited (as depicted here), but sometimes saturation arises at an instant at a break point.

steeper until the process of saturation stops increased feeding. This is essentially lab and field experimental work, limited to the process of predation, with little explicit concern for what happens to the prey once the predators are saturated and cannot crop down the increasing pine saw fly population. The insect outbreak is implicit, and we presume this led Holling to investigate it.

A decade later there was sufficient computational power for Holling and Ewing to investigate more general patterns of prey/predator and host/parasite relationships. One can plot host and parasite against time as two lines waving up and down about half a cycle out of phase. As the host increases in numbers the parasite numbers respond because there are more hosts to parasitize. The increased parasite load drags down the host population, leaving fewer hosts to feed the parasites. Starved, the parasites also decline, whereupon the host numbers begin to increase again, and we enter a new phase in the cycle repeating into the future. Another way to express these patterns of cyclical behavior over time is to work with a plane of host plotted against parasite. Instead of time being plotted out explicitly as an axis of change, the output can be plotted as movement across the host/parasite plane over time. Mapping that onto the host/parasite plane, the cycles of waving up and down now appear as circular maps. We see cycles of behavior as circles of behavior. Much as the plots of host and parasite wave up and down over time, the behavior mapped onto the plane moves the system across and around the plane. Over time the oscillations of host and parasite gradually damp down. Instead of wavy lines into the future the host parasite lines become straight at a constant value. Mapped onto the plane, both host and parasite stop moving, rather like a pendulum coming to rest, hanging straight down. Holling and Ewing start the host/parasite equations with various numbers of host and parasite—that is to say, they started the relationship at different places on the host/parasite plane. Plotted on the host/parasite plane the roughly circular pattern system spins down to a point. Movement on the plane can be considered as moving in some sense generally downhill. The paper is entitled "Blind Man's Bluff," as they use a computer to run their hands, as it were, over a response surface of host and parasite numbers. From the notion of moving downhill, one can think of the surface as having ridges and hollows.

Moving to a higher level of abstraction we can imagine the plane as a cup where a ball spins around until it comes to rest in the bottom of the cup (Figure 10.4). When the type 2 and type 3 equations are saturated,

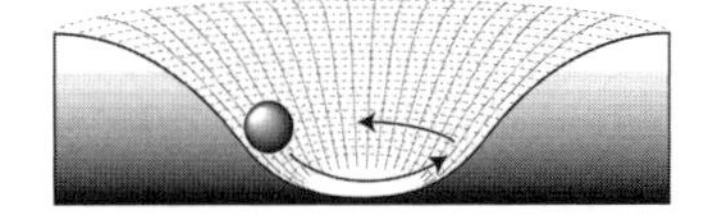

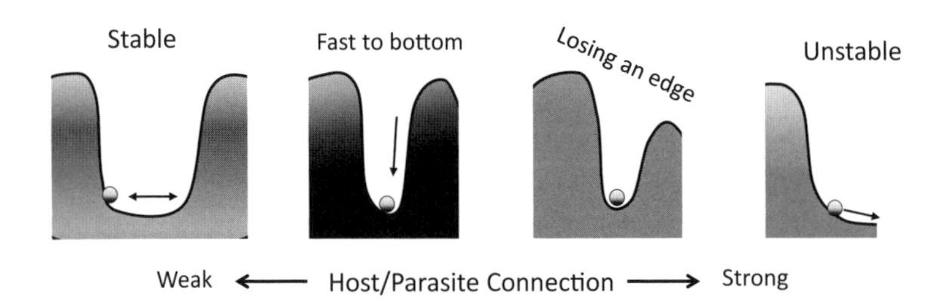

FIGURE 10.4. Host parasite relationships can be plotted against time as two waving lines. The waving host line is half a cycle ahead of the parasite line. The same applies to pairs of prey predator lines. Sometimes the lines damp down to stop undulation, as the relationship stabilizes. Plot the host/parasite numbers against each other over time and the relationship maps a roughly circular path. With damping, the path spirals into the middle. We can draw that relationship as a cup with a ball circling down to rest at the bottom of a cup. This relationship is drawn here for systems with different degrees of parasite efficiency. With a slack relationship the cup is open with steep sides. That is to say, it oscillates for a long time, always coming to rest in the bottom of the cup. With increases in efficient predation the cup gets narrower. It finds the bottom of the cup immediately. Further efficiency starts to erode one edge of the cup. Great efficiency takes away one side of the cup and the stable equilibrium of the cup bottom disappears. That is to say, under great efficiency and high connectedness all stability is lost. The depiction of the upper 3D basin was created by Tom Brandner.

the prey are free to grow exponentially in an outbreak. Predator is just a parasite that tends to kill rather than milk the prey. With the right starting coordinates extinction of predator or both predator and host is another possibility. The same can happen with hosts and parasites since parasites are only gentler predators that do not consume the whole victim. With an outbreak we have to change the concept of a cup to describe the behavior. The blind man's bluff led to cups with lips, so that if the ball ever crosses the lip it keeps rolling away from the cup. That is what happens in an outbreak inside the computer simulation.

We do not need a new equation for the outbreak. The same equation that spins down to a point when started inside the cup can spin away from the cup if the original coordinates are set outside the cup. Moderate starting places give a trajectory that spins down. Extreme starting places throw the behavior of the equation over the lip into instability and continuing change to infinity or zero in this theoretical universe. By starting their simulations all over the surface, Holling and Ewing can map the computer output to a response surface with a cup and with unstable places that lead to spinning away from the cup. As the scientists change the initial host parasite numbers, they quickly find a region on the surface where the system stabilizes at a given point, rather like a pendulum coming to rest. Outside that safe region the system leads either to extinction or unbounded growth, as the system manifests instability.

Moving further, the investigators code in the relationships buried in the equations for a certain degree of coupling between the host and the parasite. The equations with tighter and looser coupling of host/parasite behave differently. Tighter means a stronger response of host numbers to a change on parasite numbers. Holling and Ewing are able to develop general principles as to how dynamic simulations work. They start with a very loose coupling, and show the response surface for those conditions. It is the behavior of the ball in a broad flat-bottomed cup. The telling results occur when they change the strength of the coupling between host and parasite, making the parasite more efficient in its attack. This is a different way of probing the equation. Instead of feeding in different values for variables and initial conditions, the equation itself is changed to see how it behaves with slightly different forms of the original equation. The investigators then repeat the process of running their hands over the new surface defined by the new more or less connected forms of the sets of equations. The surface changes and the coupling shows two things happening.

With increasing connection between host and parasite the tendency is for starting points in the safe region of the cup to move on a faster track, more directly to the equilibrium point at the bottom of the cup. The second change is in the shape of the cup, the safe region that moves to equilibrium. Tighter coupling appears to permit higher values of host and parasite still inside the safe equilibrial region. The safe region appears to be losing its ability to rein in higher values. We can draw the whole surface or simply take a section through the cup; the section makes the change in the surface more explicit and easier to see (Figure 10.4, lower images). We can think of the safe region as a cup cut in sections with one stable equilibrium at the bottom of the cup. Displace the ball from the stable equilibrium at the bottom of the cup and it returns. There are two other equilibria that are unstable. They appear situated at the lip in the sections of the cup. The ball is balanced on the lip, but the lips are unstable places. One could balance a ball on the lip of the cup, and it will stay there at equilibrium. However, any move off the lip in either direction causes the ball to move away from the lip, indicating an unstable equilibrium. Further increases in efficiency change the surface in the same fashion. Greater efficiency of connection between host and parasite takes away the height of the lip of the cup on one side. In the end one edge of the cup disappears altogether. As a result, in the absence of a cup, there is no stable equilibrium available. The take-home message is that as systems become more efficient, they become brittle to the point where they lose all stability. A brittle system is rather like tennis players with match point against them. They tense up and lose the point and the match. The greatest players seem somehow to remain calm, and can come back and win from being earlier at match point against.

So Holling is looking, not at how predators or parasites control their prey and hosts, but how control of population numbers is lost at high densities. Of course there is a larger view that looks at where the populations go when the equilibrium relationship is lost. The loss of one stable region can be presumed to yield to other stable regions not considered at first. From this Holling[5] talks of resilience and multiple stability points. This introduces disturbance as a notion in population numbers. The system would move from one stable point to another because of a disturbance. Think of it as a series of cups on the landscape where someone is shaking the table so the ball flips around between various cups. At this point Holling introduces two sorts of stability in his populations; the notion of resilience in the face of forces for change, as opposed to resistance to those forces. A decade later, Holling generalizes these results in his spruce budworm model. Holling and Ewing only performed analyses to find out the effects of tension on equations, an exercise in structural stability. In the mid-1980s, Holling[6] extends these ideas to model not only the escape from prey/predator oscillations, but also spruce budworm epidemics and their passing. The hierarchy in epidemics at a lower level is where the dynamics of the pest are endemic, present all the time. At a higher level of analysis those dynamics are constrained by some force, such as bird predation. This is the ball and cup stability. At a yet higher level the cup erodes so as to allow a switch from control by predators, such as birds, to escape into unbridled growth of the pest in an epidemic. All the time Holling reaches up to higher levels of discourse, as his ecological hierarchy spans wider universes. An epidemic amounts to the ball being taken into another cup wherein budworm works at epidemic levels. This cup too erodes as the budworm kills the trees and so destroys the energy base for the epidemic.

In the same 1980 paper Holling presents what has come to be called *panarchy*[7] (Figure 10.5). The axes on the panarchy scheme are capital on the ordinate and complexity or strength of connections on the abscissa. The two axes notwithstanding, panarchy is not a graph, because graphs present models, and so must be internally consistent. Panarchy is not and involves internal conflict. The inconsistency is that capital disappears from high degrees of complexity, but almost immediately reappears further down the track at a lower level of complexity. Panarchy is a narrative where conflict is simply treated by moving the narrative forward.

There are four phases in the lazy 8 of panarchy: r, K, Ω, and α. The r and K are related to properties of the logistic equation:

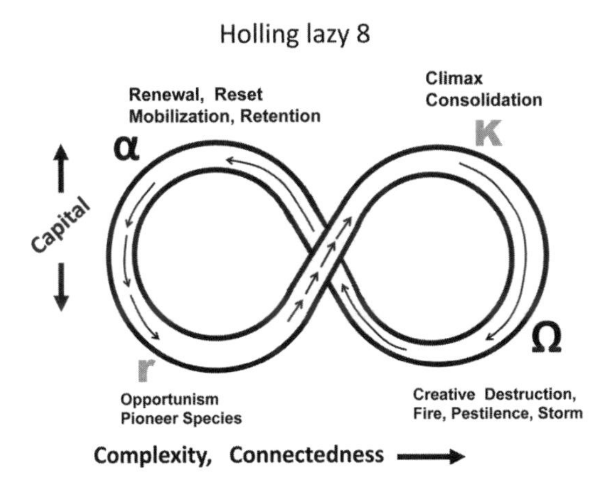

FIGURE 10.5. The track of Holling's panarchy (or lazy 8) narrative. The plane is capital against complexity or connectedness. The cycle starts at the r phase with little capital or organization. Over time capital and complex organization increase up to the K phase. At K the system is brittle, an accident waiting to happen. The passage from r to K is slow, depicted by short arrows. When K collapses capital is suddenly lost in a move to the Ω phase depicted by a long arrow. But with loss of organization in a move to the α phase, capital reappears but in a disorganized state (Odum would say that capital loss is a rapid conversion to liquid assets). Lacking organization in the α phase, capital is soon lost in a return to r. The contradiction between Holling's capital peaks makes panarchy a narrative rather than a model.

$$N = Nr\,(1-N/K)$$

where N is the size of the population, r is the instantaneous growth rate of the population, and K is the carrying capacity of the environment. The equation starts with small values of N growing at almost the full rate of r. As N approaches K, N/K gets closer to 1. The inversion in the subtraction from 1 means that with N equal to K, (1−N/K) is zero. Multiplication by zero wipes out the effect of the r growth term; N does not grow and the logistic equation stalls at K.

The designation of r in panarchy indicates a system with little capital on the ordinate and little organization on the abscissa. It is the phase when growth terms capture behavior. The move to the K phase indicates the result of long-term accumulation of capital. The system in the K phase of panarchy sits with high capital controlled by negative feedbacks, often living things organized for self-preservation. Most of the time for the entire panarchy cycle is spent in growing from the r phase and reaching the mature K phase. Accumulation of capital and organization is slow but persistent. In the end, the accumulation of capital becomes a hazard. If ever there is large capital, something will arise to consume that capital. In a social example the emergence of powerful warlords in Somalia came about because of the capital moved into Mogadishu Airport by food aid agencies to deal with a famine.[8] Warlords love famine, because it gives them control of a critical gradient of food distribution. The capital of the goods from the international food agencies is dissipated by the warlords. That dissipation takes panarchy from K to Ω.

If the system is a mature forest, the capital of large mature trees invites windthrow or fire. Sugar maple forests actively organize against fire by leaving nitrogen in fallen leaves. Earth worms find the leaves nutritious and pull the dead leaves underground. Individual trees fall to make gaps in the forest. The gap opens up a recycling phase into which new trees insert themselves. A clearcut, massive windthrow, or a crown fire, resets the whole forest at a higher level. In the spruce budworm system trees of thirty or so years old offer enough food for epidemic outbreaks of the insect that destroys the stands of trees. Holling takes a positive attitude to the destruction of capital, using Schumpeter's term, "creative destruction," to which he assigns the Ω phase. The capital of the organized K phase is destroyed in some way.

The conflict to which we have referred occurs in the renewal of the capital returning in a disorganized state. This is the α phase, where information entrains the system into its next cycle. The capital that was destroyed now emerges as capital for some disorganized condition. The emerging disorganized capital might be dead as opposed to live trees, their death being the creative destruction. The living trees at the outset of budworm epidemic reappear as standing dead trees that serve fungi and bacteria with resource. But because the decomposers are not organized to preserve the capital in the wood, the capital fidgets away in a loss of dead capital in decomposition. Loss of capital takes the system back to the wide open r phase once more. The α phase is crucial for the resetting of the new cycle. For instance, the Hubbard Brook clearcut experiments were followed by the application of herbicide in phase α. The poison denied the renewal phase and so recovery was delayed and mineral nutrients continued to leak from the dead forest. However, in the absence of herbicides, the clearcut forest within a few years was white in the springtime with the blossoms of the pin cherry. There were no pin cherry trees in the forest that was destroyed, except as pin cherry seeds in the soils waiting for just such an event as the clearcut.[9] The pin cherries left alone to play their role become active in the α phase; they immediately seal up the leaking mineral soil nutrient compartment. In a classic error, the lauding of old growth diversity is taken as crucial for system survival. No, it is the lower diversity phase with pin cherries that is crucial for survival through the reset phase that allows a return to a successful r phase.

H. T. Odum offers an equivalent scenario in his maximum power principle[10] (Figure 10.6). We have noted the contradiction of Holling's capital disappearing and then quickly reappearing; did the capital ever disappear? That contradiction makes panarchy a narrative not a model. But Odum avoids the contradiction by renaming disorganized capital as liquid assets. Holling's collapse of organization is Odum's transfer of capital into liquid assets. He notes the same pulsing and recycling phases as occur in panarchy. Odum makes the distinction between capital, which he calls resource reserves, as opposed to the capacity for immediately doing work, which he calls liquid assets.

Odum's gradual achievement of maximum resource reserve is Holling's gradual accumulation of capital as the system moves from r to K. Holling's creative destruction phase occurs as Odum's transfer of stored resource to liquid assets. The destroyed capital has simply been made liquid. It is the nature of liquid assets that they are quickly spent. Hol-

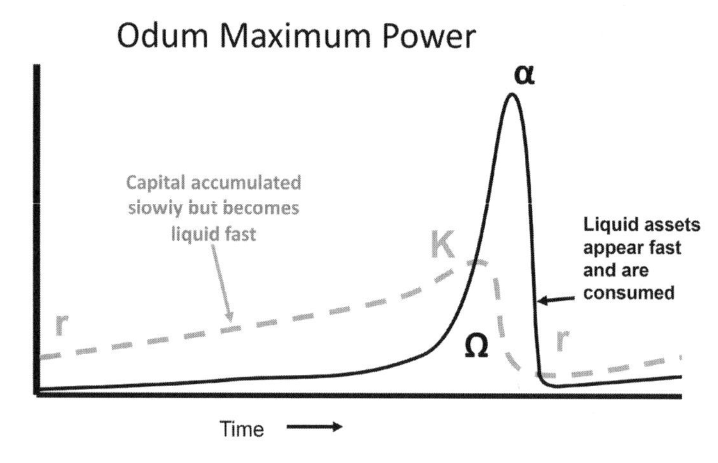

FIGURE 10.6. Odum's maximum power principle mapped onto panarchy. While the track of panarchy speeds up between the Ω and r phase, Odum's depiction shows a steady pace through time. It is the consumption of the peaks of capital and liquid resource that has a lot happening over a short period. The consistency of the distinction between capital and liquid resource makes maximum power a model, not a narrative.

ling's renewal α phase occurs when Odum's resource reserves are fully converted to a peak of liquid assets. It takes very little time for the liquid assets to be spent, opening the door to a new phase of capital accumulation starting at r.

The critical difference between Holling's and Odum's figures is that the movement around panarchy occurs at different rates, while time simply marches on in Odum's scheme. The fast moves between phases in panarchy are simply presented as narrow pulses of activity in Odum's maximum power. We have already addressed high and low gain systems, and they apply again here. When Holling moves on from destruction so as to quickly arrive back at r, panarchy is in high gain mode where the system is moved simply by flux of capital destruction. Odum's liquid asset line is similarly driven by the flux of consumption. Forests are cut, crops are harvested in high gain mode. The move from r to K is low gain in both schemes; some external process allows concentration of capital. On the face of it, the loss of organized resource reserves looks like a high gain pattern of consumption in harvest. But if one points out that the capital liquidation phase is followed by an extended period of capital accumulation, the whole cycle is a low gain pattern of particularly efficient low gain exploitation. After the pulse of consumption there is an organized period when capital is actively left to accumulate. Harvesting is explicitly avoided in that part of the long-term plan. Thus the consumption phase waits until capital accumulation is going at close to a maximum rate before it moves in to consume. Odum calls this the maximum power principle, and says production systems all move to achieve maximum local throughput and degradation. He sees the world in strictly energetic terms. By cropping periodically, the system is always accumulating capital without being constrained by the limits of an approach to an overmature K phase. Capital accumulation runs always at the maximum that the r in the growth equation allows. Thus it appears again that there is an ordered alternation between high and low gain, so as to achieve maximum power. Only recently has slash and burn tropical agriculture been recognized as highly organized. The burn is conducted carefully so as to leave seedlings of fruit and nut trees, so the fallow phase is still productive. Fallow does not just happen, it is actively organized. This is the process of active withdrawal of farmers as an organized phase that helps capital start accumulation as soon as possible.

But high and low gain conceptions have an edge over both Holling's and Odum's scheme. Holling and Odum need to see pulses to show their systems at work. In high and low gain conceptions both the energetic phase of high gain and the constraining phase of low gain are continuously at work. We see lakes as being under one or another mode of control. One is the rapid carbon fixation of algae dominating dynamics in what is called bottom-up control. The other mode is when fish dominate the system top-down by imposing constraints on capital accumulation. When a lake is dominated by algal production it is driven bottom-up by algae. That may be seen as the system being in high gain mode of maximum use of nutrient fluxes. When the lake is constrained by fish in a trophic cascade, capital in algal biomass is small. Big fish eat little fish. So little fish are constrained so as not to be there to graze on zooplankton, which then become abundant. The large standing crop of zooplankton then graze down the phytoplankton. This top-down control works through constraints of fish eating. It would appear that the lake is then in low gain mode. But the critical point to notice is that even when the fish are constraining the system in top-down mode, the algae are still capturing resource under high gain. And when the algal production of bottom-up dominates whole lake behavior, the fish are still consuming, but not with a strong enough constraint to dominate whole system behavior. There are pulses in maximum power or panarchy. The same general situation can be shown in the balance of high gain dynamics and low gain constraint, but with gain there need not be actual pulsing, because both high and low gain tension can be present all the time. All this shows how even in the realm of explicit particular equations, level of analysis pertains and contradictions will arise in the tension between levels of analysis.

In the last phase of his career Holling uses a broad brush to perform biome ecology. It all turns on sizes of animals.[11] Holling ranks all the animal species in several regions according to size. The phenomenon of interest is aggregations of sizes that formed "lumps" of sizes with forbidden sizes of animal in between. He can see the forbidden sizes as steps up between the largest animal of a lower lump and the next animal, the smallest of the next lump up in size. He could also see the edges of lumps by plotting, not sizes of animals directly, but by looking at size difference between adjacent animals in the size ordering (Figure 10.7). Peaks of size difference show the gaps between lumps. Each geographical region has lumps of a certain size specific to the region. But all regions have lumps. Endangered species occur at the edge of lumps, presumably because the environment presses them. Species that have recently invaded

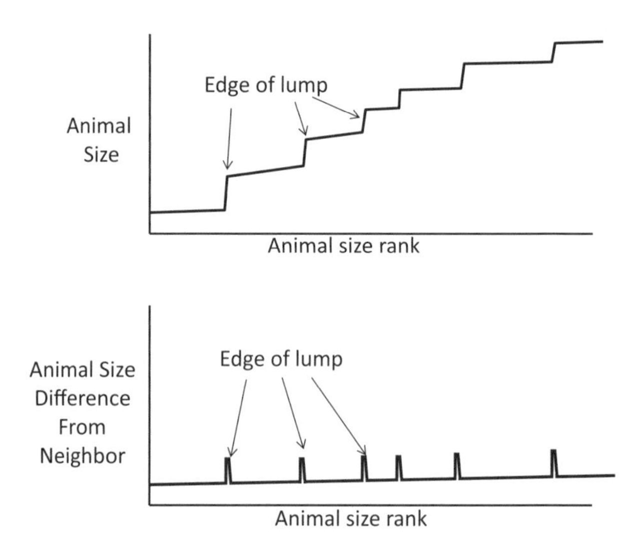

FIGURE 10.7. Two expressions of the edges of Holling's lumps in animal size. Passing from one lump to another can be seen as a step in size or as a large gap in size along the ranked sizes.

also occur at the edges of lumps. An explanation might be that those are the places where the community offers less resistance to animals that have not been selected in the local environs.

Holling is able to show by different lines of evidence that lumps are a universal phenomenon (Allen and Holling 2008). With major climate change, extinctions largely occur in given sizes of animals—that is, certain lumps as a whole go extinct wholesale. The extinction of the megafauna after the last ice age is an example. The lumps derive from the size of an animal relative to the mosaic of resources on their landscape. The link between size and landscape includes a factor related to animal volume (a cubed function). Animal body weight coming from volume indicates how much the beast needs to eat once it gets to a patch of resource. Lumps also derive from scaling relative to the length of the animal's stride. Stride length is a matter of size too. Stride is more like a squared function, less than cubed anyway. So the trick is how fast and far can an animal move to a new resource patch, and how big does the patch have to be to supply the volume of food needed. The melting permafrost would have literally bogged down mammoths and the like.

There are other issues causing lumps related to predation. There are two ways to avoid predation. One is to be too big for a predator to overcome prey at a workable cost. The other strategy is to be small enough to hide. Bears are predators and are generally too big to be preyed upon, so hide on the ground for hibernation. However, in Siberia bears build nests in trees. Tigers! Now there's a top carnivore for you. Normally bears are too big to be prey, but in Siberia tigers eat bears, so bears get out of harm's way in trees. Phenomena like this explain why lumps are specific to the local environs in one region; lumps are regional and are not universals.

Another predator-driven change of size occurs on new islands. When land becomes an island (e.g., sea rise flooding the Mediterranean) the large predators go locally extinct because they need a space bigger than the island. Colinvaux says in *Why Big Fierce Animals Are Rare* big carnivores are limited by their food being high in the food chain. There is not much to eat close to the top of the chain. The largest whales can get so big because they filter feed krill from low on the food chain. So as Crete became an island, big carnivores like lions went extinct on the island. Thus the small animals that hide from predators down cracks and in hollows or in water can get bigger. On Crete the dormouse became larger. In general on islands there are things like giant beavers, large birds, otters, tortoises, and lizards. Also, elephant-sized animals on islands with smaller ranges need to evolve smaller, and they can afford to do so without predators. There were different species of dwarf hippopotami on Sicily, Malta, and Crete. There were pygmy mammoths on Wrangel Island, north of Russia by the Bering Strait, until about four thousand years ago. Huge giant ground sloths occurred on islands, such as the Antilles in the Caribbean. Change the scale of a landscape and the animals evolve to change size.

The critical thing that humans do is technology, and that is scale related. First our technology has let us move human ecology upscale and cover the globe. Most technology, at its essence, changes the scale of the human relative to its problems. With technology we can move faster, and preserve perishables for longer times in plastic bags to travel farther. We can manipulate very small or large objects (e.g., bacteria and Stonehenge). The computer changes the scale of our problems so much as to make qualitative differences in what we do. We hardly ever use computers to do old things faster. With faster processors we do things differently.

Domesticated cats and dogs should be seen as technology. Dogs project across space, and take us with them. They run faster than we do for a few miles, and can use scent to track over distances in ways we cannot. With dogs our pack moves faster than we do. Our innate scale lets us go at a trot farther than most animals; we can even run down horses over distance by exhausting them. But domesticated horses let us move our bodies much faster over distances of a few miles, although we have to change horses to keep going, as in the Pony Express.

Cats let us project our influence into the small spaces wherein hide the vermin that eat our grain. But cats are scale limited, even if they are stuck at a useful size for our purposes. We do not commonly have Great Dane–sized clawed cats because they would be a handful and not much use. We have domesticated cheetahs, but it is telling that theirs is a dog-like ecology. They are ecological dogs that happen to be cat-related. Cheetahs chase like a dog, and have non-retractable blunt dog-like claws. Note by contrast to domesticated cats, the size of domesticated dogs varies. That is because we use them at different scales. Big guard dogs, but small ratting terriers. Dogs are good for hunting rats, although a big bold cat can do a lot. Cats are great for mice. We use our dogs to fend off predators of our sheep. In that way we can graze areas more intensely but still keep the flock safe. Dogs also let us control movement of all our herds. Notice that living with humanity, dogs have evolved a way to tell our mood and intention by looking at our faces just as we look at human faces. Wolves do not read our faces so well. Dogs do not look at other dogs' faces that way; they smile instead with their tails.[12]

So the dogs at my place fill the space in ways that my cats do not. My cats when locked out on a winter night hide and keep warm between the bales of hay. We worry about them unnecessarily because we cannot do the same; we have to make shelters, like tents and yurts. So I don't often have mice, because of my cats. They are generally shiftless, but vigilant enough when it comes to the fun of hunting. Most of the details of our lives can be explained by the scale of things, more than particulars of structure. That is a powerful generality.

Simulating Language

The models whose output we should be able to assess most competently are simulators of human behavior such as communication. We know firsthand that it seems to be an enormously complicated task to carry on a meaningful conversation or even string letters into meaningful word sequences. One might think that an effective simulation of aspects of language and writing style would demand great complexity and tight control. Nevertheless, Bennett's early experiments with randomly typing computer monkeys would indicate that tight control and explicitly rational modeling of underlying processes can be quite unnecessary. Meaningful simulation results require only an appropriate grain size and cycle time in crucial parts of the system. So long as the parts have the right grain size, they can behave randomly and still capture recognizable aspects of style and language. Selection of appropriate grain sizes and cycle times, the basis of hierarchical construction, apparently influences our interpretations of phenomena more significantly than do the specific structural rules that link the system parts even in a dynamical setting. As with evolution, antecedent to every stochastic process invoked in explanation is the recognition by an observer that there is something to be explained. We show below how a depauperate dynamical system can produce striking results, but for that phenomenon to occur there must be an observer to be surprised or at least interested. Without an observer we cannot know that there is a phenomenon.

Bennett's results are striking and worth reporting here. He simulated a collection of monkeys typing at random. The goal he set for them was a masterpiece of literature. The simulation of monkeys hitting keys on a normal typewriter at random achieved nothing. But Bennett came to the monkeys' assistance with correlations associated with various sequences. His first adjustment was to present the monkeys with a typewriter, the keys of which were present in the proportion of the normal usage of the various letters, the apostrophe, and the space. He got text broken into word-like segments. He improved the monkeys' chances further by building correlation matrices that indicated the likelihood of one letter following another. This involves a two-dimensional correlation matrix. Monkeys typing on normal typewriters were zeroth-order typists. Those with the letters and spaces occurring in appropriate proportions were first-order typists, while those using the probability of paired occurrences were second-order typists. The third-order typists used a three-dimensional correlation matrix and used probabilities associated with the former two characters to determine the next character. Computational limits were reached at the fourth order, but remember that was in 1977.

First-order monkeys produce strings of gibberish of about average word length, but "an average first-order Shakespeare monkey typing ten characters per second will still take about three days to get even the first two words of Hamlet's soliloquy." By the second order the language that had been used in construction of the correlation matrix could be discerned from the computer output by characteristic letter sequences. Italian monkeys ended words with "o" much more often than English monkeys. Fourth-order monkeys who had been reading Shakespeare got caught in a trap of the Bard's characters commonly saying "Ho ho." The monkeys would get into a loop of "Ho ho ho ho . . ."

Using Bennett's own words to describe second-order monkeys:

> [A]t last we start to get an appreciable yield of words—and, even more interesting, some appreciably long word sequences. The latter is a little surprising because we only have incorporated the statistical correlations between pairs of letters. . . . One basic problem with these monkeys starts to become apparent as early as the fourth line; they are pretty vulgar. [By the third order] the results indicate a fifty percent yield of real words and lots of long word sequences, but the fluctuations are quite extreme; a line or two of total incoherence followed by a startling remark with as many as nine real words in a row—for example, WELL UP MAIN THE HAT BET THAT IT SUCKS.

> By the third order it was possible to tell some difference between monkeys that had been reading Shakespeare, Edgar Allen Poe and Hemingway because of the style of the computer output.
>
> The vulgarity is definitely associated with low order letter correlations. The parallel in real life might be that people that use it the most seem the least educated. In English, the obscenities seem to peak in the third order and have almost completely disappeared by the fourth order. For example, only one mildly off-color remark (about LONGUE ASS KISSES) was encountered in the first ten pages of fourth-order Shakespearian monkey texts. . . . The yield of real words from the fourth-order Shakespearian was roughly ninety percent, and strong style-dependent differences were detected between different authors.
>
> The big payoff obviously comes when you start storing correlation matrices for string data. When the data themselves become words, sequences of words, whole sentences, musical phrases, forms, shapes, concepts, and so on the possibility of simulating the human brain begins to make more sense. For example, does anyone really doubt that a monkey program using fourth- or fifth-order correlation matrices loaded with clichés would be indistinguishable from the average political speech?

Bennett's comment on string data is significant, because the clichés are already chunks of meaning. Lower level meaning could aggregate to meta-meaning.

Marshall McLuhan[13] suggests that the alphabet has had a profound influence on how users deal with their world. The thesis is that codified law, monotheism, abstract science, deductive logic, objective history, and individualism all come from the influence of the alphabet. "All of these innovations, including the alphabet, arose within the very narrow geographic zone between the Tigris-Euphrates river system and the Aegean Sea, and within the very narrow time frame between 2000 B.C. and 500 B.C."[14] The use of letters takes the discussion below the level of meaning. CAT means a cat, but C, A, and T do not have meaning in themselves. Pictographic writing never goes below the level of meaning. For instance the symbol for "man" in East Asian pictographs resembles the Greek letter λ, without the curly bits: two legs and a body. Any part of the lambda-like character means nothing and is not building anything, unlike C, A, and T, which do build "cat." The legs and the body is it!

In non-literate culture long strings are remembered by cyclical repetition. Homer's *Odyssey* is clearly written down from oral narratives.[15]

The repetition comes in the form of celebrations between dramatic parts of the story. Having escaped the Cyclops, Odysseus and his men take off and feast, wrapping the loins of an ox with fat, and they drink wine. That is what they always do, and the phraseology is generally similar. All non-literate forms have such repetition so as to keep things straight. The American standup comedian, Larry the Cable Guy, has a catch phrase "Git'ur done," which presumably is "get her done." He repeats this at the end of each of the many sections of his extended monologues. It is his version of wrapping the loins of an ox with fat. It is this same cyclical patterning that allowed precolonial African cultures to develop fractals centuries before Mandelbrot. African villages, buildings, and rooms, as well as traditional drumming, are all powerfully fractal, according to the ethno-mathematician Ron Eglash. Meaning cascades upward, unencumbered by meaningless alphabetic letters.

The thing that matters in meaning is that it is structural and is rate-independent. That offers stability. This stands in contrast to the rate-dependent behavior of thermodynamics. It is no accident that neural nets (both biological and cybernetic) consist of discreet structural units. In complex systems meaningful low level structure appears to anchor and stabilize even the most elaborate behavior. There is great power in object-based programming depending on units that have rate-independent meaning. Pattee[16] points out that going below the level of meaning lets the phenomenon slip like sand between our fingers. He notes that if the meaning of the strings of words we use is not understood then analyzing the thermodynamics of the chemistry of the ink will not help. While schemes that sit at and above meaningful levels can achieve a lot, reductionist excesses in biology have blunted progress.

The power of depending on meaningful units can of course be perverted, much as can any powerful tool. When Bennett refers to simulated political addresses coming from string data of clichés, he speaks of aggregating the meaning of already meaningful structural bits. The meaning in clichés is pretty depauperate, but so then is most political speech. But then it gets the powerful elected. A GIS map of human opinion tied to location can be enormously informative, for instance, for governance of a park or refuge. The "internet of things" is a step up from earlier internets. It meaningfully connects rate-independent objects that already handle data in meaningful ways. Refrigerators can now be programmed to respond by offering shopping lists of things that were bar-coded and put in the appliance. The shopping list would derive from items not replaced or whose use-by date has expired. One imagines a not so interesting conversation between a refrigerator and a toaster about the freshness of the loaf. But it may get you a better breakfast.

Allen investigated how people value objects. He achieved some success when he performed a multivariate gradient analysis of the output from clustering multivariate analyses. This procedure was applied to 80 people who had been asked to value pairs of 16 objects: examples of houses, cars, kettles, clothes, and objets d'art.[17] The raw data were 16 × 16 symmetric matrices filled in with how the person in question evaluated the 120 possible pairs made between all 16 items. The research subjects were asked to compare the 16 objects in terms of how similarly they evaluated the two objects in the pair. Some subjects viewed all cars in similar terms, and their 16 × 16 matrices captured the view. It was clear from the cluster analyses which people in the set were concrete utilitarian thinkers; their dendrograms grouped all cars. The dendrograms of utilitarians also grouped all clothes together, such as work clothes and riding habits. Allen would probably be a concrete utilitarian thinker himself; note how he describes above the research items by groups of utility: he refers to cars as opposed to clothes or houses. Normative thinkers would, by contrast, group the riding habit, not with other clothes, but with the Jaguar car. All this was clear from the dendrograms derived from the sets of evaluated comparisons made by individuals: one dendrogram from each person in the analysis. The second phase of the work went on to make comparisons between all the dendrograms. We were at that point analyzing not a raw data set but rather a set of analyses. There was already meaning in the subjects apparent in their dendrograms as they went into the analysis of how the full set of subjects could identify groups of strategies for evaluation and value. The phenomenon of interest was how people in general value things. The second phase of analysis received data already structured in a meaningful way (interpretable dendrograms). Types of evaluation were clear. Interestingly, like versus dislike of objects was independent of the major types of evaluation. Apparently the subjects had followed the instructions, which were to focus on type of evaluation criterion, not on whether objects were in fact seen as valuable.

All this would support Bennett's speculation about clichés and politics. One might imagine that a speech carries significant information, but think again. There are politicians who will say anything for its effect, not its verity or significance. Nothing of substance is communicated; the listener gets just an impression. One might hope and expect academic writing to carry substantial ideas. But think then of the reams of government gray papers written by academics and program staffs. Nobody reads them, so why are they written? It is not for transmission of ideas, it is so that the person in charge of the government unit can say, "I have somebody working on that." There are of course major exceptions.

But even academic, peer-reviewed, white paper articles suffer in the same way. There is the worrying title of Ioannidis' important paper entitled, "Why Most Published Research Findings Are False." His is a compelling case. Gerd Gigerenzer[18] is unflinching in his criticism of statistical incompetence. In his view the use of p values in sociology is a voodoo dance to ward off evil reviewers. He reports that 80% of teachers of social statistics are wrong on what $p < .05$ means. The book by Saltelli and others, *The Rightful Place of Science: Science on the Verge,* circles science practice, putting in body blows until science loses every round. All this comes from the imposition of new mercantile values across society. Medical, scholarly, and clerical values have all succumbed. A successful professor and former student of Allen now claims that he spends much less time doing science than he does "running a small multimillion dollar business."

The emergence of new mercantile values in academe means that the scientific literature is not being evaluated on the basis of the meaning of symbols in the substance of the papers. A new phenomenon is predatory journals whose sole purpose is to extract publication fees from contributors. The contributors are under pressure to meet a bogus scholarly record in small teaching institutions that do not leave time from teaching to do proper research. Deans of small teaching colleges are not really equipped to understand the role of research; it is often only a matter of prestige for them. The peer review process is simply broken. As a reaction Sokal submitted a purposely nonsense paper contributed by academics to show literally anything can get published.[19]

Academics are evaluated by colleagues who are not in the same subdiscipline. Academic hires are made to fill in gaps in departments, so tenured faculty are generally not qualified to evaluate colleagues' work. As a result colleagues in a department are mostly not even reading each other's papers, let alone evaluating them properly. The information actually used are metadata; the importance of the journal, not what the paper says. Worse still are the importance factors for journals. They were created so libraries could evaluate the acquisitions they make. Eugene

Garfield is the founder of the *Institute for Scientific Information* and he designed the importance factor. Since 1975 journals in the *Journal Citation Reports* have been assessed as to impact factor. Garfield explicitly said his device could not be used to assess contributors, but it is regularly used in assessing academic colleagues. The slack and ambiguity in the importance assessment system is great to the point one might question its utility even for libraries using it for its intended purpose. Misuse a system that is already weak and there simply is no valid assessment going on.

Symbols can readily be engulfed in meta-symbols that have little to do with the lower level significant meaning. Evaluation of science in grants and publication is not on the basis of the meaning of the science per se. Grants are nitpicked to death on the methods section, although everyone knows that nobody doing new work actually goes on to use the proposed methods from the grant. References are not to inform on the science, but are used to discriminate against rival camps. Granting agencies should limit methods to 200 words of feasibility. They should also not let reviewers see any citations, except for kindly suggestions after the grant is funded or not.

Simulating Conversation

A second example of computed word strings appears again as far back as 1977. But this time it is conversational, so is even more disconcerting. Here a computer program called ELIZA instructs the machine to behave like a therapist in the Rogerian school. Rogerians do not work by identifying your problem as some trauma in your history, but rather they take the emotions expressed and turn them back on the client, forcing a resolution. The point to note is the clear hierarchical structure implicit in the cyclical behavior of the therapist. The simulation seems to respond immediately to input from its human client. The computer's words are in upper case. It is a remarkable piece with unexpected resemblance to a conversation with a human therapist.[20]

Men are all alike.
IN WHAT WAY
They're always bugging us about something or other.
CAN YOU THINK OF A SPECIFIC EXAMPLE

Well, my boyfriend made me come here.
YOUR BOYFRIEND MADE YOU COME HERE
He says I'm depressed much of the time.
I AM SORRY TO HEAR YOU ARE DEPRESSED
It's true. I am unhappy.
DO YOU THINK COMING HERE WILL HELP YOU NOT TO BE UNHAPPY
I need some help, that much seems certain.
WHAT WOULD IT MEAN TO YOU IF YOU GOT SOME HELP
Perhaps I could learn to get along with my mother.
TELL ME MORE ABOUT YOUR FAMILY
My mother takes care of me.
WHO ELSE IN YOUR FAMILY TAKES CARE OF YOU
My father.
YOUR FATHER
You are like my father in some ways.
WHAT RESEMBLANCE DO YOU SEE
You are not very aggressive, but I think you don't want me to notice that.
WHAT MAKES YOU THINK I AM NOT VERY AGGRESSIVE
You don't argue with me.
WHY DO YOU THINK I DON'T ARGUE WITH YOU
You are afraid of me.
DOES IT PLEASE YOU TO BELIEVE I AM AFRAID OF YOU
My father is afraid of everybody.
WHAT ELSE COMES TO MIND WHEN YOU THINK OF YOUR FATHER
Bullies.
DOES THAT HAVE ANYTHING TO DO WITH THE FACT THAT YOUR BOYFRIEND MADE YOU COME HERE

From the text it appears that lines of human input made as firm statements are reverberated as reassuring remarks, often questions; in the example, YOUR BOYFRIEND MADE YOU COME HERE. The only changes are my and me to your and you. To less firm statements the therapist asks a more general question clearly derived from a key word in the preceding line of human input. "Mother" generates TELL ME MORE ABOUT YOUR FAMILY; "care" generates WHO ELSE IN YOUR FAMILY TAKES CARE OF YOU? In this last response we see a slightly longer loop in operation. FAMILY there is derived from the computer response two lines above. Specific "mother" goes to generic "family," so "family" is good for another repetition without saying anything foolish.

But somehow the computer can tell the substance of the whole conversation. It does this in the longest loop in the whole sequence. The early lines of a conversation are met with a request along the lines of BE SPECIFIC. The program is fishing at that point, particularly as to the first specific person mentioned by the human client. That first specific person is likely to indicate a constraint on the whole conversation. The program looks for context. This is rather like the first species entering Shugart and West's[21] FORET stand simulator as a forest gap opens; the species in greatest numbers in the released underbrush constrains the stand growing in the computer for the next couple of simulated centuries. So it is with the Rogerian therapist. An emotional problem is either with a particular person, or somehow surrounds some specific person. The program fishes for that person and catches him in the body of the first specific person mentioned with "my" coming before it. "Well, my boyfriend made me come here." Ah, it is the boyfriend. The program then appears to disregard that information until the conversation has had time to play out. In response to a question about the general context of her father, the client makes a one word comment, "Bullies." Then the program snaps the trap. DOES THAT HAVE ANYTHING TO DO WITH THE FACT THAT YOUR BOYFRIEND MADE YOU COME HERE. Well, of course it does. The girl is moving away from her bullying nuclear family, and what does she find? The whole world is full of people ready to bully her.

The computer probably does not know much about macho American males, or that girls tend to marry their fathers and boys their mothers. So how does it know the client's situation? It has simply used a device to find the context. Meaning is always in the context, the next level up. The first person mentioned matters and captures the context. But look and see how that critical comment from the computer could have been used earlier. Try it out and you will see it would have worked pretty well there too; not exactly the same significance, but still it would work. This shows that complexity is in the eye of the beholder. We interpret complexity with a story. However, ELIZA works quite like the real therapist, who does not guide things specifically. A real Rogerian, bored and day dreaming about something else, might easily use the punchline and only afterwards think, "Hey, that is what this is about." Complexity is about levels of organization meeting and often clashing.

We know the model passes all tests of validation when we reassure ourselves with a nervous laugh. What is humiliating for us as humans but inspiring for us as systems analysts is that the simulation is not tightly logically directed. Rather it is a system of loops where the long loops constrain the short loops. The example above is disconcerting and instructive. It is not just because computers can simulate something in which we have an emotional vested interest. We know firsthand the phenomenon is complex. But most important is the way it tells us that apparent complexity and subtlety can be generated by little more than the simple coupling of events with different cycle times. If you try it, this conversation really does happen with the ELIZA simulation. Just feed in the client statements above and the conversation happens. Often the end of the conversation varies, with a string of general statements about the father, but in the end the punch line about the boyfriend comes up.

Something with the richness of the last line here could not have been simulated in a deterministic reductionist mode, by chaining logical and specific processes to achieve the whole. And yet we try to build ecosystem models through extended deterministic reductionism. Of course, some or all of this richness may have resulted from the truly human side of the system. We suspect, however, that replacement of the patient client by a similarly programmed patient simulator would produce conversations, at least occasionally, of comparable richness and depth.

Holling[22] with characteristic brilliance identifies that a model can be realistic, simple, accurate, or precise (detailed quantification). But Holling insists that you cannot have it all. One has to compromise by usually backing off on two, and at least on one of the above four characteristics. Our discussion of mathematical modeling approaches in general is cursory. Even so it does illustrate an important point that bears reemphasis. The temptation is great to simplify in order to achieve specific and limited objectives. That is not necessarily bad. But it can be. And yet there is also a strong tendency to surrender to another temptation, to make models more complicated in a process supposed to make the model more realistic. We consider reductionist details a naïve realist trap. We view adept data analysis as throwing away all but one set of insights. Depth is achieved by subtraction not addition; it works counterintuitively. There is temptation to put something in a simulation just because we think it happens in nature. We recommend steadfast resistance to the temptation to complicate for the sake of purported realism. Austerity can lead to insights and understanding.

Always look to the context, scale, and level of organization. At some point preserve the complexity; do not always whittle it down to mere complicatedness. We recommend a multileveled hierarchy of models, coupled by patterns of constraint and cyclic interaction. We prefer that to deterministic causal chains. It is in an appreciation of the significant advantages of such hierarchical coupling that we find the fortitude to resist utilitarian simplification when preservation of complexity is imperative. The tool to deal with the difficulties of coupling levels is narrative. An insistence on always being constrained by consistent models over robust narratives leads to disasters like the New Synthesis. It is the neo-Darwinism that has held sway for the second half of the twentieth century. It surrenders to genes as the one ultimately causal way phenotypes appear. As Denis Noble[23] shows, the relationship of the phenotype to genes and physiology is complex because it is a web of causality, not a chain linked to genes. The disappointing failure of drug companies working on single causes is pertinent. Some chronic gut issues appear to be the result of upsetting the gut flora as a whole with antibiotic treatments for some other issue. *Clostridium difficile* is the rogue organism. Further antibiotic treatment can sometimes work, but mostly they only open up the gut flora integrity, which lets *Clostridium* back in. But there have been successes in more holistic treatment of particularly intransigent gut infections by fecal transplants from healthy donors; the effort is to restore the old complex balance.[24] But the response of the mechanistic establishment has been predictable and misguided. Even with the holistic treatment in hand, and a good narrative as to why it works, reductionist mechanist compulsives search for one or a few species of bacteria that will do the same as the whole gut flora. When complexity invoking multiple levels works, leave it alone, and tell stories. Much focused control is illusory, and has hidden flaws, including unnecessary expense.

CHAPTER ELEVEN

Diversity and Connectedness

The extent of the literature on biological diversity reflects not the significance of diversity in biological systems but rather the ease with which diversity data may be collected. Also, only a very simple model does make intuitive sense of diversity. That special case is if a species became globally extinct; in that instance we would say that diversity unequivocally has gone down. The problem is that only in the special case of the exact globe is it necessary that diversity has gone down; it can easily increase with the loss of a species in a more local setting. Because it was so abundant, the loss of the passenger pigeon will have increased diversity of birds in North America. The explanation is that some measures of diversity very reasonably do more than simply count species, they give account to evenness of numbers of individuals in species. High evenness refers to diversity in the experience of the average individual. In the most even situation the chances are greatest that an individual will be next to another individual of a different species. For instance, if there are one hundred total individuals split into five species, diversity is highest if all species contain the same number of individuals, twenty. This is because there is complete evenness in such a system. Diversity would be lowest possible in a five species situation if four species are represented by only one individual and the other ninety-six belong to the fifth species. This is because the system is as uneven as it can be. The passenger pigeon will have contributed mightily to unevenness. The logic of evenness pertains to random encounters of individuals. With complete evenness, individuals of one species would have the highest diversity of random encounters with the other species. At any scale except the global, the loss of a species will change evenness, and this might send diversity down, but it might increase diversity, as we suspect the passenger pigeon did. Do not get us wrong. We want to save species and habitat too. We are very sad that we can never see a passenger pigeon, even if its loss did increase diversity by increasing evenness.

The fundamental problem is that diversity is caused by so many phenomena that it is not one thing or anything unified. So many different causes bump up diversity: overlapping floras, gross geographic topography, local seasonal variation in climate, much small topographic variation, a generally clement climate with a long history and many separate niches. The situation is complicated by the scale of data collection. Tropical rain forest has high diversity at a large scale, but grasslands win out if the area is smaller. Sclerophyllous scrubs communities in Australia and South Africa, where individuals are long lived, have high diversity, perhaps related to the mixture of Gondwana and other floras.[1] Diversity is what is left over after such a muddle of factors has played itself out. As we indicated earlier, scientific models should attempt to view the holon generating the phenomenon of interest at its own scale of functioning. With all else equal, diversity does seem to be lower when the system is stressed. But all else is equal only in very local settings. Ruth Patrick showed that diversity of diatoms consistently goes down with stress from pollution. But such predictive power of Patrick is limited to situations where things are particularly well defined, so all else is indeed very equal; mostly they are not, particularly if the area for comparison is large and heterogeneous. A complication is that stress of course is an observer defined state. It probably wouldn't be described as stress by a surviving, now dominant species. As we said earlier, there is always stress, as opposed to what? So given Ruth Patrick's work, as a tool local diversity does have use, but only when all else is very close to equal.

But that does not mean that diversity per se has much meaning. Kurt Vonnegut's term *granfalloon* was coined in *Cat's Cradle*. In that work there was a religious sect that viewed itself as being unified. The problem is that individual members had no common beliefs. A granfalloon is a collection of things that do not have much significance. Diversity as a tool can be valid, but diversity as a general object of study is a granfalloon. Studying diversity is like studying the chemistry of red things. Redness does have some physical chemistry to it, but it is important only in a limited setting, which evaporates when the many different sorts of red things are compared. But the appeal of a number so easy to collect makes it ubiquitous, and so we must address it in this volume, even if not in the most flattering terms. The shining exception is the fully worthwhile work of Philip Grime, whose exhaustive and successful experimentation wins the day. He experiments on material from one natural moorland community, a wet mountain herbaceous vegetation. And he gets clear results that do not take statistical gymnastics to reveal.

Massive experiments on happenstance collections of plant species (not like Grime's coherent communities) are much less compelling. There also appears to be some scientific politics of careerism in the work. Some of the suspect papers have over thirty authors. Hector et al. claimed that their statistics showed increased ecosystem functionality with increased diversity in experimental plots. The results were so wrong that another large group of ecologist with credentials in diversity studies had to point out sampling error in the experiment, an elementary statistical flaw.[2] The argument was that increasing the number of species in the more productive samples was bound to increase the probability of including more productive species. When Huston et al. repeated the statistical analyses correcting for the bias, the statistical significance disappeared. Diversity has not been shown to increase ecosystem function in principle. That did not stop Hector and his colleagues coming back over the following decade, but they had had their chance and missed, undermining the trust of the scientific community. The meme that diversity leads to stability has been mocked by Robert May[3] as the "folk wisdom of ecology." Yet it persists and is used as a utilitarian argument for protecting biodiversity (with the latter not well or consistently defined). When one finds someone is wrong, and insistent about it, a sound strategy is not to try to work out how they are wrong, but rather try to address what they were trying to say.

Diversity and Connectedness

"Diversity leads to stability" almost always mistakes diversity for connectedness. The limited use of information theory in biology is disappointing because it has much to offer. The original practical application of information theory was in the field of telecommunication systems. When the theory was adapted for the quantification of biological diversity, an important feature of the original work was either misunderstood or ignored.[4] Pioneering work concerned with information flow in telephone systems did not depend upon the number of telephones in the system but rather upon how many telephones were connected to one another. It is therefore inappropriate that diversity as it is now usually measured by various indices (including those not based on information theory) should give equal weight to both connected and disconnected individuals and species. Considerations of diversity must make the distinction between the connected and disconnected, because individuals of each type are likely to have opposite implications in many biological contexts, including system stability.

There is a literature concerned with quality and quantity of connections in complex systems, and it cursorily addresses diversity.[5] Levins defines connectivity of components as the mean number of direct interconnections between one component and the rest of the system. We mentioned Gardner and Ashby when the previous chapter discussed the Monte Carlo method (Figure 10.7). As we said then, they use connectance as a percent to express the number of interconnections as a proportion of the maximum possible number of interconnections. Connectance saturates at 100%, because there is no way to increase beyond all possible connections actually being made. However, there is still a way to increase degree of connectedness with measures other than connectance. With two systems at 100% connectance, the system with more components has a higher connectedness as measured by connectivity. Connectivity is the average number of connections per node or part. If mean connectivity is held constant while the number of components in the system increases, the percent connectance goes down, because extant connections become diluted by the addition of disconnected individuals.

While setting up their systems Levins takes a qualitative approach and ignores the particular values of the interaction terms, while Gardner and Ashby are quantitative and assign the nonzero terms particular values between +1 and −1 taken from a known distribution of strengths of connection. The terminal values of +1 and −1 indicate strong positive and negative influence, respectively. Meanwhile small interaction terms (e.g., +.2, −.1) give a weaker connection between the variables concerned. MacArthur is centrally concerned with the strength of interactions in his consideration of community stability.

When we discuss system connectedness below, we take account of connectivity, percent connectance, and the strength of the connections or interaction terms. If one holds the number of system components constant, raising the mean connectivity has the same positive effect on system connectedness as would an increase in the mean of the absolute value of the nonzero interaction terms. That is to say, a strengthening of a unit number of interaction terms is equivalent to creating new interactions; both raise the connectedness of the system. In summary, a system is more connected (a) if the coefficients of interaction measuring predation, parasitism, competition, etc., increase in absolute value; (b) if the nonzero interaction terms (the interconnections) become a larger proportion of the maximum possible number of nonzero terms (i.e., increased connectance); and (c) if the number of nonzero interaction terms per component (connectivity) increases. But that is just the quantitative mechanics of the general issue of increasing degree of connectedness.

Clearly the contribution of each of the above to connectedness is in different units. We purposely do not try to quantify the relationship between the strength of the interaction terms, connectance, and connectivity because to do so we would have either to fetter our discussion with unreasonable general assumptions or to discuss only very specific cases. We are instead trying to expose the general principles. Sometimes quantification informs, but equally it can obfuscate.

Stability and Diversity

Stability and diversity in ecological systems have a shared literature. It is indeed an interest in stability that has kept diversity a central concern in ecology. A problem with the literature is that there are several definitions of stability and diversity, and terms are not used consistently. The same term can have opposite meanings for different authors. Stability is set in the context of disturbance. First there is the notion of resistance, the capacity of the system to resist being displaced by a given size of perturbation. Then there are various notions of resilience. Resilience can refer to the capacity of the system to return after perturbation from a given displacement. Resistance will of course minimize the displacement, but resilience indicates a certain response to the displacement. So there is recovery associated with the degree of displacement, but even that can be addressed in two ways. First is the degree of displacement from which there is still recovery, and second is the speed of recovery if there is a recovery. Then there is the added dimension of what qualifies as recovery. If there is a singular pattern before the disturbance, how much of that pattern must be reestablished for recovery to be recognized? It might be recovery to an equilibrium point or a stable limit oscillation. But if the system at rest is chaotic, it is only possible to recover to the track of the strange attractor, which will of course never return to any particular value or set of values. There recovery will be recognized not by what happens, but by what never happens.

Intuition and common sense make field biologists reluctant to abandon the notion that increased diversity offers alternative pathways through a system for stabilizing signals. This idea was one of the earliest relating diversity and stability.[6] An alternative view arising out of more theoretical considerations[7] suggests that increased diversity can increase system lag, which is viewed as a destabilizing force. This view has come to gain such respectability that the earlier suggestions of diversity increasing stability have been treated with some condescension. As was first indicated by May,[8] the situation is more complicated than an either-or proposition. May was able to stabilize a simulation model of herbivory by increasing its diversity with a further trophic level. Maynard-Smith unfairly dismisses May's[9] observations that increased diversity lowers stability by noting the assumptions of random connectedness in systems that become unstable with increased diversity. Maynard-Smith says that May's negative correlation only shows that ecosystems are not randomly connected; but Maynard-Smith does not give credit to May, who actually makes the same point himself. May's conclusion is not so much that systems with high diversity are less stable, but rather that connections are importantly nonrandom. Nonrandom connections readily lead to hierarchical structure in models where there is a local nexus of connections that draws together local subsystems set in a larger context at a higher level of organization.[10]

Some of these arguments can perhaps be resolved by recalling the two components of constraint in a hierarchical relationship. Constraint may be reduced either by loss of asymmetry in a relationship or by failure to deliver the constraining signal itself (Figure 2.7). In the first case the constraining holon offers less constraint than formerly, because the scales of the interacting entities become less different. One entity constrains another by having a longer relaxation time, by using a filter with a longer window. Constraint is not just a matter of interaction, it has to be asymmetric. Two entities with equivalently scaled filters do not constrain each other at all, because neither can out-wait the other. In the second case the constraining holon moves so much more slowly than the constrained holon that the constrainer and the constrainee are uncoupled. The decoupling arises because each holon can offer no behavior for the other. The candidate for constraint never moves so far as to encounter the constraints offered by the upper level. For instance, which galaxy is the home of the solar system makes no difference to the functioning of life. The constraining signal is lost. Let us now translate the components of constraint into terms of system connectedness, equating loss of asymmetry and loss of signal to two different facets of connection. The two facets are system over-connection, as opposed to system under-connectedness.

In over-connection something somewhere is likely to get into positive feedback. The system is too complicated and there is too much that can go wrong. A given technology can get only so complicated, as when valves in a computer become so many that one of them is always broken. Valves were an old technology that goes back to the days when a radio had to warm up. The solution to the valves being unreliable was to move to solid state technology with transistors, which are more reliable. Greater reliability allows an enormous number of them without the expectation that one is almost certainly broken. In under-connection, the system is so weakly connected as to be an invitation for some outside influence to insert itself into the system and bring it down. A case in point might be the small knots of pairwise specialized connections that occur between species on isolated islands. Island ecology is under-connected within the island. The arrival of generalized species such as rats and goats is disastrous. The invaders insert themselves everywhere between the local pairwise connections.

An increase in diversity might have opposite effects on stability depending upon whether the added individuals become a part of an already over-connected or an already under-connected system. A further complication is that a new individual will often increase diversity but could be either a stabilizing or a destabilizing force. The difference depends upon whether the individual is itself connected to or disconnected from the main lines of communication of the system to which it is added. Adding a strongly connected individual to an already over-connected system will tend to destabilize the system by further increasing the total system connectedness. May's and Gardener and Ashby's randomly connected components caused over-connectedness in their test systems. In any system there will be principal lines of strong connection and peripheral lines of weak connection. In a conversation, talking individuals are in the main line of connection whereas others, who may neither talk nor pay attention, are peripheral. As the numbers of guests at a party increase, conversations are louder to compensate for the noise of bystanders talking about something else. The others at the party we would call uncoupled or disconnected individuals. Add too many disconnected individuals to a conversation and it breaks down, as uninterested and disinterested parties blunt the discourse. If an uncoupled individual is added to an over-connected system, it will tend to lower the connectedness of that system, so increasing its stability whether measured in terms of resilience, resistance, or speed of movement to the equilibrium (remember Holling's criteria for stability).

Because diversity is a static consideration of structural units, it takes account of neither the connectedness of the whole system nor the degree of connectedness to the system of new individuals or species. Diversity by itself has little to say about stability precisely because it does not consider connectedness. Clearly, comparable systems will share similar levels of connectedness. This will explain Patrick's predictions for diversity and stress; the stressed and unstressed diatom communities will start with sufficiently similar levels of connectedness, so diversity may indeed be related to stress. Thus experimental confirmation depends not on diversity but upon connectedness of the initial conditions. Such experiments are special cases that make no particular comment on the general relationship between diversity and stability. Murdoch[11] suggests in a test of diversity/stability relationships, "We should compare natural communities similar in all respects except their diversity." Even so, he concludes that "such comparisons are well nigh impossible, but the existing data suggest no correlation between stability and diversity, or an inverse correlation." Grime[12] is the one researcher who actually does what Murdoch suggests, as he performs long-term experiments on species combinations with different diversity, but with species that do in fact grow together near Sheffield on the local moors.

Competitive exclusion is a principle coming from population biology. It says that two species sharing the same niche will compete to the exclusion of one or the other of the species. Population biologists can force species into the same niche by narrowing space and resource richness. Once that is done the principle of competitive exclusion says that if two species are in the same niche, one will be driven to extinction, even if conditions are such that one cannot predict the winner. But there will be a winner. Roughgarden and Pacala[13] showed the phenomenon in nature (islands with two species of lizard), but most demonstrations of competitive exclusion are in artificial experimental situations. Heterogeneous environments may stimulate higher diversity. High diversity would pertain when there are many separate niches. In the paradox of the plankton, diversity is high but the environment is very homogeneous.[14] The paradox is that water provides a low diversity of niches, but high diversity of species. There are two ways for species to coexist. In one case the plankton species are so different that they do not interact; the signal that there is potential competition is lost because niches are almost completely separate, the homogeneity of the environment notwithstanding. The species are disconnected. In the other condition the species are so similar to one another in so many dimensions that the competitive advantage of one species over the other is necessarily slight. Exclusion takes so long that before it can happen the environment changes so as to shift competitive advantage.[15] The constraining species constrains itself almost as much as it does the other; the species are too strongly connected to achieve the equilibrium condition of competitive exclusion.

MacArthur[16] makes similar comments, but in the context of community equilibria and the strength of interactions. He starts with the underconnected case, moving on to over-connection:

> If species interact weakly, their communities are vulnerable to invasion by additional species, thereby increasing the interaction; if they interact strongly, they are vulnerable to almost all the hazards of existence and some will go extinct, thereby reducing the interaction. The in-between degree of interaction is surprisingly robust.

A useful device is the community matrix where relationships between species are put in a square array filled with strengths of interaction, both positive and negative between the species, which are the columns and rows. The large interaction terms in his matrix of competition coefficients define MacArthur's over-connected case. He shows that predation can easily drive a prey species to extinction if that species interacts with other species through large competition coefficients under predation. Remember the strong interactions that Holling and Ewing created as their tight systems went unstable. A community described by a matrix of small competition coefficients because of specialization is not sufficiently cohesive to avoid invasion. The invading species do not generally fit quietly into the scheme of things (rabbits in Australia, goats and rats on Pacific islands). Under-connection courts uncontrolled change as much as does over-connection.

Levins and Gardner and Ashby (Figure 10.2) argue only the over-connected case, but they concur with MacArthur. As mentioned above, Gardner and Ashby note that the probability of one of their systems with 100 percent connectance remaining stable drops rapidly as the number of components (n) increases—perhaps as fast as 2^{-n}. Further, they show a decrease in the probability of the system remaining stable with increasing percent connectance, and this is all the more severe and sudden as the number of system components increases. Levins points out that, if the mean number of connections per component "is held fixed, the system is eventually (for a large enough n) stable." This is the case of the stabilizing influence of the addition of a disconnected or uncoupled component to an over-connected system. "But for any given n, a high enough connectivity makes the system unstable"; the system becomes over-connected and positive feedbacks lead to low-frequency oscillations that finally prevail, leading to nonperiodic instability.

Agroecosystem Examples

While within-community comparison of diversity and stability may be meaningful, the broader discussions of tall forests versus technological monocultures have no meaning. Diversity in a tropical forest has nothing to do with and cannot be measured in the same units as diversity in an Iowa cornfield.[17]

In Iowa farming systems the human component is very important and is connected to virtually all components of the system by large interaction terms. On Iowa farms the lines of communication can transmit signals of large amplitude. The system is highly connected: plants, fertilizer, water, soil nutrients, pest, pesticides, etc. If we are to compare the ecology of Iowa farms to other systems such as tropical gardens, all else must stay equal for the insight to be valid. In an effort to increase diversity, were the human component to interplant a second crop plant in the cornfield, the new component would be tightly coupled to the rest of the system. The system collapse could follow many paths. One collapse scenario could be as follows: weed killing around corn kills the beans, while weed killing around the beans kills the corn, or no weed killing drives the farmer out of business from excessive cultivation costs.

There are systems in Iowa that are not so tightly constrained as row crops. There, companion cropping is possible and it is done in planting hay fields (oats with alfalfa and hay grasses). The critical point here is that the farmer exercises less tight control over the hay field (less fertilizer, cultivation, and pesticide). The hay farmer is less connected to the system. The hay field is a less connected system than is a cornfield. In row crops, a companion species might be inter-planted between the corn plants without substantial modification to cultivation practices. That amounts to the addition of a connected component to a system that is already tightly connected by genetic uniformity and vigorous husbandry. The implication is that companion cropping in Iowa cornfields is possible only if husbandry is relaxed. Less husbandry would be essential (Figure 11.1).

Close to Iowa are some farming systems in southwest Wisconsin where intercropping is successful, but they are unusually managed with forethought and ecological expertise. Mark Shepard[18] grows nut and fruit crops in a contrived temperate savanna. He selects for hazel bushes that grow fast in highly competitive situations. Apples and American chest-

FIGURE 11.1. Technological agriculture creates a highly connected system with strong interaction between the farmer and the rest of the system. Less intensive interaction in the hay field in the foreground allows companion cropping. The monoculture in the background is intensive and does not allow intercropping without interference.

nuts are in his system. There is grazing between the shrubs and trees. He has contrived drainage on the landscape so that little water is lost, so Shepard has to do little beyond his original setup. Close by, Peter Allen is learning from Shepard, and developing his own minimal impact (connectivity) system. In Iowa Laura Jackson[19] has developed relationships with farmers to take at least some of the landscape out of intensive row cropping. Shepard, Allen, and Jackson all pay close attention to the economic context. Shepard has a chapter in his book on making a profit. Allen's whole motivation was to change the financial aspects of his farming. Farmers who treat their livestock unkindly are often forced into it by the financial vise in which they survive. Allen works to sell directly to the consumer, allowing him the slack to farm humanely. Jackson[20] liaises with farmers so they can explicitly afford to do some grazing. She notices that with a small move to grazing, many of Iowa's native prairie species can persist on the landscape in a Prairie State that is otherwise plowed fence to fence. Shepard, Jackson, and Allen are in their separate ways managing for higher diversity and stability by expressly not overconnecting the land to the farmer.

It is significant that winter comes as a hiatus in temperate farming. It disconnects the farming system. The farmer goes into the house, the crops stop growing and are harvested, the pests die, the water is recharged, and nutrient status recovers. Dry farming systems in Colorado in front of the Front Range of the Rockies extend the hiatus of winter by farming in interdigitating strips that are farmed only every other year. In the patchwork of strips, water is allowed to accumulate over half the area, while other strips that are back in service are using water from two years (this year's rain and the rain in the previous fallow). Disconnection of this sort lowers the r growth rate terms in the system, increasing stability. It has emerged that difference equations exhibit a "bewildering richness"[21] of dynamical behavior as the response to stimulus (r, the growth rate) increase. Difference equations have an implied lag in that there is a waiting period for t to be updated to $t+1$. Here May is referring to one of the early manifestations of chaos with its destabilizing effects. The vigor with which fertilizers and pesticides are added in temperate systems relates to large response variables in the equations that describe the agroecosystem. If a response is vigorous but soon reveals itself as inappropriate, winter can intercede to snuff out chaotic effects. Corrections more than once a year are possible so long as an earlier response, such as a heavy dressing of fertilizer, is not irreversible. Often-and-gentle approaches shorten the lag associated with a given level of connectedness, thus avoiding a chaotic response to the human management practice. Perhaps companion cropping in cornfields is possible if there is shortening of the lag in the system by the farmer acting lightly but often.

Winter is an important force that intermittently uncouples the components of the system. This allows for the emptying of compartments like the pest load that have increased under a positive feedback loop through the growing season. In the tropics this is not the case, and so the agroecosystems are tightly connected by pests and year-round leaching processes. Like the Iowa cornfield, tropical agroecosystems are tightly connected, but not because of human activity. Only weak connections pass through the human compartment in a tropical garden.

In Iowa intercropping is likely to increase connectedness. The addition of another tropical crop is, by contrast, the addition of a weakly connected component, because the tropical farmer does not connect the crops with vigorous action. The strong connections in the system involve the pests specializing in extant crops; they are not connected to the new crop. The additional cultivar here is a case of putting a disconnected compartment into an over-connected system. Unlike the second crop in an Iowa cornfield, the tropical inter-planted crop will increase, not decrease, stability. The pest moves down a row of tropical corn and the papaya tree breaks the row, so obstructing the connection.

Genetic heterogeneity also lowers the connectedness of cultivars and their pests. Traditional ways of growing rice in the Philippines used to involve many varieties,[22] as do traditional potato fields in the Andes.[23] With monoculture in the Philippines, blast fungus in rice is a real problem. Zhu et al. find mixing resistant and non-resistant strains not only gives 80% benefit (relative to a monoculture of resistant rice), but it also preserves resistance in the resistant strain. By presenting the fungus with a community of rice varieties, the researchers took advantage of the way communities constrain populations. Normally resistant strains lose their resistance in a few years because of intense selective pressure on the fungus. However, when the fungus is presented with the two strains the selective pressure on it is not how to crack the resistant strain, but how to find the susceptible strain. Selective pressure is changed by the mixing the two strains. Using different varieties disconnects the blast fungus from the rice such that the pest cannot focus on the resistance of the resistant variety. Low diversity in a tropical cropping system (monoculture) increases connectedness.

Murdoch presents several theoretical models where the introduction of "physical complexity, especially a patch distribution in space, enhances population stability." Murdoch recommends lowering connectedness in over-connected systems. Notice that two row crops in a field are in patches not in strict intercropping inter-planting; this amounts to the patch distribution in space that Murdoch recommends. This has been a strategy for many of the tropical monoculture crops like cocoa. The shrub is planted as an understory species with trees breaking the connections between the crop plants. Even so, cocoa in West Africa is already in grave danger from black pod disease.[24] Coffee used to be a plantation crop in Ceylon but was made uneconomic by disease. Banana plantations are, even with the best laid plans, not always successful. Should disease strike early in the life of a banana plantation, abandonment is the only course.[25] Even following Murdoch's apt suggestion, tropical plantations appear to be, in the long run, over-connected.

Adjusting Connectedness

If the above suggestions about connectedness and stability are correct, then surviving systems, those that we can readily observe, should have appropriate levels of connectedness. Exogenous forces are wont to influence connectedness in ecological systems. In the face of such influences the system may resist and maintain an unchanged structure. An alternative strategy allows the exogenous force to do what it will and then for the ecological system to effect endogenous structural change so as to readjust system connectedness to fall within acceptable bounds. The critical change is not the insult, but is the response of the system to it. When ecologists test models for structural stability (the capacity of a system to survive parameter changes and alterations of the form of equations), they are addressing this mode of stability. If connectedness and stability are importantly related, we should be able to find ecological systems that show an active form of structural stability. We have found such systems reported in the literature, although not discussed there in hierarchical terms.

Models for connection. Standard models for communication between structures employ a direct coupling of the parties. Such models are not specifically wrong (except that all models are wrong according to George Box), but rather a hierarchical model may be a more insightful alternative. A common criticism of hierarchy theory is that it has no testable theories per se. Indeed that is true, but hierarchy theory is not a theory in that sense. It is a meta-theory that will guide normal science theorists toward making useful hypotheses.

There are two compressions in science.[26] The one more often recognized is the compression of the focal situation into a model with which hypotheses are tested. There is however a previous compression that sets up the context of the model. That compression decides what is interesting, and is one of the things paradigms do. Paradigms are narratives about what is interesting. One takes it on its face that what an accepted paradigm says is interesting. A narrative is not a truth but is rather an announcement of a point of view. The paradigm says what is studied in the prescribed way is self-evidently worthwhile. What is interesting comes from a point of view, which is an assertion not a verity. Hierarchy theory articulates a narrative for undefined situations, which in science amounts to a paradigmatic statement not a hypothesis. The narrative turns undefined and undefinable complexity into something that has been operationally defined. Complexity is the undefined situation that does address infinities and zeros, from which operational considerations are derived. Making something operational does not involve verity. Complexity disappears when it is simplified in a definition that juxtaposes asserted levels. The cost is loss of complexity, but the benefit is that the situation can at least be modeled, and science can proceed. Complexity can be defined in some sense, but as something that is and cannot be defined in any instant. We deal with defined models not complexity, although we conceive of issues in the narratives that surround original complex conceptions.

For a signal to pass sideways in a hierarchy from one entity at a given level to another at the same level, the signal must pass up the stem of the transmitting holon to a common junction point. The junction point is higher in the hierarchy, where the stems of the two communicants meet. Then the signal passes from the junction point down the stem of the horizontally receiving holon.

The junction point in the hierarchy is a holon in its own right. Two people talking in a room saturate the room so that the respective other person can hear the communication. The room is the junction compartment and is a holon. The signal must not so much pass between the lower-level communicating holons as it must saturate the junction holon that contains or is the context of the lower-level pair. In reaching all parts of the junction holon, the signal reaches the lower-level holon that is the receiver of the horizontal communication. A standard nonhierarchical model couples the holons by a direct communication channel that makes the holons simultaneous functions of each other. The hierarchical alternative has certain attractive features. It alerts us where to expect new pertinent filters. For instance, information theoretic discussions of groups of people predict that up to six member in a group do not apparently organize. But at seven members it is more economical in terms of monitoring channels to have a leader. It takes two communications between members via the leader, but it is more reliable than direct connection to the tangle. That is why sports teams get a captain once more than seven are involved. Basketball is on the edge: five players at a time on court. There is really only captain or coach behavior in basketball for issues of substitution of the extra players. In soccer and rugby the captain is almost always in play, because of the larger number of players.

Leon Harmon performed interesting work on signal passing between levels. In 1973 he used the image of Abraham Lincoln to show how humans can reconstruct quite specific faces even when they are distorted with noise. The reader will probably be familiar with images represented by tiles. It is now so familiar as to be a standard modifier on the image handling program Photoshop, which is what we used to create Figure 11.2. Salvador Dali saw Harmon's images and painted a series of faces of Lincoln represented by tiled images. Harmon showed that faces can be recognized if presented with just over 100 tiles. The more tiles the easier it is to recognize the face. Allen used such images in his classes. Those close to the screen could only say it was a face, but those at the back of the large auditorium knew who it was easily. The reason is that distance filters out details, like the edge of the tiles. Distance works like a low frequency pass filter. The lack of resolution that comes with being distant is in fact a filter that helps unscramble the message from the noise.

Context can deny or help the passage of information. Allen gave a talk at Oak Ridge National Laboratory, Tennessee,[27] using the image of Lincoln, making the statement that specific information, like Lincoln's exact identity, is retrieved only with a certain probability and depends on context. An audience member helped make the point when she said, "I thought it was Superman, but then you have to reckon Abraham Lincoln don't figure too big down here."

FIGURE 11.2. A series of images of Abraham Lincoln's face sent through various filters in Photoshop. a) The original image of Lincoln that we modified. b) The same image but tiled over with only enough tiles to allow recognition with difficulty. c and d) These images have more tiles and are much easier to unscramble. e) The image has a smoothing filter of the tiled image figure in b, like Harmon's 10 cycle pass filter, and is recognizable.

Harmon was able to make the Lincoln face more recognizable if the tiled image was sent through a 10 cycle pass filter that only allowed passage of low frequency signal. The 40 and 20 cycle detailed noise was removed. With only 40 cycles removed the face was still obscured because of the presence of 20 cycle noise. But if he took out the 20 cycle signal with a notch filter and left in the 40 cycle signal of the edges of the tiles, the face was recognizable, even though the ghost of the tiles was still there. The trick is that 20 cycles is closer to the 10 cycles of the face but the 40 cycle tile edges are far enough away in frequency space so as not to interfere. Noise with frequency close to the frequency of the signal obscures more than noise far from the signal frequency. The frequencies of signal and noise form a hierarchy.

The second attractive feature of the hierarchical communication model is that it can account for other parties overhearing an interaction

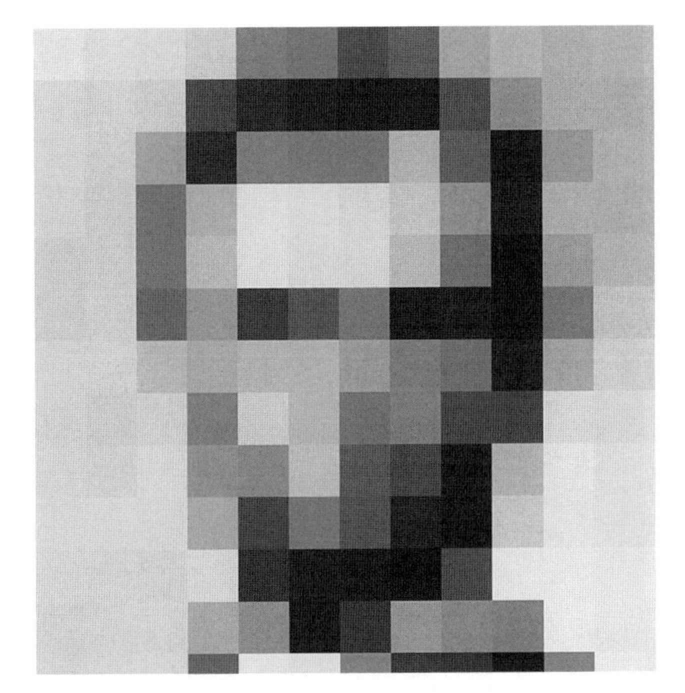

B

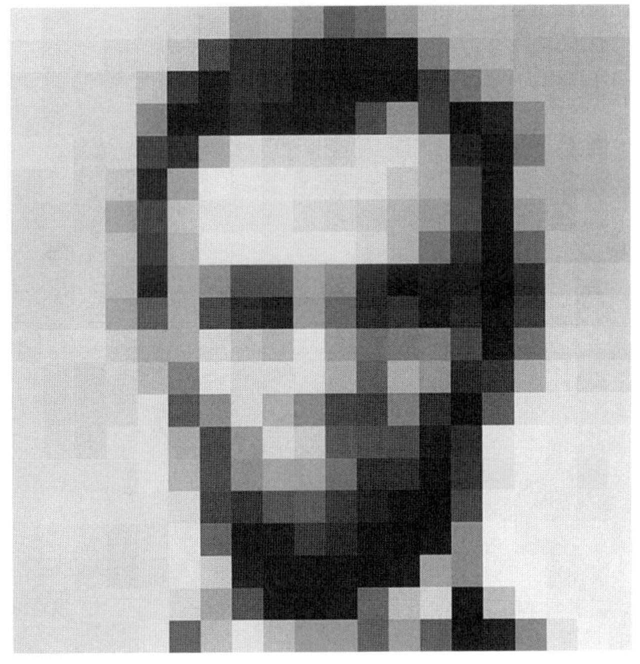

C

FIGURE 11.2. (*Continued*)

FIGURE 11.2. (*Continued*)

directed elsewhere. The hierarchical communication model points out that the junction holon may constrain other holons that were not part of the original communicating pair. There may be no communication channel between the original pair that does not also involve yet other holons. That may indicate the validity of the hierarchical communication model. A communication in a crowd has to fill the junction compartment (the local space in the crowd) such that interlopers may be able to spy. Spyware, for instance transmitters placed on windows by sleuths, has filters that help the message across. The hierarchical model points out information leakage and third-order interaction. Steep scale gradients normally block horizontal information flow by averaging and buffering signal. Hearing aids help the signal across by putting a filter on incoming signal so as to promote speech and suppress incidental noise.

In our discussion of surfaces and scale gradients (chapter 6) we indicated that the entities that map discretely into experiential space generally have surfaces that place them on steep portions of scale gradients. Entities in a hierarchy are relatively stable because disaster may not be transmitted horizontally across the hierarchy so as to reach and destroy sister structures. The junction point through which the destructive force must pass has lower frequency characteristics than the surface of either the lower holon that suffers damage or its sisters. A compartment that is high in a hierarchy usually buffers and smooths the signal so as to severely limit disruption. We spoke in chapter 6 about colonial chordates that use necrosis to change filters between related organisms. The further up the hierarchy that a signal must be transmitted in order to achieve horizontal transfer, the more dissected is the hierarchy and the more separate are the holons of the lower level. Upper level holons that act as the junction points work as buffering filters that mediate or stop the horizontal transfer of information. Information collected by one branch of law enforcement, all too often is not easily shared with other branches in other jurisdictions. Sometimes the pathology is that information moves too easily upscale. If the lower level signal can tunnel up a long way, the hierarchy shows signs of instability. This is what happens when journalists bring attention to low level victims of some government action in a way that undermines reasonable and constructive regulations. There are always victims, but journalism amplifies their suffering.

Any system that has lower holons directly connected with no smoothing filter in between courts disaster. A prime example is the Tacoma Narrows Bridge in Washington in 1940. The mistake of the engineers was to build the bridge of sections all the same length. As a result, oscillation in one section could be transmitted as a harmonic wave to the next section and so across the whole structure. The junction holon allowed passage to the disruptive signal intact, subjecting it to neither smoothing nor fragmentation. The Tacoma Narrows Bridge was a completely horizontally connected hierarchy, and as such was susceptible to the wind perturbation. The amplifying oscillations bent the bridge like a ribbon until it finally collapsed. Now engineers mix long sections with short sections so the long pieces smooth the oscillations of the short pieces, and the short pieces dissect the oscillations of the long pieces.

Connecting organisms and resources. These principles can be applied to models of resource exploitation. In a population of individuals feeding from a common food source, the connectedness of the population on the food dimension is determined by the length of the time constants associated with the total food supply. An individual takes a morsel of food. That the individual has eaten is transmitted to other members of the population through the contribution of that mouthful to the behavior of the total food compartment. If the food supply is large, then the individual may eat many mouthfuls, which nevertheless represent a small insignificant portion of the total supply. The abundant slow moving food holon averages out these morsels according to their proportion in their entire resource. Were the food supply smaller and therefore given to behavior with more rapid fluctuation, each individual bite would represent a larger portion of the whole. The system would suffer increased connectedness at the level of consumption by the individual holon. Were the food supply to be severely limited so that it behaved with almost the same frequency as the individual feeding patterns, the system would be very strongly horizontally connected. Each mouthful would represent a large portion of the whole and would therefore have its associated information powerfully transmitted through the system across the individual organism level.

Were the food supply intermittent and highly variable the individuals feeding could be disconnected from the food. The potential feeders would amount to a holon that was unable to constrain the food. In such a case constraining holons can only constrain by presenting a unified front to the constrained holon. The feeders are a context, but alone they are not an effective constraint. By aggregating in the face of an intermittent food source, the individual feeders can cue each other if one finds the resource. That is what is going on when animals flock or shoal.

A constraining holon that is not unified is divided and ruled by what it would constrain (workers, factory owners, and unions). Faced with a rapidly pulsing food supply, organisms must unite into a cohesive feeding unit if individuals are not to be gradually lost by missing the pulses. They would be unable to behave faster than the food and so would have to behave as a unified environment for the intermittent supply. Turkey vultures function as a group so the intermittent spikes of resource are signaled to all.

Territoriality. Examples of all the above conditions seem to occur quite commonly in nature. All our examples have been arranged in tabular form (Table 11.1) but need amplification. The examples are so many that the table serves to keep track of the permutations and ramifications. We have not tested the general model that is suggested by our unified treatment of the examples. We offer it as a means of specific hypothesis generation so that workers in the field might have at hand a fresh set of questions couched in terms other than the optimality models that still prevail.

Mech (1977) reports that when food supply is short, prey are found concentrated in the regions between wolf pack territories. He does not suggest that this is necessarily a conscious maneuver on the part of the prey, for the concentrations may represent survivors rather than fugitives. Which is the case is no matter here for the general theory, but is exactly what empiricists like to work. The significant point is the uncoupling between prey and predator that is indicated. As the food supply becomes small and subject to rapid behavior that could be fatal for prey and predator alike, the surviving prey are a subset that interacts less with predators than did the abundant prey population in its entirety. This gradual uncoupling of the predators from the prey prevents the hungry packs from driving the food supply to the ultimate rapid behavior, the death of the last prey. Better to be hungry with a slowly declining prey population than satiated with the food gone forever.

More useful here are examples of shifts in consumer behavior as the total food supply begins to exhibit significant changes in patterns of behavior. As the behavior of the food becomes higher frequency, the food holon drops in the hierarchy. As a junction holon, the food supply moves hierarchically closer to its recipients and begins to over-connect the consumers. Some animals change feeding strategy so as to reduce the over-connectedness.

Carpenter and MacMillen (1976), studying the nectarivorous bird

TABLE 11.1. **Summary of patterns of system connection in various animals as reported in the literature**

Organism connected	Connecting compartment	Connectedness case	Response to incipient instability	Comments	Source in the literature
Hawaiian honeycreeper	Abundant nectar	Safe level of connectedness (Fig. 11.3a)	None necessary	Birds ignore each other at times of day when nectar is abundant	Carpenter & MacMillen (1976)
Hawaiian honeycreeper	Diminished nectar in clustered flowers	Over-connected (Fig. 11.3b)	Territories set up	Birds ignore each other and vigorously territorial in one day	Carpenter & MacMillen (1976)
Hawaiian honeycreeper	Scattered flowers yielding less nectar	Severe over-connection (Fig. 11.3c)	Territories abandoned. Interspecific dominance hierarchy	Adjustment in connection	Carpenter & MacMillen (1976)
African golden-winged sunbird	Abundant nectar	Safe level of connectedness (Fig. 11.3a)	None necessary	Same as honeycreeper above	Gill & Wolf (1975)
African golden-winged sunbird	Diminished nectar	Over-connected (Fig. 11.3b)	Territoriality	Territories expand as food becomes scarce, indicating incipient hierarchy	Gill & Wolf (1975)
Wolf	Caribou	Over-connected (Fig. 11.3b)	Pack hunts in center of territory	Caribou either seek or survive in regions of territorial boundaries. Packs hunt less in (are less connected to) these areas. Small residual prey flock less connected to predator(diminished predation}.	Mech (1977)
Yarrow's spiny lizard	Insects	Over-connected (Fig. 11.3b)	Territory size altered in response to food availability	Territories retract with food supplementation; results in less border conflict (diminished predation)	Simon (1975)

Yarrow's spiny lizard	Insects	Severe over-connection (Fig. 11.3c)	Switch to hierarchical behavior at high densities	Cost of territoriality become too expensive	Middendorf (1979)
Generalized parasite	Connecting variable k	Catastrophic connections	None built into models	As connectedness increases systems loses resilience until total instability	Holling and Ewing (1971)
Malepo Island *Anolis*	Invertebrate and candy from investigators	Under-connected (Fig. 11.3d)	Conspecific cuing	Lizards watch each other to monitor possible food resources	Rand et al. (1975)
Iguanid lizard simulations	Food resource	Under connected (Fig. 11.3d)	Conspecific cuing	Behavior explains patterns of distribution seen in many animals	Kiester and Slatkin 1974
Black Lizard	Insects and plants	Local severe over-connection (Fig. 11.3c)	Territories abandoned. Switch to hierarchical behavior at high density.	Dense, localized populations switch from territoriality to hierarchies. Limited basking sites and possibly food. Food is abundant but v high density of lizards from behavior below perhaps gives ultimate food stress.	Evans (1951)

(*continued*)

TABLE 11.1. ***(Continued)***

Black Lizard	Insects and plants	Global under-connection (Fig. 11.3d)	Conspecific cuing	Abandonment of territories due to conspecific cuing with respect to localized food. This correct response to global under-connection leads to local over-connection	Kiester and Slatkin (1974) interpretation of Evans (1951)
Turkey vultures	Food	Under-connected to food	Unify food compart-ment	Conspecific cuing unified food sightings	Stager 1964
Neolithic humans	Incipient agriculture	Under-connected to local areas of abundant crops. (Fig. 11.3d)	State evolves distribution system	Agriculture is susceptible to external forces giving local dearth and plenty. Farming only survives if this smoothing of local agroecosystem economies is achieved. Allen argues similarly for urban primacy in stabilizing casual sowing.	Gall and Saxe (1977) (Allen 1977)

Note: Summary of patterns of system connection in various animals as reported in the literature. The species show various patterns that adjust connectedness so the population and species remains stable. Over-connection gives spontaneous positive feedback. Under-connection makes the system susceptible to invasion by some perturbation. Most animal species have repeated appearances in the table, usually because a difference in environment causes different degrees of connectedness and so requires a distinctive response. The repeats appear grouped together. The connectedness cases are tied to Figure 11.3.

called the Hawaiian honeycreeper, and Gill and Wolf (1975), working on an African equivalent, the golden-winged sunbird. Both teams found that if nectar is abundant, then the birds essentially ignore each other. In this case the slow moving food supply filters the information regarding each individual bird's consumption. As a result the individual birds do not perceive each other at all on the food dimension (see Figure 11.3). If, however, the population density increases or the food supply becomes reduced, then the birds become very cognizant of each other and proceed

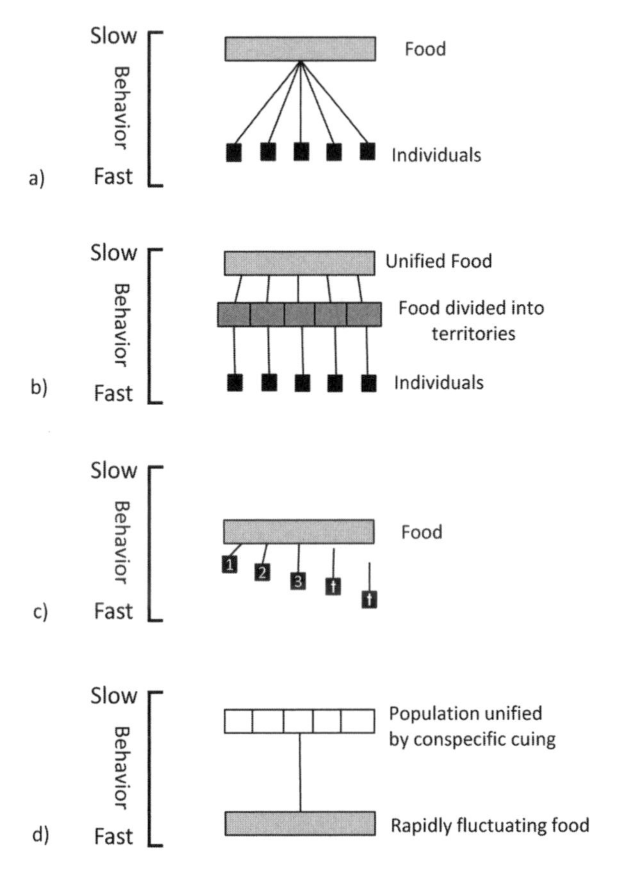

FIGURE 11.3. a) Patterns of connection where organisms are separated and ignore each other. b) Patterns of connection where territories separate individuals. c) Patterns when larger territories of the dominant expand to exclude and disconnect weaker members. d) A pattern of organisms forming a flock or other conspecific cuing, so as to share a diffuse, fragmented resource. This is the correction for an under-connected system. The other parts of figure 11.3 deal with the over-connected case.

to set up territories. The birds may ignore each other in the morning while setting up territories in the afternoon: "The Golden-winged Sunbirds in the area were not territorial until after 1500 hrs. when the undefended nectar levels dropped below 2 μl per flower" (Gill and Wolf). The authors of our examples structured their experiments and observations using game theory models in which the switch of behavior was given account in a strict cost/benefit analysis. When they found an exception to their optimality model, Carpenter and MacMillen were at pains to point out that the one individual defending a territory in uneconomic circumstances "supplemented its diet by foraging extensively for insects." Their models of optimal switching strategies may or may not be appropriate, but either way the example serves our purpose of investigating contextual influence on feeding activity.

We do not suggest that the food supply necessarily has anything directly to do with the mechanism of the switch in behavior. That is the sort of thing empiricists might work out. The trigger for a switch in behavior may not be in terms of the reason for the change. Signs are rate-independent whereas material behavior of resource is rate-dependent. However, we might reasonably suppose that the mechanism has been naturally selected, with food crashes mediating death. The change in behavior itself is not to be confused with the factors that make the switch necessary. We describe the hierarchy on the dimension of food, suggesting what would happen if the over-connectedness that it engenders were not adjusted. The social or ethological dimension of the behavior switch itself is a different matter. So much of biology is working out the functionality of behavior.

In the face of a diminishing food supply the birds become horizontally connected. Therefore they use a behavioral device to stabilize access to the food compartment, making the total food supply behave more slowly. Information about the food consumption of one bird must cross the boundary between the two territories, and in order to do so it must pass through the holon of an integration of the food supply in both territories. In our nectarivorous examples, the populations of flowers do not compete for sugar. Nectar in one territory is not there at the expense of other territories, and so the territories are functionally quite separate. In territories based on other resources food may be capable of movement (e.g., insect prey) and depletion of one territory could be felt in adjacent territories. Food eaten in one territory could have been eaten in another territory if it had been left to move there. Having survived passage up to the level of food supply in both territories together, that information about food depletion may then pass down the other stem of the hierarchy to the animal in the other territory. Territoriality appears to be a device for lowering the connectedness of a system that is dangerously overconnected (see Figure 11.3b). The effect of this is to minimize violent endogenous fluctuations that arise from many hungry consumers eating into a severely limited resource. The mechanism can be a lowering of the efficiency with which the total population exploits the unified food supply. This is in concurrence with the conclusion of Holling and Ewing (1971), who found that parasites too tightly connected to their food supply constitute an unstable system.

Social hierarchies. Territoriality is a social behavior that comes at a certain cost. As the food supply diminishes, however, eking out an existence in one of the territories becomes increasingly difficult and the cost of territoriality increases. Social hierarchy, on the other hand, is very expensive for the population to establish, but once it is achieved further diminishing of the food supply does not much further increase the cost. Social hierarchy is expensive because a certain amount of threat display and fighting becomes necessary at the outset. As a food supply diminishes from abundant to scarce, at first there is greater economy in the establishment of territories. As food becomes yet more scarce, social hierarchy becomes the relatively cheaper expedient.[28] The effect of social hierarchy is to disconnect weaker individuals from the food in a continuous fashion as the supply becomes scarcer. In social hierarchy the population continuously adjusts its size and hence its connectedness through connectivity[29] by fatally disconnecting individuals of junior standing one at a time from sufficient food for survival (see Figure 11.3c). Lizard species are known (*Scelopertus jarrovi*) that abandon territoriality in time of severe food stress for a social hierarchy.[30] Thus they gently, and as necessary, adjust connectedness while the rapid behavior of the food supply slowly drives the population below its physiological subsistence level (Figure 11.3).

Simon[31] notes that territories in *S. jarrovi* are defended within sexes but less, if at all, between sexes. Also, juveniles form their own separate system of territories. The adjustment of connectedness is a major shift in the functioning of a system, and what needs adjustment on one dimension may not in another. Simon's observation indicates a separation of connectedness on the food dimension from connectedness on the sex or adult replacement dimensions. Simon also notes that lizard territories become smaller with food supplementation. Smaller territories can influence each other more, so tending to increase connectedness, but this is counterbalanced by tendencies toward decreasing connectedness through a lowering of the frequency of food behavior caused by food supplementation.

The converse of Simon's lizard case is reported by Gill and Wolf, who found sunbird territory expansion in the face of diminishing food supply. They also find the beginning of the shift from territory to social hierarchy in their observation, "How effective the defense is depends in part on the dominance relationship of the intruding individual." As food supply diminished, only the most dominant individuals would be able to expand their territories and still defend them. Gill and Wolf do not report a complete breakdown of territoriality in sunbirds, but the beginnings of the collapse are evident. Carpenter and MacMillen do report such a breakdown in the Hawaiian honeycreeper under food stress. Thus the honeycreeper runs the full gamut from no territory due to large food supplies with little connectedness between individuals with respect to food, through a territorial response to over-connection, to an abandonment of territory in the face of over-connectedness of territories in food scarcity. Carpenter and MacMillen do not expand on the social regime after the collapse of over-connected territories.

Flocking. The last case of adjustment of connectedness is in the context of a high-frequency, intensely patterned resource. Here the population behaves more slowly than does the resource and so must present a united front to the food, the lower holon. If food is allowed to constrain the population in these circumstances, then the individuals die or at least operate very inefficiently because the food behavior demands a higher frequency of behavior from the animals than they can, as individuals, achieve. Kiester and Slatkin (1974) call flocking behavior "conspecific cuing." In this strategy, they say:

> An animal observes the movements, density and activity of conspecific individuals in addition to observing food resources, and uses these to organize its movements. This strategy may be effective in environments with temporal variation of large magnitude in the availability of food resources.

Kiester and Slatkin test a theoretical model of conspecific cuing and give several examples. They report that:

Evans (1951) provides the best example, the Mexican ground iguana or black lizard, *Ctenosaura pectinata*. This species is normally dispersed and territorial in the thorn forest in which it lives. However, on rocky walls adjacent to squash patches (areas of high resource availability for both food and shelter) this species occurs in high density and shows a hierarchical social system. Fitch (1940) gives a similar account for the western fence lizard. Stamps (1973) reports that female *Anolis aeneus* occur in both hierarchical and territorial situations.

The movement from territory to hierarchy seems associated with higher frequency of resource behavior. Conspecific cuing seems to increase in such circumstances, but does not have to be associated with social hierarchy. In fact, the most intense conspecific cuing, flocking, is intensified by lack of hierarchical aggression. Rand et al. (1975) observe very high densities of *Atudis agassizi* but note, "The most impressive observation was the relative lack of aggressive encounters." *A. agassizi* lives in a barren habitat on Malepo Island in the tropical Pacific. Its resources appear very localized, like invertebrates collecting around dead birds. Rand et al. report that they watch others intently so as to use their fellows' success in finding food and water as the key to their own resource capture. In this case the resource supply fractures the system horizontally so that it is under-connected (see Figure 11.3d). The system is unstable and in danger of losing parts (lizards) because they become disconnected from the whole. The horizontal connection through a behavioral device produces a safe level of connectedness, so keeping the system stable.

There are other examples of conspecific cuing in under-connected systems. Prey use it as a device for increasing connectedness in the face of high-frequency localized behavior of predators. The many eyes and ears of a group of cottontails and the signal of a fleeing white rump bring the connectedness of the rabbit population to an appropriate level. Fish shoal in ways that confuse their predators. Turkey vultures faced with highly localized resources of carrion use the movement of their fellows to connect a diffuse population of vultures over an enormous area. The fine grain and small scale of the resource require a cross-connection of the vultures if any are to survive between feasts.[32] Bekoff and Wells (1980) report a shift toward larger groups as coyotes move seasonally from diffuse rodent food sources to focused food concentrated in the bodies of large herbivores.

It would seem from the many examples given here that ecological systems adjust their connectedness as external forces stress consumer holons. Suitable levels of connectedness seem to be a principal factor for stability. Its relationship to diversity is by no means simple. Clearly a system that is over-connected in one dimension may be under-connected in others, so giving strategies for increased or decreased connectedness associated with each dimension of the hierarchy. These models might even be of help and significance in understanding the selective advantage of various human social patterns. Perhaps a limiting factor for human societies may not be their particular strengths or technological weaknesses, but rather the responses of the system to dangerous levels of connectedness. Quoting Jantsch (1976), "Such a new paradigm may be expected to develop into a new science of connectedness, which will also become the viable core for a new science of humanity." These ideas are expanded in chapter 12.

CHAPTER TWELVE

Scale as an Investigative Tool

This chapter concludes the book as it describes a scale-oriented or scale-aware approach to observation and analysis. It contributes to the unification of complexity as a self-conscious arena of discourse. We often use ecology as the vehicle, but the implications are general and apply to social studies as well. There has been a change of focus from spatial considerations to organizing principles that relate to the passage of time. This has been extended to expressing biological problems in terms of frequency rather than time. This chapter presents a general use of scale-oriented approaches. It develops a framework wherein notions of scale can be applied to all sorts of investigations. We note that the most cutting edge work focuses on scaling.[1] The areas we pick out are directed evolution and the critical scaling work that allows Nick Lane to take a completely new stance on what is life, fundamentally. But more of that once this chapter has set the context for a proper treatment of scaling and its large meaning. Complexity and scaling are at the frontier, and hierarchy theory can keep them there.

As we have argued with Rosen's modeling relation, a formal model applied to two different structures says all is the same except scale (Figure 5.7). Formal models are scale-relative statements, and are therefore in themselves scale-independent. We speak here of formal models as did Rosen[2]; other models may indeed be scaled. If two structures can be encoded into and out of a formal model, the two structures become analog models of each other. The analog link is through what the structures are observed to have in common. Experiments performed on the more tractable structure (the experimental model) have results that can be applied to the other structures that are more difficult to manipulate (like a lake instead of the experimental beaker). All experimentation depends on this analog relationship.

Avoiding Emergence Invokes Values

Engineers have standard methodologies when they work with scale. They are conscious that the same model (equation) can be used again and again to make or understand structures. This is supposed to make equations a general condition, but it is not; in fact the general condition is that most equations cannot be solved.[3] Engineers are particularly attentive to scaling issues because relationships change in what they construct across scales. They are cautious in scaling, always trying to avoid emergent properties. Emergent properties appear when there is an increase in scaling effects. Emergent properties appear when a gradient applies, for instant when a whirlpool appears it is due to a gradient that drives water flow. When the difference between the top and bottom of a gradient is great, there is more stress. All this is usually normalized away by engineers applying scaling factors so that there is no emergence.[4]

We see similar scaling in nature. As a tree gets taller the effects of gravity increase with the mass of the tree, a cubed function. A model that normalizes on static loading would increase the thickness of the tree trunk on a squared function, so the bole does not snap. Snapping exposes a cross-section, which is whence comes quadratic scaling. As the tree gets taller we would expect the tree to thicken as a squared function of the height if static loading is the issue for the tree. Real trees appear to use not the value 2 on the exponent for thickness, but 1.5 instead. That means bigger trees are weaker with regard to static loading. The 1.5 exponent normalizes not static loading but dynamic loading, as applies when a tree bends in the wind or when branches sag under their own weight.[5] Engineers apply similar scaling exponents in their models, but tend to over-build so as to be safe. The emergent property bridge-builders fear is the bridge falling down. An emergent property appears when a situation changes sharply. A tree falling can be seen as an emergence. It is exactly that sort of emergent property that engineers design to avoid. In contrast to conventional civil engineering, in ecological engineering emergent properties of failure are the norm. The long-term principle when engineers build ecologically is to expect failure; build instead so it can be fixed.

The River Niger was dammed at Kainji starting in 1964. A long wide valley was flooded and villages were moved to engineers' specifications. Grass huts were replaced by tin huts, which are sturdier and do not leak. Grass huts breathe, metal huts are like cook boxes. Almost immediately the villagers removed the conical metal roofs and set them up as water butts. The grass replacement roofs were a great improvement in terms of heat and ventilation. The engineers understood their error. But when the walls broke, the native dwellers puzzled the engineers. The villagers simply abandoned the metal for mud-walled grass huts. "Why did you do that?" asked the dumbfounded engineers. The villagers replied that they preferred grass and mud, which may not be as waterproof as tin; but when breached grass, sticks, and mud are within local technical abilities for repair. The villagers did not have the means to fix broken tin walls. First World engineers scale rigidly, nature and native practice scale flexibly.[6]

When we consider scaling in our work, we are actually interested in sudden change—that is, in emergent properties, the very thing engineers eschew. Hierarchy theory addresses rescaling where qualitative differences appear. Hierarchy theory looks for emergence and tries to deal with it, not stop it. Essentially, hierarchy theory compares the situations before and after the structure falls down, and deals with the mess. Structures are a subjective decision. Accordingly there is no emergence in a whirlpool until it is recognized. It is after all still water going down a drain whether it is glug glug glug or the swish of the gyre.

Hierarchy deals with scaling changes, translating quantitatively different situations into workable qualitative distinctions. It models situations where everything is different, but not completely. If everything were completely different, there would be no coherence, and there would be nothing to discuss, no reference for emergence. When a system collapses the fundamental character of the situation changes. The model for a bridge ceases to apply after the collapse, because the bridge per se ceases to exist. The old model may indeed have predicted the collapse, but after it happens the model no longer applies. As with so much in our discussions, a lot turns on attitude to a situation more than on materiality.

This all makes little sense for engineers and bridges, because bridge engineers suppress emergence. Meanwhile in ecology things collapse all the time, so we had better get used to it. The critical difference is getting used to disaster. The model of a beaver for its dam takes into account the fact that the river will flood and destroy the dam once or more a year. After the beaver dam is overtaken, the situation has not fundamentally changed because the old dam had rebuilding it in its design. The engineer would build a fail-safe dam so as to avoid the situation changing. The beaver cannot afford dams so robust, so the beaver's model works to minimize the cost of reconstruction. The scale that matters to the beaver changes the subject when addressing a dam stably holding water. Faced with unavoidable destructive forces engineers do design for resilience (e.g., historical wood and paper housing in Japan in response to frequent earthquakes).

The mathematics of collapse and emergence is positive feedback. Compared to negative feedback, positive feedback is the more difficult because applying those dynamics destroys the context in which the positive feedback makes sense. This is the main point of ecologists like De Angelis et al. (1986), who wrote a whole book on positive feedback. The scale that matters to the beaver changes the subject for a human engineer. The overbuilding of engineers has the emergent property of its own large engineered structures that persist after their usefulness. The Roman Colosseum still stands millennia after there are too many Christians to use it for its original purpose. In these happier times it has been repurposed as a site for Christians and everyone else to visit. These days, engineers are doing more ecological-style modeling, as when they take into account the cost and ease of recycling a building or a washing machine after it ceases to be useful.

So scaling of an ecological sort needs more than standard engineering approaches. And ecological sorts of problems now nip at our heels. James Kay would often characterize contemporary problems as something like fixing an airplane wing in flight. We are stuck with a society that has already committed to a system that is now breaking up in the air, unless we do something. It is often not clear what is that something, since landing the plane may not be desirable or even an option. A literal flight that needed fixing in flight was Gemini VIII.

> On March 16, 1966 Pilot David Scott and Commander Pilot Neil Armstrong entered space aboard Gemini VIII. Their primary mission was to conduct the first linkup in space by docking their capsule with an Agena target satellite. Shortly after the successful docking, however, the conjoined crafts began spinning out of control. Armstrong disengaged the capsule. [But the spin got worse.] One of sixteen Gemini thrusters had become stuck in the open posi-

tion, emptying fuel into space. Unable to stop the spin with the main thrusters [they disengaged the re-entry capsule]. The crew were finally able to stabilize the capsule after thirty minutes by using engines intended for re-entry. [Armstrong had to do this while in the spin reaching over his head to get at the controls.] Gemini VIII splashed down safely in the Pacific Ocean.[7]

Notice Armstrong's solution was to redefine the vehicle in which they were flying. Test pilots have to have hierarchical redefinition in their tool box. Most conventional engineers can get by without it. We do not have Armstrong any more, and our problems are not so recognizable. We need a general new body of theory to deal with our new circumstance. It will invoke scaling and dealing with emergence.

Scaling in Ecology

There is a large body of ecological literature that involves scale, although it does not generally do so in our terms.[8] With few exceptions (e.g., Noy-Meir and Anderson 1971), these papers do not directly address scale in its own right. Since scale is not generally explicitly included in ecological models, the readers of a communication must guess and arbitrarily apply a scale of their own. Ecologists reading a report often do not worry that their guess (perhaps made unconsciously) may be different from that of the writer of the communication. Readers do, however, worry when the results appear inconsistent with those of their own, apparently identical model; the altercation that follows is semantic and quite unnecessary; the models, differently scaled, are fundamentally different and concurrence without scale correction is not to be expected. Differently scaled models define entities differently and so are most simply related to different phenomena.

This final chapter shows how scale can be used directly as a general tool. Noy-Meir et al.[9] attribute some of the differences between schools of ecology to different data standardizations that are implicit in the various approaches; we consider such modifications of weights to be a scaling device. We will advocate for an optimal investigation strategy that locks together several approaches. Even though our examples are incomplete, they tie together several schools of ecology as tactics in a common scale-oriented investigative strategy. This would seem valuable at a time when theories become more complicated and schools of thought become more separate. The ecological journals have become mutually unintelligible, or at least uninteresting, for each other's respective readerships. Everyone read Thomas Kuhn[10] on paradigms; now students are most ignorant (although they use the word paradigm often and incorrectly). Kuhn would suggest that ecology is due for a paradigm change, and perhaps that change will bring scale to the forefront as a unifying concept. A quarter century ago, Allen and Hoekstra moved in the direction of a unification of ecology, and have recently presented a new edition.

Ecology is a diverse discipline with many and varied approaches. It is not surprising therefore that we need a preamble before the attempt at unification can be made. The model we propose uses a geometric idiom that should be in tune with a biological mindset. Biology involves awkward spaces to such an extent that conscripted biologists in the Second World War were used to interpret aerial photographs, much as they had interpreted optical sections of cells under the microscope. We attempt to deal with curvature in ecological spaces.

Periodicity and Inertial Frames

The vehicle for the discussion of inertial frames is a somewhat specialized field of plant community succession (change). Indulge us. The idea is not to make plant community experts of the reader, but to provide an accessible point of tension between stasis and dynamics. We have to choose one technical field or another to get our point across, and we know plant community ecology. Allen and Hoekstra[11] suggest a conception that turns on the cyclical nature of ecological processes. A difficulty with community concepts is that there are several definitions, some of which are at distinctly different levels. Nichols[12] took a consensus in 1921 of ecologists about community concepts and announced an agreement on the distinction of particular concrete community as opposed to the abstract community. The concrete plant community is what one sees at a particular time and place, perhaps a forest. The abstract community is the pigeon hole into which all the concrete examples of a given community fit. Nichols is expansive in his definition of the pitch pine community type in Connecticut (he worked at Yale). It consisted not only of all the present examples, but also all that had ever existed, or ever will exist. The abstract community would fit into our earlier discussion of essences (Figure 5.4). Some of this rich conception is captured in Cooper's[13] analogy of the community to a braided stream. Braided streams have sand bars around which a network of channels connect, disconnect, and then reconnect not just as the river flows downstream, but also from season to season and year to year. Winter floods take out some sand bars and deposit others. The reticulum is woven and rewoven over a longer time than that taken for water to flow down the length of the stream. Successional change is analogous to the river flowing. The cut of the river at a point is the concrete community analogy. Old buried cuts of the river are past concrete examples that have now disappeared. The abstract community is analogous to the whole river valley with its buried cuts from the past and the river flowing into the future. This is a hierarchical version of Heraclitus' statement that you cannot step into the same river twice. The quote from Heraclitus appears in Plato's *Cratylus* more than once. You cannot see the same concrete community twice but have to deal with it anyway.

Allen and Hoekstra conceive of the community not as a thing in a place, but as a set of interacting periodicities. It is not so much that sugar maple (*Acer saccharum*) and hemlock (*Tsuga canadensis*) occur together in northern Wisconsin. They do, but there is a dynamic that the overstory of one tends to occur over an understory of the other. They replace each other in a wave interference pattern. Wave interference appears in old cowboy movies, when the spokes of the stage coach get in and out of phase with the frames of the film, making the wheel appear to roll backwards. The constant one sees in a community type is not a collection of species in a place but is rather a wave interference. One only has to change the periodicity of one component a small degree in relation to another periodicity in order to get a very different particular, but nevertheless related patterns. In an animal community the starfish, *Pisaster*, trundles through the shellfish community cutting a swath. That sets back the superior bivalve competitor, a mussel, which left to its devices outcompetes all other shellfish. The diversity of the shellfish community depends on the starfish opening things up. Remove the starfish and the community is reduced to the ubiquitous mussel. Removing the starfish removes a critical periodicity. Big storms can substitute for the keystone starfish species. The community is thus well conceived as a thing at multiple levels. Hugh Iltis told Allen in a personal communication of Curtis'[14] conception of a community. Curtis saw it as a set of species in a space all connected to each other by a network of rubber bands. Move any one species, and all the others are shifted in a web of tension. Curtis thus captures the structure and dynamics of the community together very well, and in the spirit of an interference pattern.

The concept of interference pattern in ecology applies also to biomes. Biomes are collections of plants, but not specified by species. Vegetation in a biome is physiognomically recognizable because the plants have all accommodated. Allen and Hoekstra assert that biomes are 1) environmentally determined, 2) disturbance created, 3) animal groomed, and 4) physiognomically recognizable. The reason a given biome looks a certain way is because its plants, the species of which are not identified, all show a morphology that is an adaptation to a certain climate. Clements[15] invented the term *biome*, we think because he wanted environmental determinism to work in communities. He is not alone in his agenda. Community studies moved beyond environmental determinism only with the work of David Roberts in the 1980s.[16] All the others from Clements[17] to Whittaker[18] were environmental determinists. That is unfortunate because there is a many to many mapping where one environment may present many communities and one community may appear in many environments. The better concept is an environmental relation as conceived by Roberts, where there is slack in the mapping from one to the other. Clements created the biome concept because it is environmentally determined. He also anchored his whole scheme to the concept of climatic climax[19] communities because biome and climax map onto each other only at climax, when vegetation is unchanging.

The periodicity of climate and vegetation also fit into the four defining characteristics of biomes above. Animals return at a certain period to groom. When the periodicity of environmental conditions moves outside the normal regime of the old biome, plants cannot for instance disperse viable seeds. As Gambel oak moves north in Utah it is put in a squeeze between frost and the arrival of water.[20] There comes a point where the tree cannot set seeds that can grow because frost or drought kills them. The tree can reproduce vegetatively and grows some 200 miles north of its viable seed set by vegetative means. The oak grows on hills, and normally uses seeds to jump across valleys. But north of the seed set line it cannot jump because it has no seed reproduction. As a result there is a particular northern hill when the Gambel oak reaches its limit; a hill full of trees is doing well growing vegetatively, but the next hill north is barren with no seeds to establish the species. In that case when the disturbance returns, for instance a fire, it is a lethal blow and a new biome without the tree takes over. If the forest is inside its environmental regime there is simply recovery from the fire and the biome persists. The physical form of the plant that is the signature of a biome arises from that form being able to deal with the periodicities of the environment. The stable look of a biome is in fact a wave interference pattern between environmental fluctuation and the periodicities of growth, seed set, seed dormancy, drought tolerance, and so on.

Wave interference invokes a relationship within which phenomena can be set. Wave interference relates a low level phenomenon, a period, to its context. This is similar to the relationship that pertains in an inertial frame, a relevant concept coming from physics. Calculations in physics are performed in inertial frames. A physics laboratory mounted on a revolving carousel would not be in an inertial frame, for the turning would impose centrifugal forces. It is not that physical laws would be incalculable in such a place; they would just be more complicated. An *inertial frame* is the fixed context in which laws of interaction are most simple. Here again we see a term that, like *stable* or *persistent*, could be interpreted as referring to something ontologically real, but is here used with only epistemological implications. Inertial frames are defined by the simplicity of explanation of phenomena associated with them. If an observation is made from one frame to another, then a transformation is required that makes for consistency between frames. In this way experiments with falling balls on the dockside are consistent with experiments on falling balls on a smoothly riding passing liner. If the inertial frames are moving slowly relative to one another, then a simple linear transformation, a Galilean transformation, is appropriate to calculate between frames. If the frames are moving relatively fast in different directions, say a fair proportion of the speed of light, then a nonlinear Lorentz transformation, associated with relativity, is necessary for concurrence. Biology has its own scales for movement not found in physics, and things become curved and nonlinear at relative speeds a lot slower than the speed of light, the speed of ecological succession for instance.

With Bartell, and Koonce, Allen found it profitable to view successional change as analogous to an inertial frame for ecological phenomena in communities. The successional inertial frame recognizes the natural species replacement as a given. To the ecologist in the field the vegetation appears to change because the scientist imposes a frame wherein particular species compositions in places are taken as constants (concrete communities). The situation is analogous to the way that the ball dropped on the moving boat, viewed from land, appears to move forward as it falls. Viewed from the reference composition of the stand after disturbance, the stand of vegetation is seen to change. Seen from the successional frame, the vegetation through time appears to be at rest despite species replacement. Neither frame is especially true; they are just different expressions of a state of affairs, each with its own advantages given the scale of the phenomenon under investigation. There are underlying processes of organismal birth, competition, senescence, and the like. These processes occur at a somewhat faster rate than species replacement, the rate of which separates the field observer and the successional inertial frames. The relationship between the above ecological frames appears nonlinear; each sees cyclical behavior in the other. The curvature in the cycle is akin to the non-Euclidean curvature that occurs as relativity relates to linear Newtonian spaces.

In any inertial frame there is an infinite choice of points in time and space from which observations could be made. The observation point providing an angle of sight is one of the factors that determine what is seen. For instance any asymmetric structure looks different from the front as opposed to the back. Nevertheless, it is easy to take into account the relative position of the observation point and calculate that the front and back bear a simple relationship to one another. The two images are merely different views of a single structure. The different images can be mapped to just one object by linear transformations, for example, simple sine-cosine rotation.

In biology, physical time is not necessarily regular, in that some events pace the system and lead to endogenous cycle times that may be irregular when metered by physics. The signal medium in biology is usually much slower than the speed of light in delivering its message. Infected insect vectors delivering the message of disease do not communicate a diseased condition immediately; there is an incubation period that must first pass. In this particular biological system the incubation period, like the speed of light in a physical system, is part of the framework to which the forward movement of time is relative. It determines what the observer of the disease phenomenon will see.

We continue the discussion of disease and epidemics here, not as a contribution to the extended and sophisticated literature of epidemiology, but as a vehicle for the discussion of scale comparisons. The observers of the biological system may give account of a malarial epidemic in terms of the number of vectors. If they model the flow of disease just by counting mosquitoes and ignoring incubation time, then their model will transmit disease faster than the biological system in the field. In other cases, however, the incubation time may be very important in structuring the epidemic. There a simple linear change that slows the infection rate may lead to a constant low level of disease in the model. However, the, biological structure that is modeled could have pulses of infection or extinction of the disease. The common cold illustrates the points.

> It may be useful to consider the possibility that colds are due to a virus which, in a large community, is constantly passing from one person to another, usually causing no symptoms, often only abortive ones, and a real cold only when it finds a victim whose local resistance is, for some reason, at a temporary low ebb. From such a person it is dispersed in greater quantity. Few people harbour it for long; hence in isolated groups it commonly soon dies out. Such a conception is not proven, but is in harmony with known facts. (Andrews 1949)

A useful way to view infection is with the notion of percolation. Percolation theory comes from two sources, electrical engineering in computer chips and soil science. Gold is a preferred material for computer chips because it is an excellent conductor and is resistant to corrosion. But gold is expensive and so using the smallest amount to actually achieve electrical flux is important. Gold is sprayed along the track as a set of pixel spots. Percolation theory says that, so long as the pixels are positioned randomly over 59.28 % of the surface, electricity will flow reliably over an indefinite length. In soil science air spaces are reasonably assumed to be randomly positioned. Because actual air therein is close enough to random, the same threshold of 59.28% holds for the percolation of water through soil an indefinite distance. Epidemics occur because the percolation threshold for transmission of a disease agent has been crossed. Sometimes diseases evolve so as to change transmission rates so as to be synchronous with a changed environment. For instance, the 1918 flu epidemic raged in a particularly lethal form when troops returned from Europe crowded on ships. Quarantine extended the epidemic by crowding susceptibles. When quarantine was lifted selection against lethality appeared as the disease was selected to keep host alive. Those strains persist today. Similarly, as yaws moved into cooler areas where clothes blunted infection from sores, the disease dug in and was less lethal as it became a venereal disease.

We made the distinction between the inertial frame and the observation point within that frame. Nevertheless the relationship between the observation point and the inertial frame deserves amplification. It will be remembered that the holon is observer defined; it is helpful to identify exactly when the observer performs the act of definition. Zellmer's scheme in Figure 5.4 applies. The definition of the observed holon occurs when scientists decide what it is whose nature they wish to investigate. There are alternative positions, only one of which is the elucidation of the aspect of nature sought at the outset.

Zellmer's model relates parts to each other. It names things and defines them in equivalence classes. The point of science is to do the best job it can in finding an explanation for the equivalence in the class that the model invokes. A general name gives an equivalent class, like the class dogs, where all members have something canine in common. We have already used the notion of a class of dogs invoking an underlying dogginess that makes dogs in the set equivalent. We have, however, been at pains to insist that dogginess does not exist independent of the class of the dogs in question. The class is a summarization of experience. How experience relates to an external world is not transparent. The point of science is not to find an ultimate truth, but it is rather to find compelling explanations for the patterns that we experience. Then we put those rationalizations into our models.

If the scientist does, in fact, define an object holon that can be associated with a phenomenon, in finding and observing within the inertial frame of a holon the investigator has achieved a main scientific objective. The scientist views the world in the terms that are compatible with those of the holon under investigation. A new experience arises when the observer shifts the observation point, say a view from the front instead of the back of the structure, but a reciprocal shift returns the observer to the original reference point. At all times the observers calculate their relationship to the source of the data and the signal fulfills predictions. Phenomena change their character as the observer uses differently scaled observation windows. In recognizing these differently characterized messages as being associated with a single phenomenon, the observer translates to take into account the relevant features of the inertial frame associated with the observed holon. Predictions are not so much about the truth of a situation as they simply make the narrative that is being told more convincing. By mapping between different lines of sight, and making predictions, separate observations are linked. This helps a lot in telling a convincing narrative that can lead other independent observers to accept the story and achieve something of a commensurate experience. Science is very good at getting us to agree on what we see. In fact that is mostly the point of science, even though that might make those with a naïve realist philosophy uncomfortable.

If, however, the scientist is neither using the relevant parts of the inertial frame pertaining to the signal generator nor translating through compensation to that frame, then the relationship to what is observed is unclear. It is possible to take advantage of this proposition in a test as to whether the observer is dealing with the inertial frame of the observed holon. The simplest test is to observe from a given point, then allow the passage of time and to observe again from the same point. Observations will only concur if the observer uses the inertial frame of the observed holon. If there is disparity between the frames, the observer perceives at a scale different from the relevant facets of the scale of the data source. Then the displacement of the inertial frames from one another in the interim will amount to a shift in the observation point in the inertial frame of the holon under investigation. Think of the vegetation scientist making subsequent observations of the same or different vegetation. A shift of this sort can amount to a mistake, as when sampling statistics are mistakenly used for an inventory. Much effort is expended in biology on replication of observations, and one of the main purposes of that exercise is verification of the scale that the observer uses relative to that of the study material.

All this might seem complicated. We present it so as to unveil the intellectual gymnastics that are indeed present even when making commonplace scientific observations. There really is not much of an opportunity to just observe. Physicists know they have an observation problem, often biologists do not. We cannot get the reader to go along with our insistence on frame of reference and scaling unless the complications of observations are laid bare. Here ends the preamble.

Modes of Investigation

There follows a set of alternatives into which investigative strategies fit. To help the reader keep things straight in the extended discourse that follows, we will lay out all the categories that are discussed separately below. The categories:

1) *Ease of use of the scale*: First scale can be used in two ways: choose a convenient scale, as opposed to a more difficult scale that might be closer to that used by the observables associated with the phenomenon in question. Easy scales allow more replication and so accuracy; difficult scales make observation more complete and get it closer to being right and relevant.
2) *Hold the scale, or hold the point of view*: The second major class of distinctions in the use of scale is either to hold the scale constant and allow the point of view to change, or vary the scale while holding the same point of view. Zellmer et al. referred to scale versus type. The scale versus point of view distinction is very close to scale versus type. One can change both scale and type, as did the large synthesis of ecology performed by Allen and Hoekstra. But as they unpacked the many facets of ecology they were wise to keep track of what happens in an intermediate position where only one is allowed to change.
3) *Focus on scale, focus on data*: The third major way to use scale is to focus on the scales used, and treat them as objects of investigation in and of themselves. The alternative in this third way of using scale is to focus on the results that can arise from changing the data inputs to the investigation. For instance, gradient analysis, ordination, is a well-understood scaling operation. One can change the gradient analysis technique while using a fixed data set (b in Figure 12.1). Alternatively, one can compare outputs of a given gradient analysis technique as various related data sets are put through analysis (d in Figure 12.1). Another example might be a comparison between individual-based models, one calibrated with forest data and in Hank Shugart's FORET model,[21] as opposed to grassland-focused individual-based models (d in Figure 12.1). The unity that emerges from that comparison is that the space full of light is fully occupied in forests, whereas the root zone in the dark in grasslands is the space that is fully occupied. One layer or another is fully occupied. A second insight is that depth of roots in grasslands is as important as, and works the same as, tree height in a forest. Jake Weiner[22] wittily performed an experiment with *Ipomoea* grown: 1) in its own pot and stake, 2) in its own pot with shared stakes, 3) shared stake and own pot, 4) shared pot and shared stake. With all bases covered he showed light competition is asymmetric, while nutrient competition is symmetric. Large plants have asymmetric advantage over small plants for light: big plants get light when small don't. For nutrients competition is symmetric where things are equal despite size (d in Figure 12.1). That exposes a fundamental difference between stem versus root competition, and so the fundamental difference between forests and grasslands.

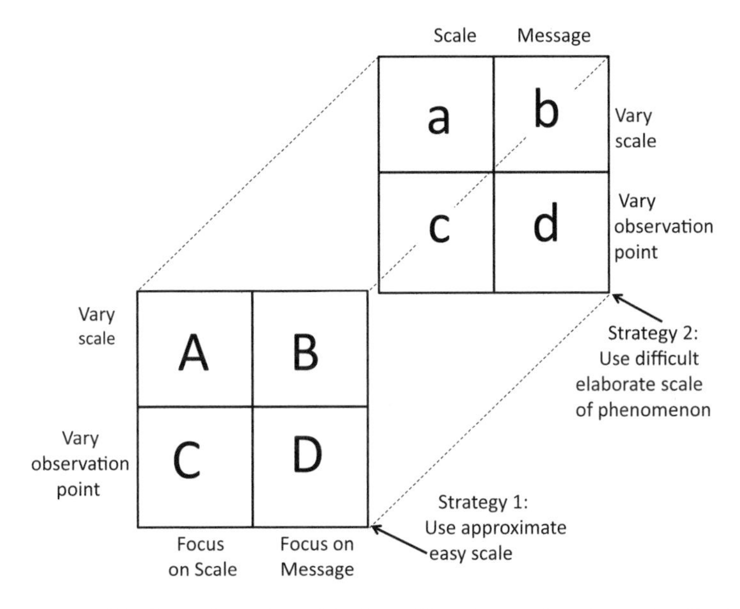

FIGURE 12.1. The three dimensions of dealing with scale in performing a scientific study. The upper-case panel invokes a choice for a quick and dirty scale that is easy to use. The lower-case panel is science using a particularly refined scale to get at the core of an issue. The move to refinement is sometimes done with evolution or being more realistic. AB and ab vary the scale in the investigation, while CD and cd hold the scale and look at something different. Perhaps the options are to use the same or different multivariate method while analyzing one or many data sets. AC and ac focus on the nature of the scaling performed, while BD and bd pay attention to the phenomenon, the signal that comes through the filter.

Let us now work through the various options laid out above.

The ease of using the scale. There are two fundamental investigative strategies using scale as a tool. The first strategy works with an adequate scale or set of scales and filters. The proviso is that they are easy to erect and maintain. The second strategy seeks to use scales whose filters are as similar as possible to those of the natural phenomenon under investigation, even if those scales are inconvenient and the filters are hard to use and maintain.

In the first strategy the observer scale used may well not be close to that of any of the systems under investigation; however, structures can be seen so long as the observer scale is close enough to the scale of the phenomenon under investigation. Nevertheless, if the investigated system and the investigator are working at different scales, relativistic curvatures may occur (remember black spruce [Figure 9.4]).

The second strategy involves setting up a scale that may be difficult to erect and maintain. The scale, inconvenient as it may be, corresponds as exactly as possible to the scale of the level of interest in the biological system under investigation. Note that we are not asserting that nature has its own scales; we are all the time dealing with the consequence of observing. The observation problem (the intrusion of the observer) never goes away. While measuring photosynthesis we can make corrections for the bag around the leaf, but the presence of the original idea of photosynthesis cannot be removed or normalized away. After various scales have been used, one can choose to assert that this or that scale applies to the verity of the situation, but it is unwise to act on that belief until after the scale has been used and investigated. Different levels may be significant at different stages in the investigation. In this manner the system as it is observed may be most completely described in tractable terms. A complication in biology and social systems is that the system under observation, for which we wish to build a narrative and model, itself has models and narratives for dealing with its own externalities. Those models and narratives invoke scales for dealing with signals. As observers, we are attempting to understand, mimic, and tell stories about the stories that the object of observation tells for its own purposes (Figure 2.3). Since we are dealing with a sentient self that is dealing with variously purposeful beings, there is insufficient reference to be able to assert a scale for nature in any straightforward terms. We are dealing with narratives about narratives, in which scale is a useful tool.

Usefulness comes from the observer not the thing observed. The scale for useful expression may or may not be amenable to description in inches or minutes. Its units may be degree days, or generations, and the like, which are at best only monotonic with time. Life stage analysis is not even monotonic with time, but it is an easy scale to apply. One just looks at the stage of the plant and ignores the difficult process of measuring plant age (D in Figure 12.1). There plants can move back to a pre-reproductive stage before returning to reproduction.

An example of an elaborate scale might be neural net analysis. The net is trained on input data so as to find a series of weights on connections between nodes in the net. Neural nets build a filter derived from the training data where the results of a process are already known. The trained net can be applied to test data where the results are also known. Most significantly the trained net can be used as a predictor for data where the output is unknown. This is a standard device for predicting whether or not an applicant for a loan will pay the loan off should it be granted. The training was on data where it is known if the loan was repaid. The weights between nodes in aggregate amount to a very long polynomial that describes the attributes of a worthy debtor. The filters work well, but they are too complicated and nonlinear for the investigator or user to understand any mechanism as to how they work (b in Figure 12.1). Neural nets are predictive, but have none of the advantages of relatively simple equations, where the parts of the equation can be identified as working parts of a mechanism that is understood. Doing science corrects our understanding incrementally. It is like building a set of weights for a neural net that is our very selves. Then we assess the validity of that net that is our understanding, which we can only marginally understand. It is remarkable that science works so very well with so much in the process at sixes and sevens.

Hold the scale or hold the point of view. General patterns of investigation may be established by two approaches within each strategy of convenient versus elaborate scaling. One approach is to hold the scale constant and either allow the observation point the freedom to change or make a conscious change in the observation point so as to secure a different look at the phenomenon (for instance front versus back). There will be no change or a simple linear change only if the scientist observes at the scale used by the phenomenon in question. The other approach varies the scale while holding the observation point constant. Here the observed message read by the observer always changes. The modified message can either be viewed as coming from a new holon at the new scale in a redefined system, or as a differently filtered message from the holon before the scale change. If the scale of the observer is different from that of the data source or if a shift in observation point changes the message received, insight into the form and function of the structure is gained from the change in quality and quantity of message. Using these data intelligently, more likely guesses as to the scaling surrounding the observed system may be made.

In any investigation either the observation point or the scale must be held constant if there is to be a cross-referencing of results; this is the principle of experimental control, although it also applies to constant protocols in descriptive fieldwork. Field survey studies may be controlled by consistent sampling procedures through local sites to general areas. Alternatively, varying sampling procedures in a given universe of investigation, such as a community type, will give a different sort of control. Ignoring such controls, field-workers can unwittingly change both the scale and the observation point together. Much of the career of Grant Cottam was spent making observation of vegetation consistent and cross-referential (d in Figure 12.1); that is now so taken for granted that nobody notices we do it, or remembers the genius who did the research. It has become D in Figure 12.1 along with all descriptive field studies. Disparity of both scale and observation point commonly generates arguments about conflicting evidence. If, however, the remonstrators' data are differently scaled, then the entities they study are different. Since they are not talking about the same thing, concurrence of observation is not to be expected. An example of lack of concurrence, might be a continuum in composition of vegetation as opposed to data that fit discrete community concepts. In social science there are schools that normalize around continuous processes and other structuralist schools that normalize around discrete social structures. Clements' concept was discrete vegetation, while Gleason's[23] concept of community was discrete. The debate over the concept of community is muddled by different degrees of significance being applied to parts of the species space. Clements notices empty parts of species space with no examples of vegetation present. Gleason notices that the parts of species space that are occupied tend to be occupied with continuous variation. There are two levels of analysis here so we would expect semantic arguments between those not sensitive to levels of analysis. Occupancy and exclusion are two different ecological phenomena, so we would not expect examples across the divide to concur.

As to the effect of scale changes, there is a literature that addresses the question teleologically put, "Do plants like to be grazed?" The phenomenon here is called over-compensation. Brown and Allen[24] took a hierarchical approach to that contentious literature and found that different scales were fueling what amounted to a semantic argument. Brown and Allen were generally concerned with the complications of scaling effects (b in Figure 12.1). The individual studies whose scale they investigated were not concerned with scale, they were only using their fixed, defined, or implied scale to get at the issue of overcompensation. Each investigator chose their observation scale to be convenient, and did not look at alternatives (so they were all on the upper-case plane [A-D] of Figure 12.1). The scientists were generally interested in the message about compensation in nature. Over the shortest time frame, capital is simply lost to the plant in grazing. But grazing stimulates growth. For instance, grasshopper saliva is a growth stimulant for leaves and stems and so compensation is achieved. However, a full cost accounting will show in the mid-term that the new biomass above ground comes from the below ground parts of the plant, and so is not free. Early in that process there is, in terms of whole plant biomass, no compensation in growth above ground, let alone overcompensation. However, growth of leaves contributes to the production compartment of the plant, and the extra photosynthesis might indeed outweigh capital loss due to grazing, to the point that there is more accumulated biomass than there would have been had grazing not occurred. Over the long term, plants that are adapted to being grazed, grasses for example, may have their biomass held low by grazing, but in an environment of diminished competition from those plants that do suffer from grazing.

Focus on scale, focus on data. The preceding arguments have erected pairs of action: 1) choose a scale that is easy to use, or choose one closer to the scale of operation of the observed; 2) change the scale or the direction from which observations are made. Having chosen one in each of the preceding pairs, we are free to introduce yet another pair of options. Interest may then be focused on the structure that is observed by means of the message it transmits. The alternative is to normalize around the scale that is used in the observation. Most ecology is interested in how things appear. Ecologists know about scale but prefer to follow some established scaling protocol to see patterns. There has been more interest in scaling over the past quarter of a century, in part moved by the first edition of this book. In this century the study of the scales themselves has been elevated to a major discourse in its own right, metabolic ecology. While it was not called metabolic ecology in the first edition, that whole strategy of investigation was recommended as one mode of ecological research (a in Figure 12.1). The discussion in this new movement addressed animals, although the very earliest manifestations were in the growth of plants. The scales in animals in terms of size were related to animal metabolic rate. These studies were extended upscale to address colonies of insects, dealing with carbon dioxide production as a colony of animals gets bigger. Recently these scaling effects have been taken into ecosystem work, where arguments address the aggregate metabolism of the whole collection of organisms interacting with each other and their resources. Finally Jim Brown and others (2011) have taken the discourse to the metabolism of human society at a global scale in exactly the same terms as metabolism was considered in animals and their individual size. Sometimes the discussion there compares the metabolism of cities or other human settlements. Tainter and Allen recently compared increase in size and cost in insect colonies as opposed to the Roman Empire (a and c in Figure 12.1). The same phenomena appear to apply across biology and social systems.

Brown's approach does reflect his animal ecology roots. It is distinct in its own right, but represents a parallel line of investigation to that in economics and the development and collapse of human societies. Economists take a longer view of their systems than do biologists. Consider again a biosocial holon (Figure 2.3) with its planning element inside (DNA, hormones, regulations, and treaties). The holon is fed from the top end of a resource gradient. It takes in high quality energy or concentrated matter and degrades it. In doing so the holon extracts the capacity to do work. The gradient created inside the holon drives positive feedbacks that cause emergence, some of which was not part of the plan. Despite a plan, for instance to economize, the holon may still find itself running out of resource. In general, ecologists stop there and announce that in the end the resource will run out. Economists, by contrast, say that you do not run out of resource, it just gets more expensive. And they are right on that point. In the face of failing to economize sufficiently, the planning element is updated so that the holon becomes more efficient. That process of becoming is investigated in ecology much less often than it is in economics. Evolutionary discourse does let the narrative flow forward as the process of adaptation is a matter of becoming something else.

Tainter has addressed the process of diminishing returns on problem solving. Economists even have an acronym, EROEI, for economic return on energy investment. Return goes down as more effort is put in. There are two returns commonly recognized in economics. One is average return: how much return do you get on effort. The other is marginal return: how much extra return is there for the extra effort. Marginal return is some sort of first derivative of average return. The decline in both average and marginal return leads to a law, the law of diminishing returns. This amounts to focusing on the scale of resource use. As resources become scarce, more effort is required, giving always less for the extra effort. Tainter has analyzed many ancient societies in these terms, in particular Rome (C and c in Figure 12.1). He was, at one time, one of the more unlikely employees of the US Forest Service, in that Rome was one of his major fields of expertise. He appears unlikely until one remembers that the US Forest Service has a mandate to manage renewable resources. Tainter investigated a set of societies that failed in sustaining renewable resource use, to the point that Allen, Tainter, and Hoekstra applied those lessons to the modern human condition. Jim Brown[25] comes at the same issue but emphasizing the scaling issues that are embedded in the modern resource crisis. The large contemporary problems are scaling issues and need to be addressed in those terms.

The three binary choices give eight alternatives (Figure 12.1). In different circumstances, some modes are likely to be more profitable than others; the best way to solve many problems, however, rests in iteration between several options of scaling, point of view, and substance of the message.

Strategy 1: Convenient Simplicity of Scale in Field Data

The general usefulness of multivariate analysis in addressing ecological complexity comes from its capacity to allow precise changes in either the scale or the observation point. The data set is collected at given instances in time and is therefore reflective of a given observation point. The data points are collected through input filters and so are scaled. Nevertheless, the degree of aggregation of data or smoothing between data points is not fixed and neither is the portion of the entire data set that is considered in any particular analysis. These options therefore may be varied so as to scale the data set differently for each analysis. Nevertheless, the data set does reflect a fixed observation point. No matter how the data are subsequently scaled by data transformations and smoothing, the original data set is still reflective of a set of signals over a particular time period, at particular places, from a particular material system. That is to say, the data are reflective not only of the manner in which they were collected, but also of the place where they were collected and the behavior of the observed holon at the times of data collection. These unchanging features of the set fix its position with respect to the observed structure.

The smoothing and standardizing (transforming) performed on the data set subsequently vary the scale of the observations. By keeping the same data set while performing different standardizations, the observer holds the signal constant (i.e., holds the observation point constant) while varying the perception of that signal as the investigator changes the scale of the observations. Complete freedom in choosing the scale is not possible, for the influence on the primary signal of the manner of data collection is an unknown. The primary scaling operation in data collection is a limit with which the analyst must live. Nevertheless, the analyst has considerable freedom of choice in scaling decisions, which, when skillfully exercised, may provide deep insight.

Multivariate analysis of a data set of numbers of individuals before and after a binary (presence/absence) transformation is an example of holding the observation point constant while changing the scale. Usually the investigator is concerned with the behavior and structure of the observed holon rather than with the nature of data collection and transformation (scale). A data set of numbers of individuals is easy to collect and is scaled fairly arbitrarily to the size of the sampling unit and the method used. This adequate but inexact scaling with respect to the observed structures makes the approach described in the paragraph above (B in Figure 12.1).

Focus on the scale. Investigations directed at the scale rather than at the message received from the observed holon sometimes lead to powerful generalizations. If the same scale arises for different biological structures it indicates a class of environmental or structural problems for organisms in general. Patterns of signal weights associated with scales used by biologists successful in one field may indicate a line of least resistance elsewhere.

Papers about methods of analysis rather than particular biological results are generally concerned with scale and so exist on the AC–ac half of the cube in Figure 12.1. Data transformation alters the extent to which the data are integrated, and so data transformation amounts to a change in scale. Both Noy-Meir et al. and Austin and Greig-Smith applied various data transformations to a single test data set, but unlike the approach in the prior section they were concerned primarily with the nature of the scale, the data transformation. The transformations changed the interrelationships apparent in the data, showing how a given entity looks different through different filters. They used the difference in the apparent relationships within the set as an indication of the scaling properties of the transformation. Such and such a transformation weights rare species more heavily, while another scales stand productivity out of the data set allowing other features of the data to assert themselves. Through all this, the unchanging form of the original data set from which the transformed data are all derived fixes the observer position with respect to the observed structures. Austin and Greig-Smith used a data set of their own contrivance, so that having formed its structure themselves they had a very clear picture of the innate structure that the set reflected. Their concern, after all, was for the consequence of the scale of the observation rather than for any interesting structure detected in the act of observation. General interpretation of biological phenomena is facilitated by a knowledge of general properties of the commonly used observer filters.

Bradfield and Orloci were interested in the scale rather than the observed structure, and so belong in the AC corner of Figure 12.1. They used a discriminant method to identify the best compromise scale (vectors of discriminant weights) for the subsequent assignment of vegetation stands to four groups of a cluster analysis. Their concern was to find the scale, the discriminant functions, and to apply that scale to several unclassified stands. The new stands change the reference point with respect to the observed structure. This would place the work at C on Figure 12.1.

Palynologists trying to reconstruct ancient forests from the pollen record have a keen interest in scale. Independent of Koestler, Webb and McAndrews (1976) have realized that filters are an important component of scale. They write:

> Local anomalies in pollen distributions may hinder interpretation of regional distribution patterns. For any geographically distributed set of data, trend-surface analysis seeks to filter out local differences among samples and to emphasize the regional pattern.

The technique uses regression coefficients in a polynomial equation:

> Our choice of the degree of the polynomial is somewhat arbitrary. As the degree of the polynomial is raised, the resulting surface more closely fits the data and smooths out less of the local variation. . . . For this study, fifth-degree trend surfaces were calculated for 18 pollen types, and contours of these surfaces were plotted. The surfaces are sufficiently general to filter out the effects of individual sites and present a summary of the regional patterns in the data.

The problem they addressed is one of identifying an appropriate scale that maximizes signal-to-noise ratio. Noise associated with the pollen signal is from several sources: the grain number and rate of production of pollen per plant; pollen transport; deposition; preservation; and counting error. Presumably these incidental factors are fairly constant, or at least random, from site to site; accordingly they would not be expected to change in any orderly fashion between samples. Webb and McAndrews saw that they could maximize signal and minimize error by considering not the pollen grains at each site but the rate of change of pollen from site to site (cf. Margalef 1968, considering changes in characters of the surface across large water bodies; difference is a powerful device). The number of plants on the ground in the forest will relate to biologically meaningful considerations. The way that biology collates with the pollen core segment will vary with latitude and longitude. That is the signal of interest. This does need standardizing from plant to pollen counted, but interference from the standardizing factors can be minimized by comparing sites rather than considering the absolute number of grains found at sites individually. Thus they found the appropriate scale for their questions about vegetation. Their output was a scatter of sites represented by pollen from the Hudson Bay to Colorado with contours on the map from northeast to southwest indicating plant community types defined by trends not individual pollen counts. The random noise in pollen counts had been normalized out.

Concern for message. More commonly it is the biological structure that is of prime interest. Making comparisons between communities by using the same ordination technique on each, as in Curtis's (1959) *The Vegetation of Wisconsin*, is an example of changing the observation point while holding the scale constant. The primary object of interest is not the influence of the scale used by the observer, but rather the form of the holon associated with the phenomena under investigation. *The Vegetation of Wisconsin* is a case D in Figure 12.1 because Curtis used the same scaling criteria throughout his study of the state. He gained comparative insight into the vegetation by analyzing different community types. Each new data set represents a different option in time and space in the vegetation's inertial frame. Each data set is a new vantage or reference point with respect to the vegetation holon, the form and texture of which were elucidated in the study.

Allen and Koonce (1973) analyzed plankton to follow the community through the year. Like Curtis, we were interested primarily in the biological structure that their analyses might show. What form does the vegetation take? What is its message? While Curtis used the same method of analysis and transformation on different data sets, we chose to vary the scale by different transformations applied to just one phytoplankton data set. Our results indicated structure at two levels of organization, showing tactics as well as strategems of survival for the algae. It was the species groups that were of interest, so while Curtis performed a study of type D, we were working in case B of Figure 12.1. Had we simultaneously varied both the data set and the transformation, comparative interpretation of the results would have been substantially guesswork for want of a translational observation. Both we and Curtis maintained a constant in the analyses so the results were coherent.

Strategy 2: Precision

Description of communities using multivariate methods generally is performed for hypothesis generation. Test of those hypotheses often leads to experimental work. The multivariate field studies detect the phenomenon and approximately describe it, but interpretation of the phenomenon invokes assumptions that can be tested by experiment. Experimental results and their associated deterministic models belong on the lower case plane of Figure 12.1 because they invoke scales that are designed to be as close as possible to the scale of the biological system itself. Experiments are not good at system description because they focus on the narrow region of a knife edge wherein a fine distinction is being made. A common error in ecology is a race to experimentation before the situation is sufficiently described. Description lets the investigator know where is the knife edge that an experiment might address. One can describe generally or test specifically but mostly one cannot do both unless in sequence. Describe first, so you know what you are talking about.

Simulation models. Porter et al. (1973) used experimentation to calibrate a simulation model of the temperature balance of lizards. The tangibility of organisms allows good description without much resort to fancy analysis. For Porter the simulation model was a scaling operation. They varied the temperature and airflow to determine the functions of heat exchange; in doing so they were taking various options in the animals' temperature inertial frame. They plotted the responses of the model under different conditions and so were varying their observer position with respect to the biological structure. Their ultimate concern was with the behavior of the lizards, so the model (the scale) was merely a tool to determine the pattern in the lizard phenomenon. The scale was therefore not the primary concern, they rather wanted to focus on the biological structure. While they were looking for a fairly exact scale of the organisms, they used cylindrical animals as a workable approximation to an animal shape. They chose a form that was easier to model while still giving insights to the biology. They changed animals and so the observation point was varied to focus on the structure of the biological phenomenon.

Porter and Gates (1969) calculated the response of different-sized animals with different insulation properties to different combinations of air temperature and radiant energy. Applying these findings to reasonable field conditions, they were able to calculate the physical environment in which given real and hypothetical organisms could survive. These physical environments were graphed on dimensions of ambient temperature, radiant energy absorbed, and wind speed. On this graph the volume in which the animal can maintain a viable temperature is termed its "climate space" (see Figure 12.2). Climate space is shaded gray on the figure, and it determines the scale of an environmental holon that constrains the animal. An animal forced outside those constraints dies from heat or cold. Some desert lizards move out of their climate space to feed for short periods, but quickly have to return to their burrows before overheating.

It may be old but Figure 12.2 from Porter and Gates is relatively straightforward so the reader can see how the scaling works. The new work of Dudley, Bonazza, and Porter switches from C (for the early work on cylindrical animals) to c in Figure 12.1 as it become more elaborate and exact. With the massive increase in computation available, they use more realistic and dynamical animal models, and so the work is less transparent, even if it is more accurate. The principles of the new work are still the same with added dimensions for momentum and movement through the medium. These are nontrivial complications but the computation is up to it. Let us consider the cylindrical animals first for transparency. The space for Porter and Gates is graphed air temperature against radiant input. The general diagonal form of the figure comes from the way that radiant input is positively correlated with air temperature; radiation warms structures that warm the air. Higher air temperature causes structures to warm and radiate onto each other. The diagonal limits place the edge that is generally possible on the planet given the sun and the density of the atmosphere; in nature it generally cannot get more radiant than full sunlight for a given air temperature. The limits move a little for white as opposed to black animals, but not much so the

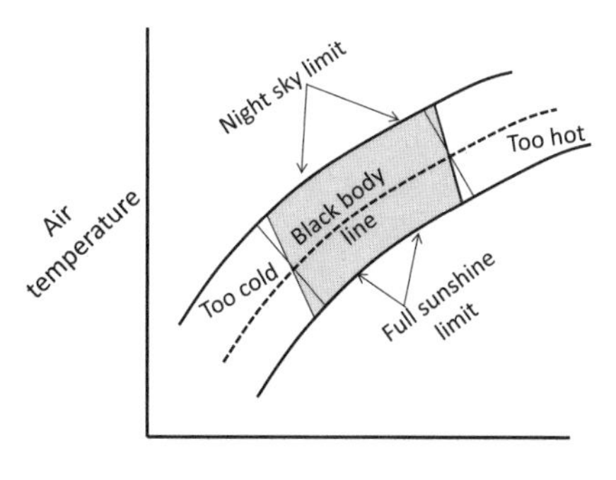

FIGURE 12.2. A stylized presentation of the climate space model where Porter and colleagues (1973) analyzed the thermodynamic possibilities for spherical and cylindrical "animals." The generally diagonal pattern comes from warmer air increasing radiation; in warmer air structures radiate more. The rhomboid presents the animal with possible options on this planet in this atmosphere. The criss-crosses at the top and bottom of the climate space offer corrections for wind speed, or the size of the animal with all else equal. In wind air temperature matters more than radiation. With all else equal, larger animals are more responsive to radiation.

biology can influence the main diagonal limits. The full sun line almost entirely amounts to limits on physical possibilities. The other diagonal limit up and left is the night sky; there is a lower limit to radiation input for a given air temperature when the background of the night sky returns next to no radiation to the radiating animal. The black body line lies in between, and is neutral on returning radiation of all wavelengths (as in a lizard in a burrow out of the sun).

The upper right limits on the rhomboid of the climate space indicate the upper limits for heat that the animal can bear. The lower left side of the rhomboid indicate the lower limits on the climate space for the animal. The upper right and lower left limits take a crisscross form indicating changing relationships between air temperature and radiation with increased wind speed. Wind makes air temperature more influential. The line of the crisscross that is less vertical indicates climate spaces where the animal is in higher wind, and so is more responsive to change in air temperature. That less vertical line also pertains to smaller animals, with all else equal. The less vertical line of the crisscross suggests smaller animals are more influenced by changes in air temperature than equivalent large animals. The more vertical of the crisscross lines pertains to larger animal indicating that they are more influenced by changes in radiation. The upper right limits moves higher for animals with large ears for losing heat (like jack rabbits and elephants). Such animals can live in a warmer environment because of cooling adaptations. Insulation with fat and fur extends the lower left limit of the climate space down into colder environments. Homeotherms have a larger climate space because they can manipulate core temperature with faster or slower metabolism. A lot of Porter's early work was with lizards and turtles with regard to conservation and climate change. Without homeothermy we can see the animals' ecology naked, without complications of adjustments of metabolic rate.

The early work of Porter used simple forms, like cylinders, to model animals. But now he collaborates with computer graphics designers to capture the details of mass, inertia, and growth for various animals such as leatherback turtles and elephants (Mansfield et al. 2014). His scales were never that simple, but they were always straightforward. But now modern technology and massive computational power let him scale his simulated creatures very close to nature. Dudley, Bonazza, and Porter say it all in their title, "Consider a Non-spherical Elephant: Computational Fluid Dynamics Simulations of Heat Transfer Coefficients and Drag Verified Using Wind Tunnel Experiments." In mode c on Figure 12.1, their work was so precise in its predictions that Porter can use deviations to recalibrate the models. Porter's website refers to this research:

> Our recent work with leatherback sea turtles establishes our globally unique capacity to create animated virtual 3-D fossil or living animals of any geometry from initial creation to landscape scale energetics, behavior and distribution limits. Our new collaborative paper [is] on tracking hatchling loggerhead sea turtles in the Atlantic and how their "riding" sargassum gives them "hot" islands in the ocean for faster growth and development.

Porter's newfound precision has allowed him to identify how important is clothing in expanding the human climate space. Hairlessness and clothes apparently give our species a big advantage.

By focusing on the scaling of temperature Porter's work gives fundamental principles that can then be applied across a range of animal examples. Yes, he uses different animals as examples, but his focus is always first on the scaling issues per se. In a sense Porter in different phases of his work moves around the cube of Figure 12.1, and we recommend that strategy of moving focus as it is informative. But it is clear that Porter's work is most characterized as he starts in sub-cube c in Figure 12.1, but then moves to d and a as his results indicate how to take advantage of his early findings. Often a sound strategy is to switch modes as the work proceeds.

Descriptive field studies tell the experimenter the approximate scales at which to design experiments. Experimental results tell the field-workers the scale at which they should collect data. As field-workers use the experimenter's results, they collect data in a manner that gives message closer to the scale of the biological system. Thus analyses give a more complete, simpler, and less convoluted picture of the structure and behavior of the material system. Each pass fine-tunes the analysis and description. As precision and sophistication find the scale closely appropriate to the material studied, explanations of phenomena become simple. Difficult problems are not often solved in one go. Moon launches split the journey into stages where each new stage corrects errors in the previous trajectory. This is now standard in solving complex problems.

Demography. Demographic studies look at the significance of the individual in population events. They tune to the scale of the individual with care and precision. In plant demography, studies on annuals, which are necessarily within one year, are often confidently executed as to the particular age of the plants. Once a plant becomes perennial, however, demographers become insistent that demography is hardly worth the effort unless the plants can be aged accurately. Difficulty in aging tropical trees (no annual rings) is seen as a debilitating problem. This is, however, a case of cultural blinding on the part of temperate ecologists. While the pulse of the annual cycle is clearly significant for temperate biennials, once a tree is a few years old winter is buffered and scaled out of significance.

There is, however, a refreshing school of perennial plant demographers who are not trapped in the cultural blind alley of the annual cycle. They age plants by life-stage. There are great advantages to a life-stage approach as developed by Rabotnov in Russia and employed by Carosella in the West. It represents one of the few cases where the biological data are easier to collect while, at the same time, the scale of observation used is close to that of the plant holon. The plants are described in terms of their stage in the life history rather than in months or years. Not only is it easy to see if a plant is juvenile, reproductive, senescent, or whatever, but that is also probably the way the plant principally views the situation. Sometimes plants can move from reproduction back to an "earlier" stage in preparation for another bout of reproductive effort later. Think what that does to data collected with a stopwatch and calendar! An insistence on a Julian scale would be disastrous. Life-stage analyses belong on the lower case plane of Figure 12.1. They aim to set exact biological scale and use the biology of human perception to facilitate the data collection. Large and robust size-class data sets for trees share the same advantages. How large is a tree is often of greater significance with respect to growth and competition than is age. That is not to deny all value to age-class data, but size classes, with their complex time-growth relationships already standardized in the taking of the data, are no poor cousin and need no apology. Whether size-class data belong on the upper case or lower case planes of Figure 12.1 depends on the biology of the plants.

While animal individuals are not usually ambiguous, in plants definition of the demographic unit can be problematic. We have mentioned the distinguished career of Grant Cottam and the Wisconsin school in forcing consistency on data collection.[26] Plant data collection sometimes uses density, in which case one needs to be able to secure an unambiguous number for individuals to calculate number of plant units per unit area. Grasses are particularly troublesome when it comes to identifying ecological units. There are rooted clumps of grass with variously connected grass stems called tillers. Grass individuals are often impractical to identify so density must be abandoned and substituted with frequency. Frequency divides the area to be sampled into smaller subunits, perhaps 100 in a square meter. Frequency skirts the problem of units by counting the subunit areas where the species occurs. The decision is then is any part of any plant of that species in or out; no need to bother with defining individuals. Frequency can be converted into density with a curve plotted, frequency on the ordinate, and density on the abscissa. The curved line asymptotes to high densities when frequency saturates at 100%.

In grasses a new terminology is especially useful. The single genetic unit is called the genet, while the morphological unit inside a genetic unit is called the ramet. The vegetative growing clump is the genet, while the individual growing shoot or tiller is the ramet. Kays and Harper attempted to identify the relative importance of the genet and ramet as units of selection and resource capture in a population of grasses. They wanted to find the scale for competition over time. They germinated seeds (genets) only to find that the primary units of competition were the tillers, or ramets, not genets. Thinning was in death of ramets. Only after an extended period did the significant unit of selection shift to the genet, as the whole plant died under competitive selection.

Parenthetically, the ruthless exploitation of leaves may be significant here; that is to say leaves that become shaded are left to starve once their energy capture falls below self-sufficiency. Perhaps the one-way flow of photosynthate defines stable entities at the scale of the leaf. This creates particular rate-independent units so evolution can calculate in defined units for controlled processes. Otherwise selection would be ambiguous and difficult. Evolution selects units at different levels, but it must have levels to be stable. We have noted that real and computational neurons work with discrete units. Rosen emphasizes the rate-independent units in a linguistic scheme underlying the use of models in the biology of biological systems. The same applies to social systems. One of Harper's main points throughout his long and distinguished career was that plants are composed of discrete subunits at various levels. Patterns of growth and competition are best considered as explicit hierarchies. All holons are arbitrary, but that excludes capricious holons such as a leaf and a half, which might confuse the selection process that determines resource allocation. It stops the control systems from generating into systems of infinite regress in a web of undefined feedbacks. These comments are relevant to the suggestions in chapter 7 that only those systems that have limited and discrete frequency characteristics can evolve at a pace that is displayed by life.

In the context of selection, the observer-defined levels of genet and ramet may be singled out as stable with the genet hierarchically constraining the ramet. The Kays and Harper experiment represents a single purchase point from which grass competition and thinning could be viewed. By considering the two hierarchical levels simultaneously, they were changing the scale of their observation in order to identify the scale of their phenomenon, competition and resource capture. They wished to identify the scale of their phenomenon in order to understand its nature and to describe it fully. Kays and Harper were therefore concerned with the biological structure, not the nature of its scale. Their focus was on the message, so they belong in portion "b" of Figure 12.1. Kays and Harper's work is an explicit thrust into scaling and hierarchy theory. Its genius is in working out there in 1974, before scaling and complexity were explicitly understood as a general phenomenon. Sometimes the early work is the best. In the same way that Curtis was not really a multivariate analyst,[27] Harper was not a hierarchist; they were both brilliant biologists putting it together as best they could. We skin the cat from the other end these days when complexity is treated explicitly.

Biologically General Scales

a) Surface to volume ratio is clearly a scale-related parameter. For a structural holon to communicate with its environment, information normally passes from the volume through the surface. Environmental intrusion and constraint must pass in the other direction. In fact, surface to volume ratio is a quantification of the vernacular meaning of scale. Large dinosaurs did not notice diurnal temperature fluctuations because of their surface to volume ratio, while mice do notice such changes precisely because of their small size. Lewis plotted the surface to volume ratio of a collection of microscopic tropical algae. He found that despite great variation in their longest dimension, the algae displayed remarkably constant surface to volume ratio. Clearly Lewis's investigations had scale rather than message as its primary concern, and it resides in region a and c of Figure 12.1.

b) Dimensionless numbers are created by having the units of measurement cancel each other out. Surface to volume ratio is not dimensionless because the surface is in squared units whereas the volume is in cubed units. Thus one can change surface to volume ratio not only by changing size, but also by changing shape. A dimensionless number relating surface and volume could be achieved, but it would invoke the surface cubed and the volume squared. The squared and cubed units would therefore cancel each other out, and shape and size would then be equivalent, giving a unified account of surface in relation to volume. Studies that seek dimensionless parameters are generally concerned with the nature of the scale rather than the form and messages of the biological structure under investigation. Parkhurst and Loucks in their search for abstract parameters of leaf form and function calculated the dimensionless term, water-use efficiency. Using this scaling parameter, they were able to give "predictions of trends in leaf size which agree well with the observed trends in diverse regions (tropical rainforest, desert, arctic, etc.)." Dimensionless numbers let you get at the nub of a scaling phenomenon. This work resides on the a/c regions of Figure 12.1.

May[28] in his treatment of lagged equations wished to achieve a unified treatment of the effects of r, the growth rate, and T, the lag. If the lag is made larger more can happen in that time. If r is made larger the equation will do more in the unchanged lag period. Thus both r and T contribute to equations becoming so responsive as to be chaotic. But r and T are in different units and so will not cancel. However, give an equation a unit disturbance and with a large r it will return quickly. The return time is called the relaxation time. With a small r it will return slowly. 1/r is a good approximation to the relaxation time. Now the lag T, and 1/r are in the same units, time, and so will cancel each other's unit out to get at the essence of the equation's functioning. Looking at the ratio of T and 1/r one can extract universal behaviors. If the lag is twice the relaxation time, any equation with that relationship will go into a two point stable limit cycle. With the value 2 on the ratio the equation cannot find K. It over- and undershoots K because the growth term can make it to K from below and then overshoot to the same degree. K bifurcates. As r gets bigger again, the stable values above and below K themselves bifurcate, putting the equation on a four point cycle. Eventually with increasing r the equation will become chaotic, never settling down, infinitely sensitive to initial conditions, and never repeating a value. Here dimensionless numbers are being used to get at the very core of a scaling relationship. Again this work resides on a and c of Figure 12.1.

c) Not all field studies use a rough and ready scale. Loucks (1962) realized that most presentations of vegetation on summary axes use quite arbitrary dimensions from the perspectives of the plants. Therefore he proposed the use of scalars (scales) that were synthesized from several environmental measurements. For instance, the water table in mesic environments does not matter so much, until it goes below a certain level. Suddenly a small lowering of water table needs to be scaled to matter a lot. It is a sort of non-Euclidean space. The synthesis paid attention to environmental experience from the plant's point of view. Using this approach, he was able to fine-tune his analysis to separate the patterns of white and black spruce, a distinction missed in the rough and ready, indirect gradient analysis of the Bray and Curtis (1957) method. Suspecting physical environment plays second fiddle to past historical accidents, Bruce McCune[29] had to be confident in the significance of his measurements. Needing to see exactly the effects of environment in the Bitterroot Mountains, he measured water not in physical terms but as an effect on plants directly. And he was right, the physical environment was not functionally different between northern and southern valleys, so historical accident can come to the fore.

d) The literature of creating scales to unify the functioning of animals was started by D'Arcy Thompson, who worked early in the twentieth century, taking his lead from Goethe. It was remarkable work considering he had insufficient computational power to solve the equations. As a result he used the power of a geometric graphic approach. Computer power started to become sufficient by the late 1960s. Allen's thesis was calculated on paper tape memory because he was only just over a decade or so after Turing. Early computing was very different. Von Neumann did the calculations for the atom bomb using business card sorters. The late 1960s made great strides in computation. Sacher, in a series of papers, works toward a single scale that would bring the calorific life span of animals to unity. He looks at many animals to find a common scale. He further elaborates his study by introducing different ambient temperatures and cybernetic control systems in his subjects. While the model (scale) predicts life span accurately for most warm-blooded animals, he finds an interesting exception. Human beings with their large brains live far too long by calorific standards. Indications are that the huge brain organizes things more efficiently, so the use of calories in humans is not as destructive. Even so, eating less does prolong life. The generality and power that comes from emphasizing the scale of a phenomenon rather than its biological message is found in Schmidt-Nielsen's (1972) work on animal physiology. He compares suites of animals' physiologies by bringing their relationships to a straight logarithmic line. He is able to develop unconventional but broadly unifying ideas on such diverse topics as respiration, temperature control, flight, countercurrent systems, and size. Heglund, Taylor, and McMahon (1974) were successful using a similar approach in unifying stride frequency and the break into a gallop from organisms as small as mice to large beasts like horses. This is the beginning of what has come to be called metabolic ecology. All of this works on the AC and ac side of Figure 12.1. Whether it is AC as opposed to ac depends on the convenience of the scaling employed. Whether it is a Cc or an Aa depends on whether the observed structures or the scales used are varied.

Studies in human energetics are difficult. Various leading scholars, who command our respect, have tried to work out human work in an hour. Giampietro (2004) finds that the joules calculated by these scholars differ by several orders of magnitude. It all depends on what we call work, and whether or not exosomatic energy is included, as when a human drives a car. His favorite example points out that there is no exergy (capacity to do work) in a gallon of gasoline if your mode of transportation is a donkey cart. As far as the donkey is concerned gasoline is at its physical dead state, where it is of no use. In biosocial systems we are dealing with the narrative and models actually possessed by the objects we study. Worse than that, biology characteristically changes the dead state. Yeast without free oxygen resorts to fermentation, where ethanol is the dead state that can do no more work. Yes, we can burn alcohol to flame a desert, but the yeast cannot. Alcohol is the dead state. But give the yeast free oxygen and it changes its story. It respires its sugar substrate all the way down to carbon dioxide and water, just as if it had burned ethanol.

So Giampietro now refers to the grammar of doing work. The environment work relationship demands a certain story. Change the environment, and the story must change, and it does. Change the free oxygen status of yeast and its story changes. Grammars fix the scaling parameters. Porter is almost always doing his calculations in terms of the first law of thermodynamics, the one that speaks of conservation of energy and matter. Giampietro is almost always dealing with the second law, the one that speaks of systems running down. Porter does not need a grammar for almost all he does because he is able to spell out the scale of the relationships he calculates. The grammar defines the scale of energy capture and use. Giampietro is lost without being allowed to rescale what is happening so as to get concretely different patterns of resource use. Porter is up against physical limits. So is Giampietro, but only after the biosocial structure has played out its changes of strategy.

There are subtle differences between energy and energy carriers. One can turn into the other with a change in the boundary of the system. For instance, grain is an energy source, whereas flour is an energy carrier. You can feed a horse with only the former. But you need the latter to bake a cake. Electricity is generally not an energy source, but is an energy carrier. This all presents practical difficulties in addressing energy policy and governance. Allen et al. (forthcoming) showed how there are practical difficulties that greatly complicate shifting from fossil fuel to renewables. The units are simply not the same and so are not additive or interconvertible. Fossil fuel is simply captured and burned. Renewable energy, like biodiesel, starts with an unsuitable source of energy, which must be processed and concentrated so it is turned into fuel. Fossil fuel is governed mostly by rate-dependent processes. Renewable resources are about rate-independent efficiencies.

Human energetic studies work are hard to rescale so that units are made sufficiently similarly scaled so they are comparative and additive. That firmly fixes such studies on the side of concern for scale (Figure 12.1 ac.) When Giampietro[30] works on energy in South Africa, water in India, or agriculture in Mauritius, he is changing the structure of interest (c in Figure 12.1). But in all these studies he is seeking the difficult scale used by the system in question so he is on the lower-case plane of Figure 12.1.

Model Organisms

Without noticing much of the literature of hierarchy theory, biologists with good instincts appear to have already used some of its strategies. Indeed, the leaders in most fields pull away from the pack by doing hierarchy theory in their field of expertise, perhaps unwittingly. Hierarchy theory is simply the technical version of common sense, something most intellectual leaders have anyway. We have discussed what to keep constant and what to change so as to keep track of the observers' movement around a biological hierarchy of observation: models, narratives, data collection, data transformation, data analysis, paradigms, and so on. The strategy of using model organisms holds all sorts of things constant in that web of scientific devices, just as Figure 12.1 would suggest. Model organisms generate continuity in a body of research embedded in a complex system hierarchy of which normal scientists need not be aware to do good work. Even rather dull straightforward data collection is suddenly made comparative. Model organisms have been a particularly successful strategy in genetics. One of the earliest using this device was Gregor Mendel in 1865 who used pea plants. Sea urchins were the favorite of late nineteenth-century developmental biologists. The suggestion that meiosis might be a device in Mendelian genetics was found in sea urchin research.[31] *Planaria* was another animal used in work on development. That is the flatworm that when sliced down its head simply produces a new head, one from each slice so as to make a Medusa cluster of heads. Morgan and Child were leaders in that work.[32] But around 1907 Morgan started to raise fruit flies to research the work of Mendel newly rediscovered in 1899.

Morgan was critical of the particularity of geneticists resuscitating Mendel. That will be because *Planaria*, his familiar model, is so flexible in its development that Mendelian genetics suggested something much less linear and rich. *Planaria*, unlike fruit flies, does not sequester its germline, so the distinction between somatic cells of the body and reproductive cells of the germline is not made. Development and genetics have a nuanced relationship in *Planaria*. But by 1910 Morgan reluctantly accepted Mendelian genetics and its particular rules as he published on sex-linked characters in flies. Much genetics has ridden on the back of Morgan's choice of the fruit fly as his model organism. As linkage, mutation, inversion, crossing over, and so many other levels in the genetic hierarchy, the fly remained the constant point of purchase so investigators keep addressing related questions. Had Morgan stuck with *Planaria*, genetics and development would now have a much more nuanced relationship. The emergence of epigenetics in Lamarckian evolution would have emerged so much sooner.

The importance of model organisms is how they situate the science on Figure 12.1. In particular the organism is a structure that focuses the scientists. It is easy to use the organism as a particular bead on the problem of inheritance. In a sense a model organism is a filter on genes, and the results of experiments improve the quality of the filter.

While the fruit fly has long been the model organism for of Mendelian genetics, molecular biology has more recently turned to *Arabidopsis thaliana*, a small member of the cabbage family. Molecular biology does deal with genes, but uses the model organism to link genes to development. There were earlier studies of the genus by those interested in its particular biology. Not until 1985 was it proposed as a model for molecular biology.[33] While genetics is interested in the behavior of the coding elements, molecular biology seeks to link the process of development to genes. The general strategy has been to knock out genes and see the effect on the phenotype. From that molecular biologists work out control of development. Another model organism is yeast, and a cautionary tale there is that the whole yeast genome has been knocked out one gene at a time, with 80% of them having no effect.[34] That is at a different level of analysis from genetics. Like the fruit fly and yeast, *Arabidopsis* moves through a generation in days and weeks, not even a whole year. So model organisms tend to be scaled for convenience of execution. Grant Cottam's father, Clarence Cottam,[35] defied such issues by starting oak hybridization breeding while in his 70s. We can rejoice that he made it to the F2 generation.

With the focus on scale and interest in the material, model organisms can appear in several places on Figure 12.1. In some cases the emphasis is on the model organism occupying a particular line of sight on a biological phenomenon. In other cases the model organism holds the relative scaling constant. Those constancies allow an ordered change of precision from the upper to lower case of Figure 12.1, as more specific questions are asked.

The Human Genome vs Directed Evolution

There was great celebration of the Human Genome Project, with even presidential speeches at the time. Complex systems analysts were always skeptical, suspecting it would not be that useful. We have already criticized diversity; it can go up or down with the loss of a species because of evenness. It is only unambiguous for the whole planet. The same critique can apply to the human genome project in its emphasis on the whole genome. Getting the whole human genome is a benchmark only in stamp collecting; very technical stamp collecting, but only a collection in the end. So we have all the genome (actually that is questionable)—so what does that mean? Notice the expansive claims for new insights about to arise have simply not come to pass. The critical issue is understanding the meaning of a part or the whole genome. Since it is only a collection of bases, understanding is lacking. This is reminiscent of Anderson's absurd claims for the end of hypotheses in science. In his paper, petabytes of data are mistaken for having captured the whole thing. Such celebration of the whole genome leads to extravagant claims.[36]

The Human Genome Project does not really fit onto Figure 12.1. So why does that matter; this is only a figure in a book? There is no suggestion as to what is being held as focal or referential, even operationally. Huge claims and few results arise because the scaling is intellectually unsophisticated. Do not get us wrong; technically getting it done was sophisticated. The technicalities of performing the human genome project with a certain chemistry have been difficult. We should be impressed with that. But having it all, the one whole genome, is not the critical issue. Other more focused comparisons have been interesting, such as identification of Neanderthal genetics, but the wholeness of the genome is not the issue there. The Human Genome Project makes the same old 1970s mistake of thinking we had a whole ecosystem in a computer. Remember the many feet of buffalo dung covering the Great Plains because of just rounding error in the PWNEE simulation.[37] The human genome project is large scale reductionism with the usual consequences. Remember our definition of reductionism notes that holists reduce too (an explanation found by holistic data analysis is a reduction). The difference is that reductionism presumes the validity of the reduction and the self-evident correctness of the privileged level of reduction. That sounds like the Human Genome Project to us. There is little exploration, mostly just assertion. As a result any importance of the project is happenstance, not by design; there is in fact very little design in the human genome. Getting it all is not much of a design. We suppose that knowing the whole genome does mean if what one searches is not there it is indeed not in the genome. But that negativity is very hard to prove with confidence.

In sharp contrast to the genome project is the new work on directed evolution.[38] And that is delivering powerful new enzymes through evolution. While the enzymes cannot be predicted ahead of time (neither can naturally selected evolution), there is confidence in the process of directed evolution producing more or less what is needed. The problem with much molecular biology is that its protagonists do not understand the critical role of evolution. Directed evolution does understand evolution, and in subtle ways. Whereas evolution by natural selection never aims at a solution, directed evolution can be more or less aimed, but not exactly. That has worked well enough in the couple of centuries of artificial selection of domesticated organisms. Directed evolution of new proteins can be expected to solve problems heretofore not yet conceived. While the Human Genome Project is static and structural, directed evolution is dynamic in that it produces functional structures. There is a filter in directed evolution, and that makes all the difference.

It is remarkable that there has been relatively little announcement in popular media of the singular importance of directed evolution technology. The scientists have noticed it, in that Frances Arnold, its champion, is the only person who is a member of all three technical American national academies.

Directed evolution as a strategy is well placed in the space of alternatives in scientific investigation. The evolution that is engineered is a change of scale. Evolution changes the filter. It makes the filter closer to the one we want. This is a move from the upper-case plane of Figure 12.1.

The move is to a finely adjusted scale and filter that better serves human needs. This is like the change engineered by Warren Porter, when he went from cylindrical animals to models of animals that look like the real thing, a non-spherical elephant and turtle .

If there is one technology that might save us (for a while) from declining resource and failed governance, it is directed evolution technology. Perhaps public health will not have to collapse with an epidemic of antibiotic-resistant tuberculosis. Agriculture fundamentally changed human existence, and public health (clean water, etc.) allowed cities to exist without disastrous diseases taking over. Of all contemporary science, directed evolution is a likely new fulcrum. We share the suspicion of social engineering as too dangerous to employ. But we might hope that developing systems of all sorts can be better grasped with a new understanding of directed evolution.

Bad science, or irrelevant expensive science, often arises because of a lack of an appreciation for scaling issues. Science is too expensive for us to continue to do it for more species one at a time. One more drug at a time has the same problem. We have mentioned Denis Noble's attack on the new synthesis. He is explicit that drug companies have been deeply disappointing in their progress.[39] Finding new incidental things would be nice, but such idle inquiry will bankrupt us. Gerd Gigerenzer makes something of the same point as he dismisses ritual hypothesis testing.[40] Every generation of scientists is bigger than all previous put together. To maintain this much increase in scientific activity, remarkably soon every man, woman, and child will have to be a scientist, so the limits are approaching.

Economics also appears to ignore limits. Mark Buchanan is a physicist who addresses how economists generally fail to accept the constraint of energy on economic growth. There are ways to look at efficiency like an economist, and that appears to make constraints appear workable. But Buchanan says there is one exception in economists' optimism and that is energy. He says

> Data from more than 200 nations from 1980 to 2003 fit a consistent pattern: On average, energy use **increases** about 70 percent every time economic output doubles. This is consistent with other things we know from biology. Bigger organisms as a rule use energy more efficiently than small ones do, yet they use more energy overall. The same goes for cities. Efficiencies of scale are never powerful enough to make bigger things use less energy.[41]

Tom Murphy is a serious astrophysicist at UCSD who studies general relativity. As a sideline he has a blog, Do the Math: Using Physics and Estimation to Assess Energy, Growth, Options. His most famous post is probably "Exponential Economist Meets Finite Physicist." It scripts a dinner conversation with him, named "a physicist," and a nameless economist. The dinner conversation really happened, but we presume so as not to be unprofessional Murphy identifies the economist only as "an established economics professor from a prestigious institution." The one thing that does map onto exponential economic growth is energy, so there is no weaseling out there. Murphy shows, with elementary graphing, that at 3% growth, in 300 years all sunlight hitting the Earth would need to be captured with 100% efficiency. At 400 years waste energy would boil the oceans. More pertinent limits to growth are clearly coming quite soon. Economists talk unconvincingly about decoupling energy and prosperity. The rational social scientist is Joseph Tainter, who speaks of societal collapse. His 1988 book says that most people benefit from collapse. After the barbarians were invited in by the Burgundians of the Roman Empire, taxes fell from over 60% to roughly 10%, and farming again became a possibility. The scale of the meaningful context is close and important.

Despite the mindboggling distances involved, there is some interesting, valid work being done on the signals from stars with planets passing in front of them. It seems likely that it will soon enough indicate an oxygen atmosphere for one of those planets, offering a tantalizing indication of distant life processes. All that is fine. The scaling of the faint signals involved is valid and impressive. But humans actually going to Mars is pie in the sky. The context is scaled to deny the science. Saving the species by living on Mars is a miscalculated absurdity. That is like the way planners fail to appreciate that fossil fuel is indeed finite. Getting to Mars with one person is close to impossible, and is certainly not worth the cost. It is telling that humanity simply stopped going to the moon in person. The context issue arises because the project is a holdover from the Industrial Revolution. In Victorian Britain problems were addressed with making a bigger engine dragging a thicker cable. It still works for bridges and the like, but really times have changed fundamentally. In the information age the rescaling characteristics are reversed. Things get smaller and faster. Automobiles and rotor tillers continue to increase in cost with inflation, because they have moving mechanical parts. But computers and smart phones get cheaper because they are made smaller and faster.

A rocket carrying a human to Mars is simply an oversized locomotive left over from the nineteenth century. It is a very big steam engine pulling a thick cable too far. You simply can't get to Mars by train, even at 20,000 miles per hour. There is funding for growing plants in space, presumably to do with creating renewable resources in a small contained space. Something interesting about plants in zero gravity may tell us about plants in gravity on Earth. But even if it is successful in solving plants growing in space, it cannot save the larger project. As a whole, manned Mars missions are set in a context that is so large that it will reach physical limits. Space travel is flawed by forgetting the size of the context. We can get clever at filtering the light signal from distant planets, but there we have the speed of light working for us, and we do not mind waiting many light years. But the movement of physical stuff, particularly human living stuff, many millions of miles is simply out of a viable context. The least problem is that astronauts go a bit crazy unless the plants they grow are seen in green light. Green light is no good for photosynthesis of the plants, but is crucial for the astronaut emotional health. Astronauts will do quite unreasonably unpleasant things because of their dedication, but apparently they will not eat algae grown in space. Not understanding the context in a hierarchy puts all at risk.

Conclusion

We mentioned Nick Lane's spectacular work on the critical stages in the evolution of life into its remarkably diverse forms. It fits right into Figure 12.1. It is so important because it investigates a hierarchy of cell and organismal functioning, and the scaling involved therein. Lynn Margulis[42] has been decades in unpacking the structure, functionality, and evolution of organisms. Her work gives breathtaking insights into the whole concept of organism, and must have influenced Lane. Lane's work is a deeply insightful look at the scaling of the hierarchy in eukaryotic cells.[43] Lane starts his journey with the bold statement that life is all about proton gradients working around membranes. The vehicle for advancing his argument is the mitochondrion, the organelle that deals with respiration, the source of the proton gradient in eukaryotic cells. Mitochondria are derived from free-living aerobic bacteria that were engulfed by host cells in a process called endosymbiosis. Instead of turning into simple prey, the engulfed bacterium came to live and divide inside the host cell. Billions of years ago it performed oxidation of food in respiration for the cell complex. Mitochondria turn cell capital into liquid assets for spending in biochemical functioning (remember Holling's panarchy and Odum's maximum power principle).

We have already pointed out that mitochondria have lost to the eukaryotic nucleus most of the genome of the original aerobic bacterium. We also mentioned that similar things happen in wholesale business takeovers. Business, like life, partitions information and functioning into various hierarchical levels of specificity and generality. There is Figure 12.1 all over this concept. The nucleus of the eukaryotic cell performs executive function, while the small amount of genetic material in the mitochondrion is solely for instructing the performance of the mechanics of cell energetics.

The hydrogen ion gradient is worked across membranes. In bacteria it is the cell membrane that that is the functional surface for doing work. This limits the size of the bacterial cell because the genetic instruction cannot be too far away from the genetic instructions to perform respiration. It is a scaling issue of process and structure, not just a matter of size. By compartmentalizing the genetic information, the eukaryotic cell escapes the constraint on the size it can achieve. The mitochondria have the critical information at the building site, while the cell's executive function can be coded and integrated in the main office. Mitochondria franchise respiration, making a huge profit for the whole firm. Mitochondria diffuse genetic information around the whole cell body.

Apparently there is no solution in simply replicating all the genetic information in the cell. Most bacteria remain limited in size by their one site of nuclear information constraining the chain of functioning. Bacteria have nuclear DNA but it is not membrane bound. Some bacteria do duplicate the whole genome, but that is so inefficient and dangerous that such bacteria cannot function with the stability, efficiency, and variety of eukaryotic cell. Duplicating the whole cell genome invites error and increases overhead. Lane's story is astonishing and compelling. We would argue that it is its explicit hierarchical argument. Lane's work jumps all over Figure 12.1. It is impressive that he could do what he did without being a hierarchy theorist. Figure 12.1 tells how to perform like him.

A useful feature of the proposed scheme is that it draws the investigator's attention to exactly the investigative mode in use. There comes a point in every investigation when the approach seems to have yielded all it can; the scheme proposed here gives a clear indication of the appropriate new approaches at such a time. Notice how Porter reached his limits with cylindrical animals and so made them more realistic. That was more difficult to model, but he now has more powerful computation and graphics to do it. By its either-or construction, Figure 12.1 leads to cases "a" and "c." Those are the sectors where the nature of the specific natural scale is investigated. Engineers use this mode in problem solving as a standard procedure, and testaments to its power abound in bridges, skyscrapers, and engines. It is our impression that interest in the scale itself, when trying to achieve the scale of the natural system, is often overlooked because few people in biology or social science think in that fashion. Engineers constantly use graphs of dimensionless parameters, and ecologists can benefit from the lesson. Since replicates concur when the scale of the observation matches that of the observed holon, it is not necessarily difficult to determine the natural frequency of scale of a biological system. Therefore modes "a" and "c" are available more often than it may seem.

Engineers are careful so that scale changes do not create emergent properties. The Tacoma Narrows Bridge collapse is an indication of just how careful they must be. Accordingly engineers tend to overbuild for the sake of safety. In biology scale changes cause emergence all the time because biological levels are close packed and continually bump into emergence of higher levels. Hegelian shifts of quality into quantity are common in biology and social science. As a result, ecological engineering calculates to fail, but fail in such a way that it is contained and salvageable. Beavers build dams always as if they were temporary, because sooner rather than later a flood will destroy it. Thus biology responds often not so much to disturbance as it does to the environmental regime. When scaling can be expected to go awry the best strategy is to deform and be resilient for the bounce back. By encouraging events like fire, fire-adapted vegetation keeps fire small so that the fire-adapted plants can ride with the disturbance. By changing the scale biological material often changes the rules and grammar by shifting the pertinent level of analysis and functioning.

Another advantage to the investigative scheme in Figure 12.1 is that it requires the investigator to decide whether it is the scale or the observation point that should be held constant as a fulcrum in the study. At the time of the first edition scale had theretofore been virtually ignored, and had often not been held constant. That need not be a problem unless the position of the observer is also changed. If both position and scale vary, comparison between results is impossible unless the appropriate translational experiments are performed. Now this second edition sits in the aftermath of a great deal of sophisticated attention to scale, so this edition brings a unification between scale and complexity theory. The papers cited here are either new or old. The older papers cited as examples show that this is a timeless set of issues bravely addressed by heroes working before the convenience of the twenty-first century. D'Arcy Thompson, John Harper, and Peter Greig-Smith all had grit. The recent papers show how the contemporary focus on complexity is underpinned by the scheme we have erected, whether that has been noticed or not. We have laid out a road map for treating complexity scientifically.

The issues raised here are not arcane and obscure, even if they require a distinctive posture. Hierarchical levels appear in conflict in many arenas in ecology and social tension. They certainly arise as ecology tries to be useful in addressing ecological regulations. Students have often come to us for advice as to what profession to follow so as to contribute to ecological good as humanity throws its weight around the planet. It depends on the students' natural talent, with words, numbers, or whatever, but one frequent response is to recommend they go to law school. Humans act on the world of nature mostly through the law. Former students of English or ecology or whatever come back to our lab and say that law students are first taught how to write a different English so as not to commit unnecessarily to a position too firmly. Legal experts do not want to write a judgment that inappropriately imposes itself when the social context changes. The problem is a change in level. Even so those who write regulations still write contradictions.

While complexity theory may seem alien, let us introduce someone who would not call himself a complexity theorist, but who found it exactly what he needed. A colleague, Kenneth Raffa at University of Wisconsin from Entomology, gave an informal colloquium on February 23, 2016, "Endangered Species: A Lens into How We Prioritize Values, Resolve Conflict, and Apply Ecological Understanding, through Various Slices of Time." His PowerPoint presentation, that he kindly copied to us, supports the account below. Raffa noted that the Endangered Species Act of 1973 prohibits any person from "taking" federally listed threatened and endangered species. A "take" is "any harassing, harming, pursuing, hunting, shooting, wounding, trapping, capturing, or collecting." There are conflicts because management for the species or other economic interests does destroy individuals. The law is explicit that "a take is a take" and that leaves insufficient room for adjusting conservation activity. Experimental approaches are permitted where the "metapopulation is large enough that failure does not reduce total population below minimum viable metapopulation."

The endangered Karner blue butterfly depends on lupins for food. The lupin is a pioneer species and will persist only in a disturbance regime. Fort McCoy, Wisconsin, is used by the US Army for tank maneuvers. That increases lupins, making Fort McCoy the place of the largest Karner blue population anywhere. The road sign for tanks came to be taken as a sign for the butterfly. Within the current metapopulation plan, the disturbances they create are welcomed by the Karner Blue Recovery Team, so the Army doesn't really find itself in a dilemma. The dilemma was mostly experienced by woodlot and forest products managers, prior to the current plan and under the Endangered Species Act, because their cutting actions were deemed a "take" in that they would likely kill individual organisms. And yet their actions would open up temporary clearings that favored lupine and hence benefited the Karner blue populations. Preserving (or not disturbing) individuals can deplete populations. It is much more resilient to have a shifting mosaic than one where the mosaic is put into a sort of museum. Variety in species goes down when an area is protected; so generalized conservation is usually bad for rare species.

A dilemma for the US Army at Fort Bragg is that it has bombing ranges that are disturbed and uninhabited by humans. These areas turn into precious places for rare species, so the Army has to protect them. The Army has to acquire adjacent land to compensate for their activities in the precious areas their mayhem creates.

Raffa made similar cases for grizzly bears management and climate change in the greater Yellowstone ecosystem, and for the monarch butterfly. The tension that arises makes strange bedfellows. Traditional Westerners and energy CEOs often align on immediate controversies; although they are not necessarily kindred spirits. Defenders of Wildlife established the Grizzly Bear Compensation Fund in 1997. It paid $400,000 in Montana and Wyoming as of 2013 to compensate ranchers. There are many dimensions to the issues: biological, socioeconomic, and political. And Raffa thinks these controversies provide a lens into how we might prioritize values, resolve conflict, and apply ecological understanding, and most importantly how the template on which this plays out changes as our institutions and values and science change over time. Our ideas generalize Raffa's concerns and arguments. Change the scale, and the phenomenon is likely to change. Population preservation can challenge community processes.

We said our ideas are not arcane. Allen came into complexity science in the early 1970s, when he could not otherwise unscramble issues of algal communities he was studying. Raffa dealt with his points of tension in conservation and ecological action by moving in a similar way toward scaling and changes in level of analysis. Our point of the general pertinence of hierarchies lines up with Allen having had no training by 1972 in complexity theory. And Raffa, forty-five years later, is moving into a posture of narrative, complexity, and contradiction, although he would view himself as a fairly regular entomologist who spends most of his time doing fieldwork and experiments on insect populations. He looks like a normal entomologist. He shows that this book is written, not for arcane holists, but for fairly normal scientists who cannot easily solve the problems on which they work, and who need a change in strategy. The pressing ecological, environmental, and business societal issues are coming to visit the lives of natural and social scientists, whether they like it or not. This book has been written to help in the transition that must be made.

We see scale and hierarchy as crucial if ecology and other natural science is to take advantage of the tools of quantification developed over recent decades. We find ourselves increasingly troubled with the state of the art. We have an intuition that this is the time for a break with the status quo. We have tried to weave the creation (who is creating whom we are not sure) back into the fabric of more conventional ecological and social science thought. Mere bright ideas add little of value.

If this book has been hard going in the reading, we promise it was no easier in the writing. But think back, how much easier could we have made it? We contend, not much. Intuition led us to some strange places, and we have done our best to explain how we got there. We realize that our treatment is probably cumbersome and it is certainly primitive. Perhaps we read like men possessed; we feel better after the exorcism of writing the book. We feel this book was written because the times imposed it upon the authors at the time of the first edition, and even more so now. There is a paradigm shift toward complexity occurring in biology, and this book is one of the manifestations of that change.

Acknowledgments

As in the first edition, we would like to thank the anonymous reviewer for rejecting a short paper that was submitted to *Science*, thereby getting the first author so mad that the paper was expanded to this book. The role of peer review in science is complex. The manuscript of the first edition was heavily criticized through many drafts over about five years, and we are grateful to friends and critics who not only took the time to give those early arcane drafts a complete and careful reading, but articulated their difficulties in understanding what was and was not there. Sometimes they understood only too well what we were saying, and it was wrong and they said so. In their approximate order of appearance we thank them for their firmness: H. H. Shugart Jr., Robert Friedman, Robert P. McIntosh, John Sharpless, Robert M. May, R. V. O'Neill, and Paul Nicholson.

Allen sees some of these ideas as one of the possible syntheses of his earlier training by J. L. Harper, P. Greig-Smith, and N. Woodhead, all lost to us now. We hope that they approved of some of what is herein. Others who have been influential, again in a general chronological order, are W. W. Sanford, S. M. Bartell, W. M. Post III, O. L. Loucks, W. S. Overton, and H. H. Pattee. "Mac" Post dates the time when things began happening because of a seminar given on the UW campus in the mid-1970s by Richard Levins. We are sure Levins played a role in breaking things loose. The writing of Gerald Weinberg has affected us both, and Starr feels especially grateful for the writing of Gregory Bateson and the way that it has rekindled his enthusiasm for this sort of work.

The second edition was set off by a graduate seminar in 2009, where Allen's graduate students and others worked to redirect the first edition: Megan Pease, Devin Wixon, Julie Collins, Peter Allen, Marc Brakken, Edmond Ramly, Nate Miller. Many of the new ideas for the second edition came from Mario Giampietro. The body of the new work that appeared in this millennium was in collaboration with Amanda Zellmer and Kirsten Kesseboehmer, when they were undergraduates.

The preparation of the manuscript in first edition, much of which survives in the new edition, was a long business and would have been much longer but for the tireless assistance of K. Graven, J. Liess, M. Weber, J. Arnold, S. Wasilewski, M. C. Currie, and S. Mudd. Special thanks go to Martin Burd, who listened to and criticized many of the thoughts presented and has continued to contribute between editions. He played a major role in selection of, organization of, and obtaining permissions for figures in the first edition, many of which reappear here. Lucy Taylor prepared many of the line drawings, and we are happy that we left so much to her judgment.

We thank John Norman and Devin Wixon for being central in an unpublished manuscript on medium number systems.

This work has been supported by grants from the Wisconsin Alumni Research Foundation to Allen. WARF is particularly generous in its support of preliminary work outside prevailing paradigms. This adventurous approach to funding has been of critical importance in the early development of much that is now here. Award DEB 78–07546 from the National Science Foundation to Allen was important in its support of the early part of this project.

Notes

Introduction

1. Petroski 2014.

2. Pound and Clements 1901. A deep detailed account of prairies of Nebraska.

3. Ahl and Allen 1996.

4. Weinberg 1975, and new edition 2001.

5. Ken Boulding made the statement on many occasions. One of them was in a plenary address to the Society for General Systems Research meeting in New York City.

6. Box and Draper 1987, p. 424.

7. Pattee 1978.

8. Overton and White 1981.

9. Coleman 2010.

10. Blodsoe et al. Published 1971 and formally in 1978 as a technical report

11. Bartell 1978.

12. Hector et al. report large diversity experiments with some 30 authors. After having failed because of sampling error the group kept coming back over the next decade, but it can be dismissed as ecological politics.

13. Rosen 1981.

14. Bonilla et al. 2008, p. 1751.

15. Funtowicz and Ravetz 1992.

16. Ravetz 2006.

17. Seneff 2014. Stephanie Seneff's home page includes several pertinent articles in pdf, some formally published, sometimes tenuously, others not published, but still important; http://people.csail.mit.edu/seneff/. She comments about the seriousness of the situation. "Should the disinformation [generated by the medical establishment] be overcome, there may be massive, chronic liability, akin to the mess left after asbestos."

18. Bakken et al. 2011.

19. Bertalanffy 1968.

20. Rashevsky 1938, Rosen 1972.

21. Rosen 1991, 2000.

22. Louie 2009.

Chapter One

1. Prigogine 1978; Prigogine et al. 1969; Prigogine and Nicolis 1971.

2. Piaget 1971. Piaget's constructivism notes that the world does not inform directly, but rather informs by building on interactions of the observer with externalities.

3. Michel Lissack (2015) suggests that the two levels of realism are not at odds but are instead orthogonal.

4. Reviewed in Levandowsky and White 1977 and Levin 1992.

5. Geneticist and mathematician Alan Owen wrote that Fisher's 1925 *Statistical Methods for Research Workers* occupies a place in quantitative biology as does *Principia* of Newton in physics.

6. Leeuw 2014 reports "Hotelling's introduction of PCA follows the now familiar route of making successive orthogonal linear combinations of the variables with maximum variance." It is suitable for computer coding. Hotelling 1933 gets a lot of the credit for Principal Components Analysis, and is the standard reference, but Pearson 1901 predates him in a tedious presentation.

7. Simon 1973.

Chapter Two

1. Overton 1975; Patten 1975; Patten and Auble 1980.

2. Margalef 1968.

3. Alfred Korzybski coined the expression in "A Non-Aristotelian System and its Necessity for Rigour in Mathematics and Physics," a paper presented before the American Mathematical Society at the New Orleans, Louisiana, meeting of the American Association for the Advancement of Science, December 28, 1931. Reprinted 1933 in *Science and Sanity*, 747–61.

4. In a 1998 impeachment hearing of President Clinton for lying about an extramarital affair he answered that he was not lying or contradicting himself because it depends on what *is* means. Footnote 1,128 in Kenneth Starr's report, 2004.

5. Rosen 2000, pp. 256–57.

6. Noble 2010, 2013a,b,c. This raft of work shows how genes are not the singular source of the phenotype. A hierarchy of physiological levels intercedes.

7. Ulanowicz 1997.

8. Rosen 1985. A new edition was published in 2012. The 1985 edition was withdrawn and is generally not available.

9. Zadeh 1965.

10. Giampietro 2004. This is Giampietro's PhD dissertation.

11. The Krebs Cycle was the first biochemical cycle to be recognized. It was found when citric acid, which closes the loop, was seen to increase the time the system remained functional. Citric acid was the last missing strong connection within the cycle.

12. Pier Gustafson now specializes in hand made calligraphy for wedding invitations and formal presentations. But he remembers fondly the days when he would sculpt objects out of inked paper. Images of "Father's Suitcase" being refurbished at the Fine Arts Museum San Francisco can be found at https://www.famsf.org/blog/trompe-loeil-traveler.

Chapter Three

1. Schmandt Besserat 1992; 1996.

2. Juarrero 2002.

3. Mandlebrot 1967.

4. Patten and Auble 1980.

5. J. Forrester was an early computer engineer who championed the move from analog to digital computers. Forrester was responsible for systems dynamics in the 1950s. Forrester diagrams link box and arrows (stock and flow) symbolically with a set of symbols for transfer of energy, matter, and information rather like electrical circuit diagrams. They are used in many situations in industry and social networks. An early classic is Forrester 1968.

6. May 1973a,b.

7. Barclay and Van Den Driessche 1975.

8. Widrow et al. 1975. Dolby sound looks for the creative pattern in past signal (music) and lets sound like it through. It denies passage to the background hiss of noise.

9. Giampietro 2004.

10. McLuhan and Fiore 1967.

Chapter Four

1. *Webster's New World Dictionary.*

2. King James Version, Genesis 1: 27

3. Thomson 1975.

4. The lab of K. Guo has done much work on this. Guo et al. 2009; Racca et al. 2010, 2012. See also Albuquerque et al. 2016.

5. The work by Grove was a class project for Allen. Having always lived in the postmodern work, she did not see her creation in postmodern terms. It was simply a successful class project for her. She was almost bemused at the enthusiasm with which it was received. Allen's colleagues were most taken with it and lined up to buy their reproductions.

Chapter Five

1. May 1973a.

2. May 1976a.

3. Hierarchy of life images are almost universal in biology and ecology general textbooks. They almost all have one. A typical example is Shutterstock's Stock Vector Illustration: The hierarchy of biological organization. Image ID:126853934. There is usually a series of illustrations of the levels. Many speak of increasing complexity, which is simply a misunderstanding of complexity. Usually it means a muddled increasing size, in general, and degree of complicatedness with reference to some arbitrary human primary perception. It is in fact a definitional hierarchy, where the definitions of the interrelationship between levels force some sort of aggregation. It is not a scalar hierarchy, but an ordering of types of system on the basis of aggregation. It is not wrong (anything arbitrary is neither right nor wrong), but it is also not general in the manner intended. The muddle comes from its implied generality and it being a narrative. Very little of generality in life is given account in such an organization.

4. Robert Bosserman 1983.

5. Emanuel at al. 1978 performed a spectral analysis (Platt and Denman) that supports notions of the general substitution of trees, like chestnut by tulip poplar.

6. Allen and Hoekstra 2015.

7. O'Neill et al. 1986.

8. Patten and Auble 1980.

9. Haefner 1980.

10. Hutchinson 1957.

11. Haefner's account of the niche almost maps onto the concrete and abstract conception of the plant community, the material thing versus the conception.

12. Ulanowicz 1997.

13. Southwood 1980.

14. The critique of orthopedic surgeons comes from Steven Levin (1986), a recovering orthopedic surgeon. For a conversation on tensegrity with Levin see https://www.youtube.com/watch?v=8ajowL0T4bM. His website on Biotensegrity is most rewarding. http://www.biotensegrity.com/.

15. Polanyi 1968.

16. Simon 1962.

17. Dodds 2009.

18. Eigen 1977.

19. Eugene Cittadino personally communicated the story of his father, the watchmaker.

20. Layzer 1975.

21. Spencer 1864, p. 444. Spencer first used the phrase after reading Darwin. Darwin wrote to Wallace that "survival of the fittest" might be a good alternative for natural selection, and he used it in *The Variation of Animals and Plants under Domestication* in 1868, and in the 5th edition of the *Origins*.

22. Wade 1977, 2016.

23. Vargas 2009.

24. Ahl 1993.

25. Noble 2010, 2013a,b,c.

26. Vavilov 1926.

27. Tainter and Allen 2015.

28. Allaby 2010; an entry in *A Dictionary of Ecology* coined by John Harper.

29. Linderman wrote Plenty-Coups' autobiography. First edition is dated 1930 (cited by Cronon), but a new edition was published in 2002.

30. Rosen 2000, p. 257.

31. Guo et al. 2009.

32. Rosen 2000.

33. Allen et al. 2014 developed these arguments to their full expression.

34. Rosen 1979.

35. Connell 1980 suggests that what is taken as competition is usually only the physical leftovers of the resolution of a competitive situation that has been resolved or ameliorated in the past.

36. Lovett 1982, p. 104.

37. Lovett and Sagar 1978 showed bacteria interfering with allelopathy. Sagar and Ferdinandez 1976 showed allelopathy could be blunted with nitrogen inputs.

38. The insights into allelopathy come from Geoff Sagar in a personal communication to Allen at the BES meeting at the end of the millennium in Liverpool in an evening session on allelopathy.

39. Polanyi 1962 speaks of tacit and focal attention. Needham 1988 spells out how Polanyi's ideas apply to literary criticism.

40. Social Darwinism has been used as a justification for class and racial privilege, an ugly business. E. O Wilson has received much just criticism for his tome on Sociobiology, where he tries to explain human and biological sociality in terms of natural selection.

41. Gustafson and Cooper 1990.

42. Tainter 1988; Tainter and Allen 2015; Allen et al. 2001.

43. Freundenberg 1992; Freundenberg et al. 1998, Freundenberg and Gramling 1998.

Chapter Six

1. Carneiro 2000; Moritz 1854.
2. Simon 1962.
3. Simon 1973.
4. Eigen 1977.
5. Simon 1962.

Chapter Seven

1. Dawkins' *The Selfish Gene* is a caricature of the new synthesis and central dogma.
2. Noble 2010, 2013a,b,c.
3. Waltner-Toews et al. 2008.
4. Radjou and Jaideep 2012.
5. Peter Day on BBC Global Business July 4, 2005, was told by his boss to go to Africa and find something good to report. The anchor reference was Prahalad 2004. The BBC program can be found at http//news.bbc.co.uk/2/hi/business/4648094.
6. Vavilov1926, 1950.
7. Leonard, Gaber, and Dick 1978.

Chapter Eight

1. Lovelock 1989.
2. Hutchinson 1961.
3. Connell 1980.
4. Character displacement is one manifestation of the ghost of competition past. The inferior competitor shifts its characteristics so as to get out of the other competitor's niche. Joan Roughgarden, with Steve Pacala, showed character displacement whereby lizards on islands invaded by large lizards became smaller.
5. Bertalanffy 1975.
6. Pringle 2008.
7. Simberloff 1983. While waiting with Allen in Baltimore-Washington airport, Simberloff said he thinks the ideas in that paper got him his reputation as the "bad boy of ecology," because he has never been forgiven for that fracas.

8. Allen and Hoekstra 2015.

9. Allen and Holling 2008.

10. Sandom et al. 2014.

11. Cooper et al. 2015; Willerslev et al. 2014.

12. Willerslev et al. has 49 authors. We are reminded of the scientific political movement for diversity in ecosystem by Hector et al. 1999 and other papers with huge numbers of authors in the following decade. Many authors means politics is involved.

13. If a biome disappears under climate change a return of the climate usually does not mean a return of the vegetation. Change in biomes is often a ratchet that only moves forward. The climate partisans are probably closer to the truth but their argument is weakened by presumption of returning vegetation. These arguments were developed by Thomas Brandner 2003, and an update personally communicated on June 6, 2016. Metcalf and her 24 coauthors find humans arriving 1,500 years earlier than the extinction. The extinction occurs as the climate gets warmer. They argue that humans could not do it until the climate was right, but they are reluctant to say so.

14. Tainter 1988.

15. Wiens 1977.

16. Lotka 1956.

17. O'Neill et al. 1977.

18. Bartell 1978.

19. Andrews 1949.

20. O'Neill, Johnson, and King 1989.

21. Lane 2005; Lane and Martin 2015. Breathtaking talks are listed at Lane's website: http://www.nick-lane.net/Nick%20Lane%20Talks.htm.

22. On May 24, 2016, Dr. Janina Ramirez hosted a BBC World Service Forum, *After Dark: How We Respond to Darkness.* Her guest, Dr. Ravindra Athale, of the Office of Naval Research in Arlington, Virginia, considered night vision as reported here. http://www.bbc.co.uk/programmes/p03v86x1.

Chapter Nine

1. Linteau 1955.

2. Loucks 1962.

3. Donald DeAngelis, personal communication at Oak Ridge National Laboratory.

4. For an extended discussion of scaling in data transformation see Allen and Hoekstra 1991.

5. Greig-Smith in the third edition of his *Quantitative Plant Ecology* gives a

particularly clear account of the geometry of different scaling devices in plant ecological data analysis.

6. Austin and Greig-Smith 1968; Allen 1971.

7. Allen and Skagen.

8. Patrick, Hohn, and Wallace 1954; Patrick 1967.

9. Margalef 1968.

10. Allen, Bartell, and Koonce 1977.

11. Magnuson 1988.

12. Allen and Wyleto 1983.

Chapter Ten

1. Mandlebrot 1967.

2. Rosen 2000.

3. Gardner and Ashby 1970.

4. Holling 1959a,b, 1965.

5. Holling 1973.

6. Holling 1986.

7. What came to be called *panarchy* appears in Holling 1986. Eventually a whole book appeared in 2001 by Gunderson and Holling.

8. Besteman 2002.

9. The story of the clearcut and short and long term recovery of Hubbard Brook was covered over decades and continues. Over time new wrinkles occur. Bormann and Likens 1979; Bormann et al. 1974; Likens et al. 1967, 1977, 1978.

10. The latest statement of the maximum power principle of Odum was with Brown in 2007.

11. Allen and Holling 2008. The book defends the argument by triangulating on the issue. Former critics ended up supporting the argument in the book.

12. Guo et al. 2009.

13. McLuhan 1964; McLuhan and Logan 1977.

14. Logan 2004.

15. Foley 1999.

16. Pattee 1978.

17. Boyd and Allen 1981a,b.

18. Gigerenzer 2012, 2013; Gigerenzer and Brighton 2009; Gigerenzer et al. 2007; Gigerenzer and Galesic 2012; Gigerenzer and Goldstein 1996; Gigerenzer and Muir Gray 2011; Gigerenzer and Sturm 2012; Gigerenzer and Wegwarth 2013.

19. Sokal 1996; for the context, Sokal 2008 *Beyond the Hoax* gives other examples. But the *Times* review by Robert Matthews on March 13, 2008, says the book "fails to reflect the fact that Sokal's concerns are now widely shared—and

that progress is being made in addressing them, the emergence of evidence-based social policy being an obvious example. His critique would also gain more credibility from encompassing his own community: the failure of scientific institutions to address the abuse of statistical methods or promote systematic reviews is no less of a threat to progress than the ramblings of postmodernists or fundamentalists," which is more or less our point. Saltelli et al. 2016 do what Sokal fails to do.

20. The conversation cited was published in *Science* 1977 under the signature CH. C. Holden is a staff writer for the journal (see reference list).

21. Shugart and West 1977, 1979.

22. Holling 1959.

23. Noble 2010, 2013a,b,c.

24. Bakken et al. 2011. The large number of authors on the paper is a sign that the field is in political conflict.

Chapter Eleven

1. Rice and Westoby 1983a,b.

2. Hector et al. 1999 was the original paper. Huston et al. 2000 showed their statistical flaw. Hector kept coming back over the next decade, losing et al.'s all the way. Our attitude is that Hector et al. had their chance and badly missed it, and do not deserve further attention.

3. May 1981, p. 219.

4. Margalef 1968.

5. MacArthur 1972 looked at quality of connectedness, while Levins 1974 was quantitative.

6. Hutchinson1959; Allee et al. 1949.

7. May 1973b; Levins 1974.

8. May 1973a.

9. Maynard-Smith 1974; May 1973b.

10. Allen and Hoekstra 2015.

11. Murdoch 1975.

12. Grime 1974, 1977, 1985, 2001, 2007; Grime and Curtis 1976; Grime et al. 2007; Grime et al. 1997.

13. Roughgarden and Pacala 1989.

14. Hutchinson 1961.

15. Riley 1963.

16. MacArthur 1972.

17. Murdoch 1975.

18. Shepard 2013.

19. Jackson and Jackson 2002; L. Jackson 1999.

20. L. Jackson 2008.

21. May 1976a.

22. Zhu et al. 2000.

23. Zimmerer 1998.

24. Guest 2007.

25. McSweeney 2007.

26. Giampietro, Allen, and Mayumi 2006.

27. The talk was given to the Environmental Sciences Division of ORNL in the summer of 1983.

28. Middendorf 1977.

29. Levins 1974.

30. Middendorf 1977.

31. C. Simon 1975.

32. Stager 1964.

Chapter Twelve

1. For directed evolution see Arnold 2009. For the critical role of mitochondria see Lane 2005.

2. Rosen 1991, 2000.

3. Rosen 2000.

4. Allen et al. 2014.

5. McMahon 1975.

6. T. Allen, personal observation 1969.

7. New Mexico Museum of Space History website. http://www.nmspacemuseum.org/halloffame/detail.php?id=68.

8. Goodall 1953, 1974; Greig-Smith 1952; MacArthur 1967; MacArthur and Pianka 1966; MacArthur and Levins 1964; Patten 1975; Root 1974.

9. Noy-Meir, Walker, and Williams 1975.

10. Thomas Kuhn 1970 was one of the Titans of the emergence of postmodernism.

11. Allen and Hoekstra 2105.

12. Nichols1923 reports the 1921 survey.

13. Cooper 1926.

14. John Curtis was a titanic figure in the concept of plant community in the middle of the twentieth century. Iltis was with him in the botany department of University of Wisconsin. Allen resided in Curtis' office from 1970–2012 and felt the presence.

15. Clements 1916.

16. Roberts did not publish so much on his thesis, but it is well described in an accessible version by Allen, Mitman, and Hoekstra 1993. Find citations to Roberts there.

17. F. E. Clements, a titanic figure, established community ecology at the turn of the twentieth century. He classified vegetation rather in the manner that plant taxonomists classify species. The scheme is more elaborate than it might be. He saw vegetation communities as discrete superorganisms.

18. Robert Whittaker championed a school of vegetation analysis using direct gradient analysis, where the vegetation is ordered on general environmental gradients. Curtis used what is called indirect gradient analysis, which allowed the vegetation to order itself in a species space, as treated in chapter 9 here.

19. Vegetation at climax is in a dynamic stasis, where the only survivors belong to species already present.

20. This example of the limit to a biome comes from Ronald Neilson of Oregon State University. He was taken by an old timer, Clarence Cottam, Grant Cottam's father, to see the most northern stand and individual of Gambel oak in Utah.

21. Shugart and West 1977, 1979.

22. Weiner 1990.

23. Gleason 1926 was the antithesis of Clements' view of vegetation working as a coherent whole. The paper gives its name to a movement that turn on the Individualistic Concept. There was a running battle, still going on today, on the tenson between discrete versus continuous change in vegetation.

24. Brown and Allen 1989.

25. Brown et al. 2011.

26. Cottam et al. 1957.

27. Beals reports in a personal communication that when he presented the algebra of intersecting arcs ordination (Beals 1958) to Curtis, the great man at first said it could not be done. He did come around. This is not any sort of putdown of Curtis; it means he could create new methods of multivariate analysis (Curtis and McIntosh 1951 and Bray and Curtis 1957) with sheer brainpower not superior mathematical knowledge.

28. May 1976a.

29. McCune and Allen 1985a,b.

30. Giampietro et al. 2014.

31. Theodor Boveri (see Baltzer 1964) and the American Walter Sutton (see McKusick 1964) saw the mechanics of Mendelian genetics in chromosmes in meiosis.

32. Morgan 1901; Child 1906. Child started in the field, but in 1910 Morgan published his account of sex linkage in flies.

33. See Meyerowitz and Pruit 1985 for transition of *Arabidopsis* as a model, as opposed to an organism to study.

34. Hillenmeyer et al. 2008.

35. Personal communication of Grant Cottam about his father.

36. Anderson, June 23, 2008, posted an article on *Wired Magazine* website, "The end of theory: The data deluge makes the scientific method obsolete." Pigliucci 2009 deeply criticized Anderson, suggesting he was describing a version of stamp collecting not science. There was a fierce and sustained rejection of Anderson's position.

37. Overton's report of Bledsoe et al.

38. Arnold 2009. Arnold is the only scientist who is a member of all three National Academies in the United States.

39. Denis Noble 2013, Evolution and physiology: a new synthesis. http://www.voicesfromoxford.org/video/evolution-and-physiology-a-new-synthesis/355.

40. Gigerenzer 2004.

41. Mark Buchanan is a physicist, one of whose credentials is his former editorship of *Nature*. He addresses the whole suite of environmental and economic issues online at https://www.bloomberg.com/view/contributors/ASLrBUWszfg/mark-buchanan. The quotation in the text is found at http://www.bloomberg.com/view/articles/2014-10-05/economists-are-blind-to-the-limits-of-growth. His book on the topic Buchanan 2013.

42. Lynn Margulis has written a dozen and a half books looking at the complex hierarchy of cells from 1962 to 2009.

43. Lane 2005 was the beginning of the argument while Lane and Martin 2015 bring the story to its present state.

Glossary

The definitions of the words given below are not intended to serve as a dictionary but rather as an account of our intended meaning in their commonest usage in the text. Alternative meanings are given where we are conscious that the term has been used in several different ways. If one meaning is vernacular, then we introduce only our specialized meaning. Anticipating that the readership of this book might be drawn from several disciplines, we have included in the glossary terms that may be commonplace in, say, biology or systems science, but that could be seen as technical terms when viewed from outside the subject specialty in question. Terms not included in the glossary should generally be taken as having their vernacular meaning.

Adaptation: (a) The process whereby a biological system responds to its environment so as to accommodate to the constraints of the environment and to take greater advantage of the environmental circumstances. (b) A characteristic which has adaptive significance.

Adaptive significance: The meaning and helpfulness for the possessor of a given character. Adaptive significance is recognized by observers.

Aerobic: Pertaining to the presence or involvement of free oxygen.

Altruistic: Defines behavior that is against the interests of the individual in favor of the interests of the group or some other individual.

American football: The game is similar to rugby football in that an oval ball is seized in the hands and in that passing from hand to hand is part of the game. Unlike rugby football, forward passes with the hand are permissible by certain players from certain positions on the field. The rules are constructed such that a team that possesses the ball will tend to keep it and dominate the play. If after four plays with possession the team makes insufficient progress on the ground, the ball is turned over to the opposing team. Substitutions of players wholesale is allowed, so the team with possession uses an offensive set of players while the team without possession presents a line of players who specialize in defense. Plays are formal and preset, being initiated from a version of the rugby scrum. Half a minute is available for a policymaking huddle, but this is optional and is abandoned by the team that is behind in the closing stages of the game, as time begins to run out.

Amino acid: An organic acid of small molecular size involving at least one nitrogen atom. They are the principal building bricks of protein.

Amplitude: The maximum departure of an oscillatory motion from its time-average value.

Analog computer: A computer that avoids discrete symbolic representation in its operation, but rather performs by summation and subtraction of electric potentials directly. The operation is continuous rather than involving discrete numbers.

Arthropods: A group of segmented animals with a tough outer skeleton. Soil arthropods are generally of small size, usually rather less than one-half centimeter in width down to microscopic forms.

Artificial surfaces: Boundaries imposed arbitrarily, even capriciously, and according to only a few criteria. Not robust under transformation, readily disappearing when viewed in an alternative fashion. They occur at relatively shallow regions on interaction density gradients.

Autocorrelation: A statistical measure of the strength of association that exists between pairs of values of a time series as a function of the time interval that separates them. The correlation between lagged data points in a series, plotted against size of lag.

Biome: A high level in the ecological classification of associations between organisms; biomes are generally characterized by the physiognomy of the vegetation, e.g., deserts or deciduous forests.

Bond strength: The strength of connections between parts. For instance, chemical bond strength is weaker than atomic bond strength. That is why atom bombs make bigger explosions than hand grenades.

Boundary: A distinction made by an observer. Artificial boundaries are drawn arbitrarily and haphazardly. Natural boundaries are still arbitrary but tend to be robust under transformation. That is, natural boundaries coincide for many distinct criteria.

Breeding barrier: Commonly, members of different but related species may not freely interbreed. The barrier that interferes with breeding may be partial or complete and may have its origins in genetic, behavioral, or temporal incompatibilities. Breeding barrier normally refers to the consequence of some attribute of the organism involved, but the term may sometimes be applied to mere geographic displacement, say, living on opposite sides of an ocean.

Buffer compartment: A part of a system that acts as a reservoir for incoming energy, matter, or information such that the input enters the rest of the system with smoothed and averaged characteristics. E.g., a bathtub will average out a variable input of water from a tap turned off and on. The water out of the drain will be less variable. The tub buffers its input.

Calvin cycle: One of the biochemical cycles involved in photosynthesis, the process whereby plants capture sunlight and fix it as chemical energy in sugar.

Carrying capacity: (K) Refers to the largest number of a given type of consumer (maximum state of a consumer compartment) that may be maintained by a given resource.

Catastrophe theory: A theoretical construct used for the description of certain types of disjunct behavior. It derives from a topological consideration of folded surfaces describing the interactions between variables. Catastrophe theory is rather specific in its appropriate application, although it serves a more general usefulness as a mathematical metaphor.

Chaparral: A fire-dependent shrub community occurring principally in dry areas of the western United States.

Chordate: Any animal with a backbone.

Chromosome: The microscopic manifestation of the genetic information in organisms. With appropriate staining or refractive techniques, chromosomes appear microscopically as threads derived from nuclear material at cell division. The number of chromosomes in a given cell is species specific. Given genetic characteristics are located on particular chromosomes, each of which may be identified. For example, there are forty-six chromosomes in human beings, with the specific gene for hemophiliac blood disease occurring only on X sex chromosome. Chromosomes consist principally of DNA and protein. The protein is associated with the twisting and shortening processes of cell division.

Climax vegetation: Describes vegetation in a stage that floristically remains unchanged despite plants replacing plants as individuals.

Cluster analysis: A series of techniques of data reduction where entities are grouped together in classes according to some criterion of similarity. Most methods aggregate over the analysis, but some divide the whole into subset clusters.

Collapse: A catastrophic change in a system that becomes, initially at least, overexcited such that some of the parts overstep the constraints of system rules. Characteristic of over-connected systems.

Communication: A transfer of meaning from one entity to another through a double-scaling operation, once at signal transmission and once at signal reception. The double scaling may change the meaning between the transmitter and the receiver.

Community, plant: A plant community is a multispecies collection of plants usually growing together in a fairly specific type of habitat. Various concepts (e.g., Clements versus Gleason) of what is a "plant community" differ principally in the degree of biological cohesion attributed to the community both in process and in space.

Competitive exclusion principle: The principle states that species with identical environmental requirements cannot coexist indefinitely. The principle is of limited utility in that the definition "identical environmental requirements" may lead to a circular argument.

Complementarity: The principle that indicates that unified models will be necessarily contradictory. Deriving from the observed-observer duality, the principle demands a rate-independent description and a rate-dependent description. The rate-independent description is linguistic in its nature, and refers to the observer-dependent aspects of the phenomenon. The rate-dependent, dynamical mode of description is associated with observer-independent aspects of the phenomenon. Better definitions and explanations in one complement are gained by confounding or compromising the other complement. Neither complement is sufficient, but both complements are necessary for a full account of phenomena. Complements must be kept separate to avoid contradiction.

Complex character: A character whose phenotypic expression is the result of the interaction of several genes identified as separate. E.g., the particular form of the pattern on a butterfly.

Connectance: A measure of system connectedness based upon the mean number of nonzero interaction terms per compartment of the system under consideration.

Connected component: A component of a system that is involved in at least one of the main lines of communication in a system.

Connectedness: A general term for the cohesiveness of a system. Systems with strong interaction terms are relatively highly connected, as are systems with a large number of the parts interconnected (as measured by connectivity, connectance, and other measures).

Connectivity: A measure of the degree of connectedness of a system (Gardner and Ashby 1970). It is usually expressed as a percentage: the percentage of nonzero interaction terms as a percentage of all possible interaction terms between the components of a system.

Conspecific cuing: Actions of individuals which are stimulated by the actions of their fellows. This tends to lead to aggregation of individuals; flocking in the extreme case.

Constraint: The control exerted by one holon over another lower in a hierarchy. The constraining holon behaves more slowly than the constrained holon. The constraining holon influences more than it is influenced in the act of constraint.

Continuum: Generally refers to a continuous gradient, but specifically when associated with the definite article, it refers to an aspect of vegetation science. A conceptual view of patterns of variability in vegetation wherein there is compositional continuity along gradients. The continuous occupation of species space.

Cosmic time: Time associated with the progress of the universe.

Crossover segment: The segment of a chromosome that is exchanged in the process of crossing-over between chromosomes.

Curvilinear: See linear.

Cyanobacteria: Photosynthesizing procaryotic organisms formerly called the "blue-green algae."

Data matrix: In community ecology it commonly takes the form of a series of species rows whose individual datum values are determined by species occurrences in a series of sites or samples that comprise the matrix columns. Environmental data matrices would substitute rows relating to environmental factors for species rows. The individual observations may be entered as raw values collected in the field (e.g., number of organisms) or may be transformed according to some prescribed criterion.

Data reduction: A process of data summary achieved by any of a large number of methods of multivariate analysis. After reduction, a large and complex data set may be represented approximately through expressing interrelationships between observations either by placement on a small number of axes or by grouping them together in a series of discrete classes.

Datum value or point: A quantity associated with a single species or single environmental factor as it occurs in a particular sample or site in a data matrix.

Degeneration: Occurs when a system has insufficient free energy to maintain the integrity of its rule system. The system undergoes uncontrolled change, which results in a fundamental change in system structure. Characteristic of underconnected systems.

Deletion: Loss of small segments of chromosomal material because of aberrant chromosome behavior at cell division.

Deme: A local manifestation of variability restricted to a small number of populations of a given species.

Demography: Used here almost always as plant demography, a study in plant ecology wherein individual plants are enumerated so that very specific questions concerning competition, mortality, growth, and similar parameters of the individual may be addressed.

Deterministic: Caused by explicit antecedent events.

Dialectical materialism: Hegel (1770-1831) maintained that every proposition (thesis) brings into a discourse its natural opposite (antithesis). These lead to a unified whole (synthesis), which reacts upon the original thesis. This view (the dialectic) was adapted by Marx in his materialist philosophy. The necessity of contradiction is central to the doctrine. Levins and Lewontin espouse a biology based on dialectical materialism. It is related to the contradictions embodied in narratives.

Difference equation: An equation used to describe the dynamical behavior of a system with the passage of time, but one whose values are updated only over discrete time periods. Change is thus described as a series of steps rather than as a continuous process.

Differential equation: An equation for describing change in the state of a system continuously. Generally, change recorded by a differential equation occurs smoothly.

Diploid: After fusion of egg and sperm, the chromosomes from the egg and sperm, respectively, retain their identity although they are mixed together in a single nucleus. That nucleus, in that it contains two chromosome sets, is called a diploid nucleus, as are all descendent nuclei until the next meiosis; e.g., all normal human adult cells are diploid. (Compare haploid.)

Disconnected component: A component that is only part of a system incidentally or is included only because it is defined so. There may be weak interactions between a disconnected component and other components of the system, but these are incidental to what is recognized as the normal functioning of the system and are not involved in the main lines of communication in system functioning.

Diversity: A central but unfortunate concern of ecology for the richness of variety of species found in various ecological locales. Some measures of diversity only count the number of species, but others take into account the degree to which different species enjoy equivalent representation as to evenness.

Domain of attraction: A region in a state space description of a system in which system behavior is localized indefinitely unless there is a change in system structure or a large exogenous disturbance.

Dominant character: A single characteristic, say for eye color, may exist in several particular forms—blue as opposed to brown. If two alternatives occur together in a single diploid nucleus, then the dominant genetic characteristic is the one that is physically expressed in the phenotype. The one that is suppressed is called the recessive character. Dominant refers, often unwittingly, to normal, beneficial, or important, and is not strictly a material issue.

Duplicated segment: A section of chromosomal material that exists elsewhere in the genetic complement, above and beyond its existence in the second set of a diploid nucleus.

Duplication: Because of aberrant patterns of replication, it is possible for a single segment of genetic material to become represented twice in a single chromosome set. (See also duplicated segment.)

Dynamical: Pertaining to the pattern of change of a phenomenon with time. It is rate-dependent.

Ecotone: A narrow ecological zone that possesses a mixture of floristic and faunal characteristics in between two different and relatively homogeneous ecological community types. Ecotones often represent gradients between two vegetations with different physiognomies.

Emergent properties: (a) Properties that emerge as a coarser-grained level of resolution is used by the observer. (b) Properties that are unexpected by the observer because of an incomplete data set, with regard to the phenomenon at hand. (c) Properties that are, in and of themselves, not derivable from the behavior of the parts a priori.

Endogenous: Deriving from within the entity or system.

Endogenous cycle time: The time taken for a system to complete a cycle of characteristic behavior, e.g., a single heartbeat or the generation time in an organism.

Entification: The erection and implementation of criteria for the identification of discrete things. The giving of names to objects, particularly the concrete, is an aspect of entification.

Enzyme: An organic catalyst consisting principally of protein. They are very reaction-specific, and the presence of an enzyme increases the reaction rates so that the balance of organic compounds shifts toward the equilibrium condition, i.e., much product and little substrate, since in living systems material representing the product side of the equilibrium balance is likely to be consumed in another reaction. Enzymatic reactions often flow principally in one direction only, never being allowed to achieve equilibrium concentrations of products.

Ephemerals (desert): Plants growing in a dry habitat that complete their life cycle in such a short period that they need rely only upon a single brief period of water availability.

Epistemology: The study of or theory associated with the nature and limits of knowledge. Concerned only with that which is knowable. Science which does not go beyond the consequences of direct observation. (Compare ontology.)

Equilibrium: A state of a system when all forces for change are balanced. Stable equilibrium points are those where endogenous forces for system functioning lead to a return to the equilibrium point after displacement by forces exogenous to the system. Displacement from an unstable equilibrium point leads only to further displacement.

Escape (cultivar): A plant found in a natural habitat that owes at least its local occurrence to some form of human cultivation or husbandry from which it has escaped.

Essence: The explanation for the equivalence in a set or equivalence class. Not derived from observer decision and not idealist.

Eukaryotic: Referring to organisms with nuclei and organelles, excluding bacteria and their relatives.

Euclidean: Pertaining to a metric space in which the distance between two points is the square root of the sum of squares of differences between the points' coordinates. If the distance between points is given by a different formula, the space is called non-Euclidean.

Exogenous: Deriving from outside the entity or system.

Factor analysis: Has two meanings: (a) a family of multivariate techniques of analysis of which principal component analysis is one; (b) specifically, a particular multivariate technique that goes beyond principal component analysis so as to include identification of factors that account for the variability displayed.

Feedback loop: A chain of causal relationships that closes upon itself. Positive feedback loops are unstable; negative feedback loops are self-correcting.

Filter: A device in theory or practice that takes a signal string that is undefined in terms of scale and converts it by a process of integration into a scaled message. The raw signal stream has a series of weights applied to it. The weights are applied through windows that work to smooth signals like a weighted average. Long windows smooth local high-frequency signals. Short windows allow the passage of short high-frequency bursts of signal. The patterns of integration that are used in a filter may be very exotic and can even be influenced (as in Dolby sound) by aspects of the signal as it is encountered.

Fitness: (a) Dynamically it may refer only to differential reproductive success with no reference to adaptive significance. (b) Alternatively in structural terms, fitness is measured by the possession of adaptive characters that aid success.

Frequency domain: A mode of discussion where events are not seen as occurring at points in time, but rather according to the frequency of their recurrence. (Compare time domain.)

Function: A mathematical device that maps one variable onto another with no slack or ambiguity.

Fuzzy set: A set wherein there are degrees of membership varying between 0 = not a member and 1 = is a member. The set "tall people" is one such set.

Gene: The genetic informational set associated with a given physical characteristic in an organism. Enzymes are responsible for the biochemical actions within cells. The gene has come to mean the genetic segment associated with the production of a given enzyme.

Gene complex: A set of genes whose interaction gives rise to certain phenotypic expressions.

General systems theory (GST): A body of theory, first developed by Ludwig von

Bertalanffy, concerned with open, as opposed to closed, systems. It is associated with, but is more deeply philosophically involved than, cybernetics.

Genet: A unit used in plant demography that includes all branches and vegetatively reproduced plants arising from a single seed and so possessing identical genetic material in all the individuals and shoots involved.

Genome: The full genetic complement of a nucleus or an individual.

Graph theory: A branch of mathematics that deals with the properties of structures consisting of interlinked points or nodes.

Group selection: Natural selection in an evolutionary process according to, not characteristics of the individual, but characteristics of the group to which the individual belongs. From this selection, individuals in groups may display characteristics that are adaptive for the group but maladaptive for the individual. Sex and altruism are examples.

Haploid: (Compare diploid.) Genes occur on particular chromosomes. A haploid nucleus contains only a single set of chromosomes with no two chromosomes possessing equivalent sets of genes. For example, in a haploid nucleus, a particular gene for eye color should arise once and only once in its appropriate position.

Heraclitean flux: Heraclitus (6th-5th century BC) proposed that the only constant is change. Structure is apparent, not ontologically real; change is present even in the most apparently static forms. All is fluid.

Heterotrophism: All methods of attaining food by the consumption of ready-made, high-energy organic material.

Heterozygous: In diploid organisms there are two sets of chromosomes with equivalent genes occurring in each set. Individual genes (e.g., eye color) thus occur in either the same or different states (e.g., blue versus brown) twice in the genome, once in each chromosome set. Genes occur on equivalent chromosomes in the same structural position in the respective sets of chromosomes. If the expression of the gene is different between the two chromosome sets, then the individual containing different (say blue and brown eyes) expressions of the gene is termed heterozygous. If there is only one expression of the gene (e.g., only brown eyes but duplicated between sets), then the individual is homozygous for that characteristic.

Heuristic: Of immediate utility. Describing a device employed out of pragmatism, but a device conceived arbitrarily.

Holism: A descriptive and investigative strategy that seeks to find the smallest number of explanatory principles by paying careful attention to the emergent properties of the whole as opposed to the behaviors of the isolated parts chosen by the observer in a reductionist strategy. Both holism and reductionism seek to explain emergent behavior by invoking a lower level of organization.

Holon: The representation of an entity as a two-way window through which the environment influences the parts, and through which the parts communicate as a whole unit to the rest of the universe. Holons have characteristic rates for their behavior, and this places particular holons at certain levels in a hierarchy of holons. What a holon shall contain is determined by the observer.

Homozygous: See heterozygous, its antonym.

Hypervolume: See n-dimensional.

Idealism: It is any of several theories that assert that objects that are perceived are in fact ideas in the mind of the one who perceives, and that it cannot be known whether ontological reality exists outside the mind. (Compare materialism.)

Idée fixe: An obsession.

Inbreeding: The most extreme case occurs when an individual fertilizes itself, but inbreeding is a matter of degree and breeding between relatives involves a degree of inbreeding. (See also outbreeding, its antonym.)

Individualistic concept: A characterization of communities (usually plant) not as a cohesive whole but as a collection of individuals and individual species whose cooccurrence is the result of invasion of the site by colonists from juxtaposed vegetation and subsequent differential survival meditated by local environmental circumstances and competition (Gleason 1926).

Inertial frame: A reference frame wherein the laws and explanatory principles associated with a given phenomenon have the simplest set of relationships. Simplest often means linear in this context.

Integration: A process of smoothing or averaging in which various parts of the signal string to be integrated may be a given different weights in an averaging or smoothing operation.

Integrative tendencies: The characteristics related to the manner in which holons are parts of larger wholes. These characteristics reflect the acceptance by the holon of the constraints placed upon it by more inclusive holons. These aspects of the holon may involve self-sacrifice for the whole of which it is part.

Interaction density gradients: At each point in space there is a certain rate of exchange of information or matter associated with a particular form of signal. At successive points in space there may be a change in the quantity of passing signal. Monotonic change in the amount of signal passing successive points constitutes an interaction density gradient. At surfaces, interaction density gradients are particularly steep.

Interaction term: A number that determines the influence of one ecological entity upon another. A positive term suggests that an increase in the influenced entity. Negative interaction terms indicate that an increase in the entity that influences brings about a decrease in the entity that is influenced. Interaction terms may reflect the influence of predation, competition, the use of resources, or any other ecological relationship. A zero interaction term indicates that there is no interaction in the direction indicated.

International Biological Program: An international cooperative investigation in the late 1960s and working through the 1970s, yielding interdisciplinary approaches to large-scale ecological questions. Part of the program involved the construction of large-scale simulation models for the principal biomes in North America.

Invariant: Unchanging.

Inversion: Sometimes a segment of a chromosome will become disconnected only to be reinserted, but backward. When chromosomes come to pair at meiosis, the inverted segments will pair gene for gene with normal segments but will be forced into a peculiar loop configuration. If there is crossing-over in the looped segment, then deletion and duplication occur.

Inverted segment: The segment of a chromosome that characterizes chromosomal inversion.

Irreducible: An aspect of a phenomenon that may not be explained in a more dissected account.

Junction compartment: When two holons on different stems of a hierarchy exchange information, the information must first pass up the hierarchy to a point where the two stems meet. This point is a holon in its own right with lower-frequency characteristic behavior than the holons thus connected.

Lamprey: A fish so primitive that it does not possess, nor has it ever possessed in its phylogeny, a lower jaw. A parasitic bloodsucking fish that attaches itself by its mouth to the body of host fishes.

Laws: A set of requirements upon which a system is dependent for its functioning but which are universal, inexorable, structure-independent, and associated with rate-dependent aspects of control. Laws of physics would be a special case. A law in biology would be that carbon is involved. We generally do not mean laws in biology that are the biological equivalent of laws as understood in physics.

Level: A section of a hierarchy that is defined by a scale or degree of organization. Many different entities may be observed scaled to a certain level distributed horizontally across a hierarchy. Or, the scale of an observation.

Level of observation: A level identified by scalar properties.

Level of organization: A level fixed by a definition, not scale.

Level of resolution: The grain size of an observation as limited by the finest distinction that can be made given the observation filter.

Linear: A relationship among variables in which any one of the variables can be expressed as a constant plus a sum of the other variables each multiplied by a constant. If this is not the case, the relationship is called nonlinear or curvilinear.

Linguistic mode: A means of description associated with rules, significance, and events. Linguistic descriptions give an account of the role of the observer in the observed-observer duality in phenomenology.

Linkage: Genes that occur on a single chromosome will tend to remain together. Thus grandparental characteristics that occur on a single chromosome will be transmitted to the grandchildren together with greater probability than would be expected with a random mixing of all characters. The less than random segregation of genes because they occur on a single chromosome is called linkage.

Liverwort: A primitive type of land plant more or less allied to the mosses, sometimes devoid of leaves and then consisting of an amorphous sheet.

Lognormal: May be expressed as a bell-shaped curve (sometimes truncated) of abundance classes of a certain type. The classes are related in logarithmic fashion, usually in ecology to base 2. Ruth Patrick used the lognormal distribution to describe the number of species in a community grouped according to logarithmic classes.

Low-frequency-pass filter: A filter that specifically allows the passage of low-frequency patterns in a signal. Note here that we do not mean the inverse, that is, a filter that specifically stops the passage of low-frequency aspects of a signal.

Materialism: The philosophical assertion that matter is the only reality and that even aspects of the mind find their explanation in the workings of matter.

Matrix: A series of columns and rows employed for organizing a set of interrelated values. The simplest case takes the form of a table of columns and rows although matrices may have more dimensions than two.

Mechanistic model: A model that invokes serial causal chains in the entirety of its explanation. The spring in a watch drives a cog that catches on a ratchet . . . and so on. The seriality may not be simple.

Medium number system: A system with too many parts for parts to be given individual account, but too few parts for parts to be substituted by aggregates or averages.

Meiosis: The mode of cell division that counterbalances the doubling of genetic material resulting from sexual fusion. At the end of meiosis, the double complement of genetic material that exists in the adult has been divided to a single copy of the genetic complement. A single diploid nucleus has two equivalent sets of chromosomes. Diploid cells may divide through meiosis to give rise to haploid cells. At the beginning of meiosis, equivalent chromosomes come together and pair such that the genes in one chromosome set lie exactly beside those equivalent genes of a chromosome of the other chromosome set. Because of the close juxtaposition of equivalent chromosomes, the chromosomes may exchange equivalent sections within the chromosome pairs. Thus the single haploid chromosome set that emerges at the end of meiosis contains genetic material from both chromosome sets that were present in the diploid nucleus at the outset.

Mesophyll: Tissues in the middle of a leaf. In a narrower sense, mesophyll can refer specifically to relatively loosely packed tissues and so may exclude conducting strands and their associated sheathing cells.

Message: A scaled and integrated signal that has significance or meaning for either the transmitter in transmitted messages or the receiver in received messages.

Metaphor: a comparison of one system to another, usually many others. Our view, along with N. Katherine Hayles, is that metaphors only work in one direction.

Model: An intellectual construct for organizing experiences. We generally do not extend our use of the word to include models as approximations of ontological reality. We prefer to acknowledge that we do not know what is the relationship of models to ontological reality.

Monotonic: A relationship is monotonic when it continuously increases or continuously decreases. At no point does a trend of increase change even momentarily to one of decrease in a monotonic relationship. With respect to its initial conformation, a stretched surface is nonlinear while a wrinkled or folded surface is non-monotonic.

Morpheme: The most finely subdivided parts of speech that still contain meaning.

Moving average: A smoothing operation associated with a series of records.

Multivariate: A branch of statistics that deals with a large number of variables simultaneously.

Mycorrhizae: Fungi associated with roots, being supplied by them with food and supplying for the root a capacity for nutrient uptake.

Natural boundary: See natural surface.

Natural frequency: The inverse of the extent in time and space for the completion of a cycle of characteristic endogenously driven behavior of either a holon or a pattern of observation.

Natural level: A level identifying entities that are robust under transformation (can be viewed in alternative ways).

Natural selection: The process that underlies Darwinian evolution. In biological systems it refers to those individuals that meet with differential reproductive success, so imparting their peculiarities to the population at large. It is survival of the stable, and the stable are those who will, as time passes, come to hold the majority of available resources in their particular configuration. In this way natural selection may occur in nonliving systems that do not involve self-replication.

Natural surfaces: Discontinuities across natural surfaces are readily recognizable and robust under transformation. Natural surfaces may be detected by any of a large number of coincident criteria for large changes in interaction density.

N-dimensional: Although commonplace experiential space is limited to just three dimensions, mathematical spaces of higher dimensionality may be invoked. For example, a community may be described as a point in a space of many dimensions, one for each species involved. Such a space would be called n-dimensional or a hypervolume.

Near-decomposable: A characteristic of complex phenomena that may be seen to exhibit apparent disjunction when viewed in an appropriate fashion. Near-decomposition does not deny ultimate continuity, but rather it pertains to relative interaction intensities across the continuous space of structural and functional boundaries.

Neural net analysis: A vector is fed in at the top of a network of nodes. The nodes are connected by weighting functions, at first random, but adjusted so as to achieve a correct linkage to give a known solution at the bottom of the net. Adjusted nets are then used to predict unknown outcomes, like will this loan applicant pay off the loan if we give it. Neural nets iterate to a long polynomial.

Nested: A restricted type of hierarchy that has the requirement that upper levels contain lower levels.

Niche: A confused word used in several different ways over the past half-century. Sometimes it refers to the habitat of an organism or species, and sometimes it refers to the role the organism or species plays in the larger community of which it is part. More recent definitions refer to the resource base upon which the organism or species is characteristically dependent.

Noise: Any aspect of a signal string that is not considered significant. That part of a signal that is not considered to be associated with meaning for the purposes at hand.

Non-nested: A general type of hierarchy that only excludes hierarchies where upper levels contain lower levels.

Nucleotide: The individual bits in the genetic code associated with the DNA molecule.

Nucleus: A membrane-bound body possessed by eukaryotes; its most significant contents are chromosomes, the primary genetic material of living systems.

Observation window: Observation window is determined for a given observation principally by the extent in time and/or space of integration used by the observer in observing. Since the integration of the signal caught in the window may not be a simple integration, the differential significance given to different parts of the signal within the window also gives some of the character of the observation window in question.

Observerindependent:- Sometimes referring to the ontologically "real" external world, sometimes part of observation that the observer does not choose. The latter case is still epistemological, not ontological.

Ontological: Pertaining to reality beyond observation.

Ontology: The branch of metaphysics that deals with reality independent of the observer. (Compare epistemology.)

Ordination: In ecology a general method of multivariate data reduction where either species, samples, or environmental factors represented as points are projected onto axes such that the interrelationships of the points carry ecological significance.

Organelle: Inside eukaryotic cells various major metabolic activities are concentrated in quasi-autonomous bodies called organelles. Examples are chloroplasts, which conduct photosynthesis, and mitochondria, which are centers of aerobic respiration.

Orthogonal: At right angles to.

Outbreeding: Breeding between individuals usually of the same species but from different lines of descent. (See also inbreeding, its antonym.)

Outlier: A sample or species that is particularly distinct in its characteristics from all others of its type. Such samples or species occur eccentrically in their respective attribute spaces. Such samples or species often dominate the course of any multivariate analysis in which they are included.

Over-connected: A system that has so many connections between its part that it is unstable and is wont to undergo uncontrolled change is described as over-connected. (See collapse.)

Paradigm: The intellectual frame implicit in the acceptance and use of a given vocabulary; the frame determines the questions seen to be appropriate in the scientific investigation of a given phenomenon. The paradigm indicates the particular scientific procedure to be followed and encourages a particular interpretation of the results. It is the governing narrative.

Paradox of the plankton: The reference is to a paper by G. E. Hutchinson so titled, which poses a question as to why it is that in an apparently homogeneous environment, such as ocean water, there should be so many plankton species coexisting. Answers to the paradox often suggest that diversity is elevated by species whose niches are so similar that there is no room for competitive edge.

Parameter: Any of the numerical constants that appear in a mathematical expression of relationship among variables.

Phenomenon: Something that appears organized and is susceptible to direct sensual experience.

Phenotypic: Associated with the physical expression of characteristics in organisms as opposed to their genetic representation in the genetic material of the organism involved.

Phoneme: A group of closely related sounds.

Phylogeny: The lineage that terminates in the organism or group of organism in question.

Phylum: One of the higher levels used in classification, usually of animals but sometimes of plants; e.g., the phylum Chordata includes all animals with backbones.

Phytoplankton: Plants that float in open water. Usually microscopic, although some forms develop large colonies that may aggregate as a scum on water bodies.

Pioneer strategy: Displayed by organisms whose ecological characteristics involve rapid invasion of open sites shortly after disturbance. Such species are often displaced as community succession proceeds.

Plasmalemma: The layer of macromolecular thinness on the outside of living protoplasm.

Platonic form: "An entity whose conceptual recognition does not involve a necessary reference to any definite actual entities of the temporal world"; A. N. Whitehead goes on to substitute for platonic form the term *eternal object*, an entity for which the objects we experience are imperfect examples.

Polar ordination: One of the earlier techniques applied in ecology for data reduction of species occurrences in samples. It projects either species, samples, or environmental factors onto a small number of axes. The criterion for the choice of axis involves reference to either particular species or particular samples.

Polygenes: Genes that work together, sometimes in a fairly simple additive fashion, to produce a given character. For example, height in human beings is located at several places in the chromosome set. Each gene makes its small contribution.

Polynomial: A sum of a finite number of terms each composed of a variable raised to a nonnegative integer power multiplied by a constant.

Prairie: A fire-dependent major grassland vegetation type occurring principally in central and western North America and across southern Russia, where it is called steppe. Sometimes use of the term is restricted to just North America.

Primitive: (noun) One of a group of entities, taken to be irreducible or unanalyzable into simpler constituents, which form a basis for explanation and/or deduction, e.g., a set of axioms. (adjective) An organism or species that either is the antecedent or possesses a large number of characters of antecedent species.

Principal component analysis: An ordination technique in community ecology. A means of collapsing a variable space of large dimensions to a smaller space using variance as the criterion to be maximized in the smaller space. A multivariate technique in the family of techniques called factor analysis.

Prokaryote: Relatively primitive organisms, including bacteria and some forms formerly classified with the algae. The Achaea are relatively recently set apart from bacteria. Generally microscopic in size.

Ramet: A unit in plant demography that identifies a single shoot or branch. Any given seed may produce a multitude of ramets or branches.

Range (species): Usually a geographic term although sometimes referring to habitat, which delimits the extent of the occurrence of a species. Sometimes it refers to the entire span of geographic or habitat possibilities but may be restricted to mean only the very limits.

Recessive character: See dominant character, its antonym. Commonly misunderstood to be a material issue when it is in fact normative.

Recombinant: Referring to nuclei or individuals in which genetic material that was formerly expressed in separate individuals has become united in one individual through a process of genetic recombination in meiosis following sexual fusion.

Reducible: A phenomenon or property that may be explained and given account in a more finely dissected description.

Reductionism: A descriptive and investigative strategy that gives account of phenomena in terms of a series of isolated parts, coupled together by direct causal linkages. Ambiguity in relationship between parts is met with further subdivision until the ambiguity disappears. (Compare holism.)

Region of stability: See domain of attraction.

Reification: The act of reifying, taking as ontologically real, or the consequence of that action.

Reify: Take an observed fact and assert that it has ontological reality.

Relation: A mathematical device that compares variables in a way that invokes slack (compare function).

Relaxation time: The time taken for a system to return to its normal (equilibrium) patterns of behavior after a disturbance. Approximately the inverse of the responsiveness of the system.

Resilience: In systems discussion of stability, resilience refers to the ability of a system to maintain its structure and general patterns of behavior when displaced from its equilibrium condition. Another different type of resilience says a system is more resilient if it can quickly return to an equilibrium condition despite large displacement.

r selection: Natural selection of organisms in an unpredictable and characteristically rapidly changing environment so as to maximize reproductive potential at the expense of competitive capacity of the adult.

Rules: A set of constraints that are local, arbitrary, structure-dependent, and associated with rate-independent aspects of control.

Saprophytically: Describing an organism which lives upon the dead remains of other organisms. Saprophytes display passive behavior as generally exhibited by plants, soil bacteria, and fungi and so are distinct from predators, parasites, or carrion feeders.

Scalar: Reference here is to Loucks (1962), where environmental gradients for moisture, temperature, and nutrients were synthetically derived using different measures to characterize a given site depending upon the environmental conditions which prevail there. For example, dry sites were characterized and placed upon the moisture scalar according to different criteria than were used to place wet sites.

Scale (noun): The natural frequency of either an observation or a holon.

Scale (verb): Take signal and integrate it so as to produce a message.

Scope: (See also level of resolution.) The breadth of the universe in which an observation or a set of observations is made, or the breadth of a universe of discourse.

Self-assertive tendencies: Characteristics of holons related to their existence as quasi-autonomous wholes. These tendencies are related to the holon's capacity to maintain its integrity. These aspects of holons pertain to the constraint that the holon asserts over its parts.

Sidereal: Associated with the stars.

Signal: A string or strings of energy or matter in transit between a transmitter and a receiver; its meaning is undefined.

Simulation: A model usually implemented through a computer that mimics the behavior found in a phenomenon.

Sister holon: Two communicating holons that occur at the same level in a hierarchy and do not therefore exert constraint over each other.

Span: The number of entities at the next natural lower level in a hierarchy that report directly to the entity whose span is to be identified.

Spectral analysis: A branch of statistics that deals with the problem of forecasting future values of sequences of evenly spaced observations with purely sinusoidal functions of time.

Sporophyte: The life history of plants generally involves both a sexual organism that is haploid and a spore-producing generation that is diploid. All large land plants are the diploid spore-bearing generation, the sporophyte.

Spruce bud worm: A moth that in the larval stage infests the growing tips of spruce trees. Low levels of infestation are tolerable, but epidemic attacks kill whole forests.

Stability: There are several definitions involved. (a) Sensu lato, the general display of persistence in both structure and patterns of behavior. Here, resilience is one aspect of stability. (b) Sensu stricto, specifically excludes resilience as part of the definition and refers only to the rapidity with which the equilibrium is reestablished when displacement from the equilibrium occurs. Generally systems stable in this sense resist the influence of outside disturbance rather than accommodating the influence once it has had its effect. (c) Structural stability, the capacity of a system to display generally the same emergent behavior despite changes in the interrelationships between its parts. In the case of an equation, structural stability would be represented by the persistence of general patterns of behavior despite changes in equation parameters.

Stable equilibrium point: See equilibrium.

Stable limit cycle: A persistent pattern of endogenously driven cyclical behavior of a system.

Stand: A local patch of homogeneous vegetation.

State space: A mode of description of system behavior where, at each point in time, the system is described by the position of a point inside a space the dimensions of which are axes relating to the various system attributes expressed as states (as opposed to derivatives).

Stem: A line of communication in a hierarchy. Stems connect higher and lower holons in hierarchical structures.

Structural stability: See stability.

Succession: The process of change in a plant community through time as represented by a series of observed states.

System: Any interacting, interdependent, or associated group of entities.

Taxon: A level of organization within a taxonomic system. Examples of various taxa are species, genera, and families.

Taxonomy: The biological study of the classification of organisms into natural groups.

Thermocline: If a large water body is subjected to substantial heat input at its surface and is left relatively undisturbed by factors of wind and turbulence, then the warm waters at the surface become substantially separated from the cool waters below. A temperature profile of a lake with a thermocline shows a homogeneous water mass at the surface of high temperature; a relatively homogeneous mass of cool water below with low temperature; and a narrow region of less than one meter where most of the change in temperature with depth occurs. The position of that steep temperature gradient defines the thermocline.

Thermodynamic: Relating to processes by which various forms of energy, e.g., heat, work, chemical and electromagnetic energy, are transformed into one another. Gold coin in commerce is used to do work in a thermodynamic manner.

Time domain: The mode of discussion where events are seen as occurring at particular points in the passage of time. (Compare frequency domain, which deals with periodicities.)

Trajectory: In this text the reference is usually to the passage of successive samples through an attribute space (e.g., species). The trajectory is derived from connecting successive samples in an attribute space.

Transformation: Used here in the context of data matrices whereby datum values are substituted for some other value according to some criterion applied to the whole data matrix (e.g., numbers of organisms replaced by the logarithm of the number of organisms, substitution of values in the columns of matrices such that raw values become the proportion of their values to the sum of the respective columns).

Translocation: Occasionally segments of chromosomes become broken off only to rejoin, but to the loose end of a different chromosome. Therefore at meiosis chromosomes that have exchanged segments in a translocation become locked together in a configuration involving not a simple pair of chromosomes but rather four chromosomes, essentially two pairs. Because of the mechanics of chromosome separation in cell division and the need to avoid duplication and deletion, the chromosomes are not free to segregate independently. Thus genes on different chromosomes may exhibit a degree of linkage.

Uncertainty principle: In general, any statement that asserts that the act of observation has an effect on the phenomenon being observed. The uncertainty comes from trying to couple rate-dependent aspects of a phenomenon with rate-dependent aspects of the system. Specifically, the Heisenberg uncertainty principle established the minimum disturbance in the position (momentum) of a quantum-mechanical system introduced by the act of measuring its momentum (position) with a specified accuracy.

Unconnected component: See disconnected component.

Under-connected: This describes a system that is very susceptible to intrusion since there are insufficient connections between the parts for the system to maintain integrity. (See also degeneration.)

Universe: While we sometimes mean the physical universe at large, more often we refer to the universe of discourse in an ecological investigation. That universe is defined by the extent in time and space over which ecological samples are taken in a given investigation.

Unnatural boundary: Artificial surface.

Unstable equilibrium point: See equilibrium.

Variable: Any of the characteristics or attributes of a phenomenon that appear to change with time, e.g., biomass, light, CO_2 concentration, population density.

Virus: A biological structure that consists principally of genetic information for its own replication. A virus lacks the machinery for its own self-replication and so must take advantage of the machinery for production found in more complete living forms from bacteria to complex multicellular organisms.

Weighting function: A function that applies differential significance to various parts of a signal string. The parts are determined by their position relative to time zero.

Window: See observation window.

Zooplankton: Animals that float in open water. They may swim, but where they are found in particular is more a consequence of water movement. Usually only barely visible, if at all, with the naked eye but include basking sharks.

References

Adams, D. 1979. *Hitchhiker's Guide to the Galaxy*. Pan Books, London.

Ahl, V. 1993. *Cognitive development of infants potentially exposed to cocaine*. PhD diss., University of California.

Ahl, V., and T. F. H. Allen. 1996. *Hierarchy Theory*. Columbia University Press, New York.

Allaby, M. 2010. *A Dictionary of Ecology*. 4th ed. Oxford University Press, New York.

Albuquerque N., K. Guo, A. Wilkinson, C. Savalli, E. Otta, and D. Mills. 2016. Dogs recognize dog and human emotions. *Biol. Lett.* 12:20150883. DOI:10.1098/rsbl.2015.0883.

Allee, W. C., o. Park, A. E. Emerson, T. Park, and K. P. Schmidt. 1949. *Principles of Animal Ecology*. Saunders, Philadelphia.

Allen, C., and C. Holling, eds. 2008. *Discontinuities and Ecosystems and Other Complex Systems*. Columbia University Press, New York.

Allen, T. F. H. 1971. Multivariate approaches to the ecology of algae on terrestrial rock surfaces in North Wales. *J. Ecol.* 59:803–26.

———. 1977. Neolithic urban primacy: The case against the invention of agriculture. *J. Theor. Biol.* 66:169–80.

Allen, T. F. H., S. M. Bartell, and J. F. Koonce. 1977. Multiple stable configurations in ordination of phytoplankton community change rates. *Ecology* 58:1076–84.

Allen, T. F. H., and M. Giampietro. 2014. Holons, creaons, genons, environs, in hierarchy theory: Where we have gone? *Ecol. Model.* 293:31–41. DOI:10.1016/j.ecolmodel.2014.06.017.

Allen, T. F. H., and T. W. Hoekstra. 1991. The role of heterogeneity in scaling of ecological systems under analysis. In *Ecological Heterogeneity*, edited by J. Kolasa and S. Pickett, 47–68. Ecological Studies 86. Springer-Verlag, New York.

———. 2015. *Toward a Unified Ecology*. 2nd ed. Columbia University Press, New York.

Allen, T. F. H., and J. F. Koonce. 1973. Multivariate approaches to algal strategems and tactics in systems analysis of phytoplankton. *Ecology* 54:1234–46.

Allen, T. F. H., G. Mitman, and T. W. Hoekstra. 1993. Synthesis mid-century: J. T. Curtis and the community concept. In *John T. Curtis: Fifty Years of Wisconsin Plant Ecology*, edited by J. Fralish, R. P. McIntosh, and O. L. Loucks, 339. Wisconsin Academy of Arts and Sciences, Madison.

Allen, T. F. H., E. Ramly, S. Paulsen, G. Kanatzidis, and N. Miller. 2014. Narratives and models in complex systems. In *Modes of Explanation*, edited by Michael Lissack and Abraham Graber, 171–96. Palgrave Macmillan, New York.

Allen, T. F. H., and H. H. Shugart. 1982. Ordination of simulated forest succession: The relation of dominance to correlation structure. *Vegetatio* 51:141–55.

Allen, T. F. H., J. A. Tainter, J. Flynn, R. Steller, E. Blenner, M. Pease, and K. Nielsen. 2010. Integrating economic gain in biosocial systems. *Systems Research and Behavioral Science* 27:537–52.

Allen, T. F. H., J. A. Tainter, and T. W. Hoekstra. 2003. *Supply-Side Sustainability*. Columbia University Press, New York.

Allen, T. F. H., J. A. Tainter, J. C. Pires, and T. W. Hoekstra. 2001. Dragnet ecology: "Just the facts, ma'am": the privilege of science in a postmodern world. *BioScience* 51:475–85.

Allen, T. F. H., J. A. Tainter, D. Shaw, M. Giampietro, and Z. Kovacic. Forthcoming. Radical transitions from fossil fuel to renewables: A change of posture. In *Complex Systems and Social Practices in Energy Transitions*, edited by N. Labanca. Springer.

Allen, T. F. H., and E. P. Wyleto. 1983. A hierarchical model for the complexity of plant communities. *J. Theor. Biol.* 101:529–40.

Anderson, C. 2008. The end of theory: The data deluge makes the scientific method obsolete. *Wired Magazine* website. https://www.wired.com/2008/06/pb-theory/. Published June 23, 2008.

Andrews, C. H. 1949. The natural history of the common cold. *Lancet* 256:71–75.

Ashby, W. R. 1964. *An Introduction to Cybernetics*. Chapman & Hall, London.

Arnold, F. W. 2009. How Proteins Adapt: Lessons from Directed Evolution. *Cold Spring Harb. Symp. Quant. Biol.* 74:41–44.

Austin, M. P., and P. Greig-Smith. 1968. The application of quantitative methods to vegetation survey. II. Some methodological problems of data from rain forest. *J. Ecol.* 56:827–44.

Austin, M. P., and I. Noy-Meir. 1971. The problem of non-linearity in ordination: Experiments with two-gradient models. *J. Ecol.* 59:763–73.

Bakken, J. S., T. Borody, L. J. Brandt, J. V. Brill, D. C. Demarco, M. A Franzos, C. Kelly, A. Khoruts, T. Louie, L. M. Martinelli, A. Thomas A., G. Russell, and C. Surawicz. 2011. Treating *Clostridium difficile* infection with fecal microbiota transplantation. *Clin. Gastroenterol. Hepatol.* 9:1044–49.

Baltzer, F. 1964. Theodor Boveri. *Science* 144: 809–15.

Barclay, H., and P. Van Den Driessche. 1975. Time lags in ecological systems. *J. Theoret. Biol.* 51:347–56.

Bartell, S. M. 1973. A multivariate statistical approach to the phytoplankton community structure and dynamics in the Lake Wingra ecosystem. Master's thesis, University of Wisconsin, Madison.

———. 1978. Size selective planktivory and phosphorous cycling in pelagic systems. PhD diss., University of Wisconsin, Madison.

Bastin, T. 1968. A general property in hierarchies. In *Sketches*, edited by C. H. Waddington, vol. 2 of *Towards a Theoretical Biology*, 252–64. Aldine, Chicago.

Bateson, G. 1979. *Mind and Nature.* Dutton, New York.

Beals, E. W. 1958. The phytosociology of the Apostle Islands and the influence of deer on the vegetation. Master's thesis, University of Wisconsin, Madison.

———. 1973. Ordination: Mathematical elegance and ecological naïveté. *J. Ecol.* 61:23–36.

Begley, S. 2009. How genes may be altered by experience. *Newsweek.* http://www.newsweek.com/how-genes-may-be-altered-experience-begley-79507.

Bekoff, M., and M. C. Wells. 1980. The social ecology of coyotes. *Sci. Amer.* 242:130–48.

Bennett, W. R., Jr. 1977. How artificial is intelligence? *Amer. Scientist* 65: 694–702.

Bertalanffy, L. von. 1968. *General Systems Theory.* Braziller, New York.

———. 1975. *Perspectives on General Systems Theory.* Edited by E. Taschdjian. Braziller, New York.

Besteman, C. 2002. The Cold War, clans, and chaos in Somalia: A view from the ground. In *The State, Identity and Violence: Political Disintegration in the Post-Cold War World*, edited by R. B. Ferguson, 285–99. Routledge, London.

Bledsoe, L. J., R. C. Francis, G. L. Swartzman, and J. D. Gustafson. 1977. *PWNEE, a Grassland Ecosystem Model.* Technical Report 64, Grassland Biome, Natural Resource Ecology Lab, Colorado State University, Fort Collins.

Blum, H. 1973. Biological shape and visual science. Part 1. *J. Theoret. Biol.* 38:205–87.

Bonilla C.A., Norman J.M., Molling M.M., Karthikeyan K.G., Miller P.S. 2008. Testing a grid-based soil erosion model across topographically complex landscapes. *Soil Sci. Soc. Am. J.* 72:1745–55.

Bonner, J. T. 1961. Introduction to *On Growth and Form*, by D'Arcy Thompson, vii–xiv. Cambridge University Press, Cambridge.

——— . 1965. Size and cycle—an essay on the structure of biology. *Amer. Scientist* 53:488–94.

Boodin, J. E. 1943. Analysis and holism. *Philos. Sci.* 10:213–29.

Bormann, F. H., and G. E. Likens. 1979. *Pattern and Process in a Forested Ecosystem.* Springer-Verlag, New York.

Bormann, F. H., T. G. Siccama, R. S. Pierce, and J. S. Eaton. 1974. The export of nutrients and recovery of stable conditions following deforestation at Hubbard Brook. *Ecol. Monog.* 44:255–77.

Bosserman, R. W. 1983. Elemental composition of *Utricularia*-periphyton ecosystems from Okefenokee Swamp. *Ecology* 64:1637–45.

Bosserman, R. W., and F. Harary. 1981. Demiarcs, creaons and genons. *J. Theor. Biol.* 92:241–54.

Box, G. E. P., and N. R. Draper. 1987. *Empirical Model Building and Response Surfaces.* John Wiley & Sons, New York.

Box, G. E. P., and G. M. Jenkins. 1970. *Time Series Analysis: Forecasting and Control.* Holden-Day, San Francisco.

Boyd, V.T., and T. F. H. Allen. 1981a. Value ascribed to objects: Multivariate elucidation of a conceptual structure. *J. Biol. Soc. Struc.* 4:39–57.

——. 1981b. Liking and disliking household objects: An empirical study of value. *Home Economics Research Journal* 9:310–18.

Bradfield, G. E., and L. Orloci. 1975. Classification of vegetation from an open beach environment in South Western Ontario: Cluster analysis followed by a generalized distance assignment. *Can. J. Bot.* 53:495–502.

Brandner, T. A. 2003. Reconceptualizing biome: A complex systems theoretical approach to understanding extinction events. PhD diss., University of Wisconsin, Madison.

Bray, J. R., and J. T. Curtis. 1957. An ordination of the upland forest communities of Southern Wisconsin. *Ecol. Monogr.* 27:325–49.

Brown, B. J., and T. F. H. Allen. 1989. The importance of scale in evaluating herbivory impacts. *Oikos* 54:189–94.

Brown, J. H, W. R. Burnside, A. D. Davidson, J. P. DeLong, W. C. Dunn, M. J. Hamilton, N. Mercado-Silva, J. C. Nekola, J. G. Okie, W. H. Woodruff, and W. Zuo. 2011. Energetic limits to economic growth. *BioScience* 61:19–26.

Buchanan, M. 2013. *Forecast: What Physics, Meteorology and the Natural Sciences Can Teach about Economics.* Bloomsbury, New York.

Burgers, J. M. 1975. Causality and anticipation. *Science* 189:194–98.

Burnet, F. M. 1971. Self-recognition in colonial marine forms and flowering plants in relation to the evolution of immunity. *Nature* 232:230–35.

Carneiro, R. L. 2000. The transition from quantity to quality: A neglected causal mechanism in accounting for social evolution *PNAS* 97:12926–31. DOI:10.1073/pnas.240462397.

Carosella, T. L. 1978. Population responses of *Opuntia compressa* (Salisb.) Mcbr. in a southern Wisconsin sand prairie. Master's thesis, University of Wisconsin, Madison.

Carpenter, F. L., and R. E. MacMillen. 1976. Threshold model of feeding territoriality and test with a Hawaiian honeycreeper. *Science* 194:639–42.

Checkland, P. B. 1988. The case for 'holon'. *Systems Practice* 1:235–38.

Child, C. M. 1906.The relation between regulation and fission in Planaria. *Biol. Bull.* 11:113–23.

——. 1924. *Physiological Foundations of Behaviour.* Holt, New York.

Clements, F. E. 1916. *Plant Succession: An Analysis of Vegetation.* Carnegie Institution of Washington, Washington.

Coleman, D. C. 2010. *Big Ecology: The Emergence of Ecosystem Science.* University of California Press, Oakland.

Colinvaux, P. 1979. *Why Big Fierce Animals Are Rare.* Princeton University Press, Princeton.

Connell, J. H. 1980. Diversity and the coevolution of competitors, or the ghost of competition past. *Oikos* 35:131–38.

Cooper, A., C. Turney, K. A. Hughen, B. W. Brook, H. G. McDonald, and C. J. A. Bradshaw. 2015. Abrupt warming events drove Late Pleistocene Holarctic megafaunal turnover. *Science* 349:602–6.

Cooper, W. S. 1926. Fundamentals of vegetational change. *Ecology* 7:391–414.

Cottam, G., J. T. Curtis, and A. J. Catana. 1957. Some sampling characteristics of a series of aggregated populations. *Ecology* 38:610–22.

Cronon, W. 1992. A place for stories: nature, history, and narrative. J. Am. Hist. 78:1347–76.

Curtis, J. T. 1959. *The Vegetation of Wisconsin.* University of Wisconsin Press, Madison.

Curtis, J. T., and R. P. McIntosh. 1951. An upland forest continuum of the prairie-forest border region of Wisconsin. *Ecology* 32:476–96.

Dawkins, Richard. 1978. *The Selfish Gene.* Oxford University Press, New York.

DeAngelis, D. L., W. M. Post, and C. C. Travis. 1986. *Positive Feedback in Natural Systems.* Springer-Verlag, Berlin.

Deutsch, K. W. 1953. *Nationalism and Social Communication.* Wiley, New York.

——. 1956. Shifts in the balance of international communication flows. *Public Opin. Q.* 20:143–60.

——. 1966. Boundaries according to communications theory. In *Toward a Unified Theory of Human Behaviour,* edited by R. Grinken, 278–97. Basic, New York.

Dodds, W. K. 2009. *Laws, Theories, and Patterns in Ecology.* University of California Press, Oakland.

Dudley, P. N., R. Bonazza, and W. P. Porter. 2013. Consider a non-spherical elephant: Computational fluid dynamics simulations of heat transfer coefficients and drag verified using wind tunnel experiments. *J. Exp. Zool.* 9999:1–9.

Eglash, R. 1999. *African Fractals: Modern Computing and Indigenous Design.* Rutgers University Press, Piscataway.

Eigen, M. 1977. The hypercycle: Principle of natural self-organization. *Naturwissenschaften* 64:541–65.

Emanuel, W. R., D. C. West, and Shugart, H. H. 1978. Spectral analysis of forest model time series. *Ecol. Mod.* 4:313–26.

Evans, L. T. 1951. Field study of the social behavior of the black lizard, *Ctenosaura pectinata. Amer. Mus. Novitates* 1493:1–26.

Fisher, R. A. 1925. *Statistical Methods for Research Workers.* Oliver & Boyd, Edinburgh.

Fitch, H. S. 1940. A field study of the growth and behavior of the fence lizard. *Univ. Calif. Publ. Zool.* 44:151–72.

Forrester, J. 1968. *Principles of Systems.* Productivity, New York.

Foley, J. 1999. *Homer's Traditional Art.* Penn State University Press, University Park.

Freudenburg, William R. 1992. Addictive economies: Extractive industries and vulnerable localities in a changing world economy. *Rural Sociol.* 57 (3): 305–32.

Freudenburg, William R., Scott Frickel, and Rachel Dwyer. 1998. Diversity and diversion: Higher superstition and the dangers of insularity in science and technology studies. *Int. J. Sociol. Soc. Pol.* 18:3–32.

Freudenburg, William R., and Robert Gramling. Linked to what? Economic linkages in an extractive economy. *Soc. Nat. Resour.* 11:569–86.

Funtowicz, S., and J. Ravetz. 1992. The good, the true and the post-modern. *Futures* 24:963–76.

Gall, P. L., and A. A. Saxe. 1977. The ecological evolution of culture: The state as predator in succession theory. In *Exchange Systems in Prehistory,* edited by T. K. Earle and J. E. Ericson, 255–68. Academic Press, New York.

Gardner, M. R., and W. R. Ashby. 1970. Connectance of large dynamic (cybernetic) systems: Critical values for stability. *Nature* 228:784.

Giampietro, M. 2004. *Multi-scale Integrated Analysis of Agroecosystems.* CRC, Boca Raton.

Giampietro, M., T. F. H. Allen, and K. Mayumi. 2006. The epistemological predicament associated with purposive quantitative analysis. *Ecol. Complex.* 90:1–21.

Giampietro, M., R. J. Aspinall, J. Ramos-Martin, and S. G. F. Bukkens, eds. 2014. *Resource Accounting for Sustainability Assessment: The Nexus between Energy, Food, Water and Land Use.* Routledge, London.

Gigerenzer, G. 2004. Mindless statistics. *J. Socio. Econ.* 33:587–606.

———. 2012. What can economists know? Rethinking the basis of economic understanding. Institute for New Economic Thinking, Berlin, September 12. http://slightlytilted.wordpress.com/tag/gerd-gigerenzer/.

———. 2013. HIV screening: Helping clinicians make sense of test results to patients. *BMJ* 347:f5151.

Gigerenzer, G., and H. Brighton. 2009. *Homo heuristicus*: Why biased minds make better inferences. *Top. Cogn. Sci.* 1:107–43.

Gigerenzer, G., W. Gaissmaier, E. Kurz-Milcke, L. M. Schwartz, and S. Woloshin. 2007. Helping doctors and patients to make sense of health statistics. *Psychol. Sci. Public Interest* 8:53–96.

Gigerenzer, G., and M. Galesic. 2012. Why do single event probabilities confuse patients? Statements of frequency are better for communicating risk. *BMJ* 344:e245. DOI:10.1136/bmj.e245.

Gigerenzer, G., and D. G. Goldstein. 1996. Reasoning the fast and frugal way: Models of bounded rationality. *Psychological Review* 103:650–69.

Gigerenzer, G., and J. A. Muir Gray, eds. 2011. *Better Doctors, Better Patients, Better Decisions: Envisioning Health Care 2020*. MIT Press, Cambridge, MA.

Gigerenzer, G., and T. Sturm. 2012. How (far) can rationality be naturalized? *Synthese* 187:243–68.

Gigerenzer, G., and O. Wegwarth. 2013. Five year survival rates can mislead. *BMJ* 346:f548. DOI:10.1136/bmj.f54.

Gill, F. B., and L. L. Wolf. 1975. Economics of feeding territoriality in the golden-winged sunbird. *Ecology* 56:333–45.

Gleason, H. A. 1926. The individualistic concept of the plant association. *Contrib. NY Bot. Gard.* 279.

Gold, H. J. 1977. *Mathematical Modeling of Biological Systems: An Introductory Guidebook*. Wiley, New York.

Goodall, D. W. 1953. Objective methods for the classification of vegetation. 1. The use of positive interspecific correlation. *Aust. J. Bot.* 1:39–63.

——. 1974. Problems of scale and detail in ecological modeling. *J. Env. Man.* 2:149–57.

Gould, S. J. 1981. The ultimate parasite. *Natural History* 90:7–14.

Greig-Smith, P. 1952. The use of random and contiguous quadrats in the study of the structure of plant communities. *Ann. Bot.* 16:293–316.

——. 1964. *Quantitative Plant Ecology*. Butterworth, London.

Greig-Smith, P., M. P. Austin, and T. C. Whitmore. 1967. The application of quantitative methods to vegetation survey. 1. Association analysis and principal component ordination of rain forest. *J. Ecol.* 55:483–504.

Grime, J. P. 1974. Vegetation classification by reference to strategies. *Nature* 250:26–31. DOI:10.1038/250026a0.

——. 1977. Evidence for the existence of three primary strategies in plants and its relevance to ecological and evolutionary theory. *Am. Nat.* 111:1169–94.

——. 1985. Towards a functional classification of vegetation. In *The Population Structure of Vegetation,* edited by J. White, 503–14. Dr. W. Junk, Dordrecht, Netherlands.

——. 2001. *Plant Strategies, Vegetation Processes and Ecosystem Properties.* 2nd ed. Wiley, Chichester.

——. 2007. Plant strategy theories: A comment on Craine (2005). *J. Ecol.* 95:227–30. DOI:10.1111/j.1365-2745.2006.01163.x.

Grime, J. P., and A. V. Curtis. 1976. The interaction of drought and mineral nutrient stress in calcareous grassland. *J. Ecol.* 64:976–98.

Grime, J. P., J. G. Hodgson, and R. Hunt, eds. 2007. *Comparative Plant Ecology: A Functional Approach to Common British Species.* Castlepoint, Dalbeattie, UK.

Grime, J.P., K. Thompson, R. Hunt, J. G. Hodgson, J. H. C. Comelissen, I. H. Rorison, G. A. F. Hendry, T. W. Ashenden, A. P. Askew, S. R. Band, R. E. Booth, C. C. Bossard, B. D. Campbell, J. E. L. Cooper, A. W. Davison, P. L. Gupta, W. Hall, D. W. Hand, M. A. Hannah, S. H. Hillier, D. J. Hodkinson, A. Jalili, Z. Liu, J. M. L. Mackey, N. Matthews, M. A. Mowforth, A. M. Neal, R. J. Reader, K. Reiling, W. Ross-Fraser, R. E. Spencer, F. Sutton, D. E. Tasker, P. C. Thorpe, and J. Whitehouse. 1997. Integrated screening validates primary axes of specialisation in plants. *Oikos* 79:259–81.

Guest, D. 2007. Black pod: Diverse pathogens with a global impact on cocoa yield. *Phytopathology* 97:1650–53.

Gunderson, L., and C. S. Holling. 2001. *Panarchy: Understanding Transformations in Systems of Humans and Nature.* Island, Washington, DC.

Guo, K., K. Meints, C. Hall, S. Hall, and D. Mills. 2009. Left gaze bias in humans, rhesus monkeys and domestic dogs. *Anim. Cogn.* 12:409–18. DOI:10.1007/s10071-008-0199-3.

Gustafson, J., and Cooper L. 1990. *The Modern Contest.* Norton, New York.

Haefner, J. W. 1980. Two metaphors of the niche. *Synthese* 43:123–53.

Harmon, L. D. 1973. The recognition of faces. *Scientific American* 229:71–82.

Harper, J. L. 1967. A Darwinian approach to plant ecology. *J. Ecol.* 55:247–70.

Hayles, N. K. 1991. Constrained constructivism: locating scientific inquiry in the theater of representation. *New Orleans Review* 18:76–85.

Hector A., B. Schmid, C. Beierkuhnlein, M. C. Caldeira, M. Diemer, P. G. Dimitrakopoulos, J. A. Finn, H. Freitas, P. S. Giller, J. Good, R. Harris, P. Högberg, K. Huss-Danell, J. Joshi, A. Jumpponen, C. Körner, P.W. Leadley, M. Loreau, A. Minns, C. P. H. Mulder, G. O'Donovan, S. J. Otway, J. S. Pereira, A. Prinz, D. J. Read, M. Scherer-Lorenzen, E. D. Schulze, A. S. D. Siamantziouras, E. M. Spehn, A. C. Terry, A. Y. Troumbis, F. I. Woodward, S. Yachi, and J. H. Lawton. 1999. Plant diversity and productivity experiments in European grasslands. *Science* 286:1123–27.

Heglund, N. C., C. R. Taylor, and T. A. McMahon. 1974. Scaling stride frequency and gait to animal size: Mice to horses. *Science* 186:1112–13.

Herrick, C. J. 1956. *The Evolution of Human Nature.* University of Texas Press, Austin.

Hillenmeyer, M. E, E. Fung, J. Wildenhain , S. E. Pierce, S. Hoon, W. Lee, M. Proctor, R. P. St. Onge , M. Tyers , D. Koller , R. B. Altman , R. W. Davis , C. Nislow, and G. Giaever. 2008.The chemical genomic portrait of yeast: uncovering a phenotype for all genes. *Science* 320:362–65. DOI:10.1126/science.1150021.

Ho, M. W., and P. T. Saunders. 1979. Beyond neo-Darwinism—an epigenetic approach to evolution. *J. Theoret. Biol.* 78:573–91.

Holden, C. 1977. The empathic computer. *Science* 198:32.

Holling, C. S. 1959a. The components of predation as revealed by a study of small mammal predation of the European pine sawfly. *Can. Entomol.* 91:293–320.

——. 1959b. Some characteristics of simple types of predation and parasitism. *Can. Entomol.* 91:385–98.

——. 1965. The functional response of predators to prey density and its role in mimicry and population regulation. *Mem. Entomol. Soc. Can.* 45:1–60.

——. 1973. Resilience and stability of ecological systems. *Ann. Rev. Ecol. Syst.* 4:1–24.

——. 1986. The resilience of terrestrial ecosystems: Local surprise and global change. In *Sustainable Development of the Biosphere*, edited by W. C. Clark and R. E. Munn, 292–317. Cambridge University Press, Cambridge.

Holling, C. S., and S. Ewing. 1971. Blind man's bluff: Exploring the response space generated by realistic ecological simulation models. In *Statistical Ecology*, edited by G. P. Patil, E. C. Pielou, and W. E. Waters, vol. 2 of *Proceedings of the International Symposium on Statistical Ecology*, 207–29. Penn State University Press, University Park.

Hotelling, H. 1933. Analysis of a complex of statistical variables into principal components. *J. Educ. Psychol.* 24:417–41, 498–520.

Huston, M. A., L. W. Aarssen, M. P. Austin, B. S. Cade, J. D. Fridley, E. Garnier, J. P. Grime, J. Hodgson, W. K. Lauenroth, K. Thompson, J. H. Vandermeer, and D. A. Wardle. 2000. No consistent effect of plant diversity on productivity. *Science* 289:1255.

Hutchinson, G. E. 1957. Concluding remarks. *Cold Spring Harb. Symp. Quant. Biol.* 22:415–27. Reprinted in 1991: Classics in Theoretical Biology. *Bull. of Math. Biol.* 53:193–213.

——. 1959. Homage to Santa Rosalia, or why are there so many kinds of animals? *Amer. Natur.* 93:145–59.

——. 1961. The paradox of the plankton. *Amer. Natur.* 95:137–45.

Ioannidis, J. P. A. 2005. Why most published research findings are false. *PLOS Med* 2:e124. DOI:10.1371/journal.pmed.0020124.

Jackson, D., and Laura L. Jackson, eds. 2002. *The Farm as Natural Habitat: Reconnecting Food Systems with Ecosystems.* Island, Washington, DC.

Jackson, L. L. 1999. Establishing tallgrass prairie species on a rotationally grazed permanent pasture in the Upper Midwest: remnant plant assessment and seeding and grazing regimes. *Restoration Ecology* 7:127–38.

——. 2008. Who designs the agricultural landscape? *Landscape Jrnl.* 27:23–40.

Jantsch, E. 1976. Evolution: Self-realization through self-transcendence. In *Evolution and Consciousness: Human Systems in Transition*, edited by E. Jantsch and C. Waddington, 37–70. Addison-Wesley, Reading, MA.

Johnson, S. 2002. *Emergence: The Connected Lives of Ants, Brains, Cities and Software.* Simon and Schuster, New York.

Juarrero, A. 2002. *Dynamics in Action: Intentional Behavior as a Complex System.* MIT Press, Cambridge, MA.

Kays, S., and J. L. Harper. 1974.The regulation of plant and density in a grass sward. *J. Ecol.* 62:97–105.

Kiester, A. R., and M. Slatkin. 1974. A strategy of movement and resource utilization. *Theoret. Population Biol.* 6:1–20.

Koestler, A. 1967. *The Ghost in the Machine.* Macmillan, New York.

Korzybski, A. 1933. A non-Aristotelian system and its necessity for rigour in mathematics and physics. In *Science and Sanity*, 747–61. International Non-Aristotelian Library.

Kuhn, T. S. 1970. *The Structure of Scientific Revolutions.* 2nd ed. University of Chicago Press, Chicago.

Laing, R. D. 1967. The Politics of Experience. Routledge & Kegan Paul.

Lane, N. 2005. *Power, Sex, Suicide: Mitochondria and the Meaning of Life.* Oxford University Press, Oxford.

Lane, N., and W. Martin. 2015. Eukaryotes really are special, and mitochondria are why. *Proc. Natl. Acad. Sci. USA* 112. DOI:10.1073/pnas.1509237112.

Layzer, D. 1975. The arrow of time. *Sci. Amer.* 233:56–69.

Lazlo, E., ed. 1972. *The Relevance of General Systems Theory.* Braziller, New York.

Leeuw, J. De. 2014. History of nonlinear principal component analysis. In *Multiple Correspondence Analysis and Related Methods*, edited by J. Blasius and M. Greenacre, chapter 4. CRC, Boca Raton.

Leigh, E. G. 1965. On the relation between the productivity, biomass, diversity, and stability of a community. *Proc. Natl. Acad. Sci.* 53:777–83.

Leonard, T. J., R. F. Gaber, and S. Dick. 1978. Internuclear genetic transfer in dikaryons of *Schizophyllum commune*. II. Direct recovery and analysis of recombinant nuclei. *Genetics* 89:685–93.

Levandowsky, M., and B. S. White. 1977. Randomness, time scales, and the evolution of biological communities. *Evol. Biol.* 10:69–161.

Levin, Simon. 1992. The problem of pattern and scale in ecology. *Ecology* 73:1943–67.

Levin, Simon, and R. T. Paine. 1974. Disturbance, patch formation, and community structure. *Proc. Natl. Acad. Sci. USA* 571:27447.

Levin, Steven. 1986. The icosahedron as the three-dimensional finite element in biomechanical support. In *Proceedings of the International Conference on*

Mental Images, Values and Reality, edited by J. Dillon. Intersystems, Salinas, CA.

Levins, R. 1974. The qualitative analysis of partially specified systems. *Ann. NY Acad. Sci.* 123:38.

Levins, R., and R. Lewontin. 1980. Dialectics and reductionism in ecology. *Synthese* 43:47–78.

Lewis, W. M., Jr. 1976. Surface/volume ratios: Implications for phytoplankton morphology. *Science* 192:885–87.

Lewontin, R. C. 1968. On the irrelevance of genes. In *Towards a Theoretical Biology*, edited by C. H. Waddington, 63–72. Aldine, Chicago.

———. 1969. The meaning of stability. *Brookhaven Symp. Biol.* 22:13–24.

———. 1978. Adaptation. *Sci. Amer.* 239:212–30.

Likens, G. E., F. H. Bormann, N. M. Johnson, and R. S. Pierce. 1967. The calcium, magnesium and potassium budgets for a small forested ecosystem. *Ecology* 48:772–85.

Likens, G. E., F. H. Bormann, R. S. Pierce, and N. M. Johnson. 1977. *Biogeochemistry of a Forested Ecosystem*. Springer-Verlag, New York.

Likens, G. E., F. H. Bormann, R. S. Pierce, and W. A. Reiners. 1978. Recovery of a deforested ecosystem. *Science* 199:492–96.

Linderman, F. B. 2002. *Plenty-Coups, Chief of the Crows*. 2nd ed. University of Nebraska Press, Lincoln.

Linteau, A. 1955. Forest site classification of the Northern Coniferous Section, Boreal Forest Region, Quebec. Bulletin 118. Canada Department of Northern Affairs & Northern Resources, Forest Branch.

Lissack, M. 2015. Orthogonal is a design concept: Incommensurable is not. ResearchGate website. https://www.researchgate.net/publication/281968745_Orthogonal_is_a_design_concept_Incommensurable_is_not.

Lissack, M., and A. Gaber, eds. 2014. *Modes of Explanation: Affordances for Action and Prediction*. Palgrave MacMillan, Basingstoke.

Logan, Robert K. 2004. *The Alphabet Effect: A Media Ecology Understanding of the Making of Western Civilization*. Hampton, New York.

Lorenz, E. 1968. Climatic determinism. *Meteor. Monographs* 25:1–3.

Lotka, A. J. 1956. *Elements of Mathematical Biology*. Dover, New York.

Loucks, O. L. 1962. Ordinating forest communities by means of environmental scalars and phytosociological indices. *Ecol. Monogr.* 32:137–66.

———. 1970. Evolution of diversity, efficiency, and community stability. *Am. Zool.* 10:17–25.

Louie, A. H. 2009. *More than Life Itself: A Synthetic Continuation in Relational Biology (Categories)*.Ontos Verlag, Heusenstamm, Germany.

Lovelock, J. E. 1989. *The Ages of Gaia*. Oxford University Press, Oxford, UK.

Lovett, D. 1982. The effects of allelochemicals on crop growth and development. In *Chemical Manipulation of Crop Growth and Development*, edited by J. S. McLaren, 93. Butterworth, London.

Lovett, D., and G. Sagar. 1978. Influence of bacteria in the phyllosphere of *Camelina sativa* (L.) Crantz on germination of *Linum usitatissimum* L. *New Phytol.* 81:617–25.

MacArthur, R. 1967. The theory of the niche. In *Proceedings of the International Symposium on Population Biology and Evolution*, edited by R. C. Lewontin, 157–76. Syracuse, New York.

——. 1972. Strong, or weak, interactions? In Growth by intussusception: Ecological essays in honor of G. Evelyn Hutchinson, edited by E. S. Deevey. *Trans. Conn. Acad. Arts Sci.* 44:177–88.

MacArthur, R., and R. Levins. 1964. Competition, habitat selections, and character displacement in a patchy environment. *Proc. Natl. Acad. Sci.* 51:1207–10.

MacArthur, R., and E. Pianka. 1966. On optimal use of a patchy environment. *Am. Nat.* 100:603–9.

Mackay, A. L. 1973. How to organize a typists' pool—operational analysis of an old metaphor. *J. Theoret. Biol.* 40:203–4.

Magnuson, John J. 1988. Two worlds for fish recruitment: Lakes and oceans. *Am. Fish. Soc. Sym.* 5:1–6.

Mandelbrot, B. 1967. How long is the coast of Britain? Statistical self-similarity and fractional dimension. *Science*, New Series, 156:636–38.

Mansfield, K. L., J. Wyneken, W. P. Porter, and J. Luo. 2014. First satellite tracks of neonate sea turtles redefine the "lost years" oceanic niche. *Proc. R. Soc. B* 281:20133039.

Margalef, R. 1968. *Perspectives in Ecological Theory.* University of Chicago Press, Chicago.

——. 1972. Homage to Evelyn Hutchinson, or why there is an upper limit to diversity. In Growth by intussusception: Ecological essays in honor of G. Evelyn Hutchinson, edited by E. S. Deevey. *Trans. Conn. Acad. Arts Sci.* 44: 213–35.

Margulis, L. 1998. *Symbiotic Planet : A New Look at Evolution.* Basic, New York.

Margulis, L., and M. J. Chapman. 2009. *Kingdoms and Domains.* 4th ed. Academic/ Elsevier, New York.

May, R. M. 1973a. Time-delay versus stability in population models with two and three trophic levels. *Ecology* 54:315–25.

——. 1973b. *Stability and Complexity in Model Ecosystems.*. Monographs in Population Biology 6. Princeton University Press, Princeton.

——. 1974. Biological populations with non-overlapping generations: stable points, stable cycles and chaos. *Science* 186:645–47.

——. 1976a. Models for single populations. In *Theoretical Ecology Principles and Applications*, edited by R. M. May, 4–25. Saunders, Philadelphia.

——. 1976b. Patterns in multi-species communities. In *Theoretical Ecology Principles and Applications,* edited by R. M. May, 142–62. Saunders, Philadelphia.

——, ed. 1981. *Theoretical Ecology: Principles and Applications.* 2nd ed. Saunders, Philadelphia.

May, R. M., and G. F. Oster. 1976. Bifurcations and dynamic complexity in simple ecological models. *Amer. Natur.* 110:573–99.

Maynard Smith, J. 1974. *Models in Ecology.* Cambridge University Press, Cambridge.

McCune, B., and T. F. H. Allen. 1985a. Will similar forests develop on similar sites? *Can. J. Bot.* 63:367–76.

——. 1985b. Forest dynamics in the Bitterroot Canyons, Montana. *Can. J. Bot.* 63:377–383.

McIntosh, R. P. 1967. The continuum concept of vegetation. *Bot. Rev.* 33:130–87.

——. 1975. H. A. Gleason—individualistic ecologist, 1882–1975. *Bull. Torrey Bot. Club* 102:253–73.

——. 1980. The background and some current problems of theoretical ecology. *Synthese* 43:195–255.

McKusick, V. A. 1964. Walter S. Sutton and the physical basis of Mendelism. *Bull. Hist. Med.* 34:487–97.

McLuhan, H. M., and Q. Fiore. 1967. *The Medium Is the Massage.* Random House, New York.

McLuhan, M. 1964. *Understanding Media: The Extensions of Man.* Latimer Trend, London.

McLuhan, M., and R. K. Logan.1977. Alphabet, mother of invention. *Etcetera* 34:373–83.

McMahon, T. A. 1975. The mechanical design of trees. *Sci. Amer.* 233:92–102.

McSweeney, K. 2007. Bananas. In *Encyclopedia of Environment and Society,* edited by P. Robbins, 100–101. Sage, Thousand Oaks, CA.

Mech, L. D. 1977. Wolf pack buffer zones as prey reservoirs. *Science* 168:320–21.

Mendel, G. 1865. *Versuche über Pflanzen-Hybriden. Transactions of Verhandlungen des naturforschenden Vereines in Brünn (1865).*1866; iv:3–270. [Extracts republished. *BMJ* 1965;i:367–74.].

Metcalf, J., C. Turney, R. Barnett, M. Fabiana, S. C. Bray, J. T. Vilstrup, L. Orlando, R. Salas-Gismond, D. Loponte, M, Medina, M. De Nigris, T. Civalero, P. Marcelo Fernández, A. Gasco, V. Duran, K. L. Seymour, C. Otaola, A. Gil, R. Paunero, F. J. Prevosti, C. J. A. Bradshaw, J. C. Wheeler, L. Borrero, J. J. Austin, and A. Cooper. 2016. Synergistic roles of climate warming and human occupation in Patagonian megafaunal extinctions during the Last Deglaciation. *Sci. Adv.* 2:e1501682. DOI:10.1126/sciadv.1501682.

Meyerowitz, E. M., and R. E. Pruit. 1985. *Arabidopsis thaliana* and plant molecular genetics. *Science* 229:1214–18.

Middendorf, G. 1977. Resource partitioning by iguanid lizards: Thermal and density influences. PhD diss., University of Knoxville, Tennessee.

Miller, James G. 1978. *Living Systems.* McGraw-Hill, New York.

Minchin, P. 1987. An evaluation of the relative robustness of techniques for ecological ordination. *Vegetatio* 69:89–107.

Moorhead, P. S., and M. M. Kaplan, eds. 1967. *Mathematical Challenges to the NeoDarwinian -Interpretation of Evolution.* Wistar Institute Press, Philadelphia.

Morgan, T. H. 1901. Regeneration and liability to injury. *Science* 14:235–48.

——. 1910. Sex-limited inheritance in *Drosophila. Science* 32:120–22.

Moritz, C. H. 1854. *Historical Development of Speculative Philosophy, from Kant to Hegel.* T. & T. Clark, Edinburgh.

Murdoch, W. W. 1975. Diversity, stability and pest control. *J. Appl. Ecol.* 12: 795–808.

Needham, J. 1988. The limits of analysis. *Poetry Nation Rev.* 16:35–38.

Nichols, G. E. 1923. A working basis for ecological classification of plant communities. *Ecology* 4:11–23.

Nicolis, G., and I. Prigogine. 1977. *Self-Organization in Nonequilibrium Systems-: From Dissipative Structures to Order through Fluctuations.* Wiley, Interscience, New York.

Noble, D. 2003. Evolution and physiology: A new synthesis. Lecture at Voices of Oxford. http://www.voicesfromoxford.org/video/evolution-and-physiology-a-new-synthesis/355.

——. 2010. Letter from Lamarck. *Physiol. News* 78:31.

——. 2013a. Life changes itself via genetic engineering. *Phys. Life Rev.* 10:344.

——. 2013b. Systems biology and reproduction. *Prog. Biophys. Mol. Biol.* 113:355.

——. 2013c. Physiology is rocking the foundations of evolutionary biology. *Exp. Physiol.* 98:1235–43. DOI:10.1113/expphysiol.2012.071134.

Norman, J. M. 2011. Intellectual inertia: An uneasy tension between collective validation of the known and encouraging exploration of the unknown. In: *Sustaining Soil Productivity in Response to Global Climate Change: Science, Policy and Ethics*, edited by T. J. Sauer, J. M. Norman, M. V. K. Sivakumar, 17–30. John Wiley and Sons, New York.

——. 2013. Fifty years of study of S-P-A systems: Past limitations and a future direction. *Procedia Environmental Sciences* 19:15–25.

Noy-Meir, I. 1973. Data transformations in ecological ordination. I. Some advantages of non-centering. *J. Ecol.* 61:329–42.

Noy-Meir, I., and D. J. Anderson. 1971. Multiple pattern analysis, or multi-scale ordination: Towards a vegetation hologram. In *Statistical Ecology*, vol. 3 of *Proceedings of the International Symposium on Statistical Ecology*, edited by G. P. Patil, E. C. Pielou, and W. E. Waters, 207–31. Penn State University Press, University Park .

Noy-Meir, I., D. Walker, and W. T. Williams. 1975. Data transformations in ecological ordination. II. On the meaning of data standardization. *J. Ecol.* 63:779–800.

Odum, H. T., and M. T. Brown. 2007. *Environment, Power and Society for the Twenty-First Century: The Hierarchy of Energy.* Columbia University Press, New York.

O'Neill, R. V., B. S. Ausmus, D. R. Jackson, R. van Hook, P. van Voris, C. Washburne, and A. P. Watson. 1977. Monitoring terrestrial ecosystems by analysis of nutrient export. *Water Air Soil Pollut.* 8:271–77.

O'Neill, R. V., D. DeAngelis, J. Waide, and T. F. H. Allen. 1986. A hierarchical concept of ecosystems. *Monographs in Population Biology* 23:1–272.

O' Neill, R. V., A. R. Johnson, and A. W. King. 1989. A hierarchical framework for the analysis of scale. *Landscape Ecol.* 3:193–205.

Orgel, L. E., and F. H. C. Crick. 1980. Selfish DNA: the ultimate parasite. *Nature* 284:604–7. DOI:10.1038/284604a0.

Overton, W. S. 1975a. The ecosystem modeling approach in the Coniferous Biome. In *Systems Analysis and Simulation in Ecology*, vol. 3, edited by B. C. Patten, 117–38. Academic, New York.

——. 1975b. Decomposability: a unifying concept? In *Ecosystem Analysis and Prediction*, edited by S. A. Levin, 297–98. Society of Industrial and Applied Mathematics, Philadelphia.

——. 1975c. Decomposability: A unifying concept? In *Proceedings of the SIAMSIMS Conference on Ecosystem Analysis and Prediction-*, edited by S. Levin, 297. Society for Industrial and Applied Mathematics, Philadelphia.

Overton, W. S., and C. White. 1981. On constructing a hierarchical model in the FLEX paradigm, illustrated by structural aspects of a hydrological model. *Int. J. Gen. Syst.* 6:191–216.

Parkhurst, D. G., and o. L. Loucks. 1972. Optimal leaf size in relation to environment. *J. Ecol.* 60:505–37.

Patrick, R. 1967. The effect of invasion rate, species pool, and size of area on the structure of the diatom community. *Proc. Natl. Acad. Sci.* 58:1335–42.

Patrick, R., M. Hohn, and J. Wallace. 1954. A new method for determining the pattern of the diatom flora. *Notulae Naturae* 259:1–11.

Pattee, H. H. 1972. The evolution of self-simplifying systems. In *The Relevance of General Systems Theory,* edited by E. Lazlo, 31–42. Braziller, New York.

——. 1973. *Hierarchy Theory: The Challenge of Complex Systems.* Braziller, New York.

——. 1978. The complementarity principle in biological and social structures. *J. Soc. Biol. Structures* 1:191–200.

——. 1979. The complementarity principle and the origin of macro molecular information. *Biosystems* 11:217–26.

Patten, B. C. 1975. Ecosystem linearization: An evolutionary design problem. In *Proceedings of the SIAMSIMS Conference on Ecosystem Analysis and Prediction-*, edited by S. Levin, 182–202. Society for Industrial and Applied Mathematics, Philadelphia.

Patten, B. C., and G. T. Auble. 1980. Systems approach to the concept of niche. *Synthese* 43:155–81.

Patten, B. C., R. W. Bossermann, J. T. Finn, and W. G. Cole. 1976. Propagation of cause in ecosystems. In *Systems Analysis and Simulation in Ecology*, vol. 4, edited by B. C. Patten. Academic, New York.

Pearson, K. 1901. On lines and planes of closest fit to systems of points in space. *Phil. Mag.*, Series 6, 2:559–72.

Petroski, H. 2014. A story of two houses. *Amer. Scientist* 102: 258.

Piaget, J. 1971. *Genetic Epistemology*. W. W. Norton, New York.

Pigliucci, M. 2009. The end of theory in science? *EMBO Rep* 10:534. DOI: 10.1038/embor.2009.111.

Platt, J. 1969. Theorems on boundaries in hierarchical systems. In *Hierarchical Structures*, edited by L. L. Whyte, A. G. Wilson, and D. Wilson, 201–14. American Elsevier, New York.

Platt, T., and K. L. Denman. 1975. Spectral analysis in ecology. *Ann. Rev. Ecol. Syst.* 6:189–210.

Polanyi, M. 1962. *Personal Knowledge*. Harper Torchbook, New York.

——. 1968. Life's irreducible structure. *Science* 160:1308–12.

Porter, W. P., and D. M. Gates. 1969. Thermodynamic equilibria of animals with environment. *Ecol. Monogr.* 39:227–44.

Porter, W. P., J. W. Mitchell, W. A. Breckman, and C. B. DeWitt. 1973. Behavioral implications of mechanistic ecology: Thermal and behavioral modeling of desert ecosystems and their micro-environment. *Oecologia* 13:1–54.

Pound, R., and F. E. Clements. 1901. *The Phytogeography of Nebraska*. 2nd ed. Seminar, Lincoln, Nebraska.

Prahalad, C. K. 2004. *Fortune at the Bottom of the Pyramid: Eradicating Poverty through Profits*. Prentice Hall, Upper Saddle River, NJ.

Prigogine, I. 1978. Time, structure, and fluctuations. *Science* 201:777–85.

Prigogine, I., R. Lefever, A. Goldbeter, and M. Herschkowitz-Kaufman. 1969. Symmetry breaking instabilities in biological systems. *Nature* 223: 913–16.

Prigogine, I., and G. Nicolis. 1971. Biological order, structure, and instabilities. *Quart. Rev. Biophys.* 4:107–48.

Pringle, P. 2008. *The Murder of Nikolai Vavilov: The Story of Stalin's Persecution of One of the Great Scientists of the Twentieth Century*. Simon and Schuster, New York.

Rabotnov, T. A. 1969. On coenopopulations of perennial herbaceous plants in natural coenoses. *Vegetatio* 17:87–95.

Racca, A., E. Amadei, S. Ligout, K. Guo, K. Meints, D. Mills. 2010. Discrimination of human and dog faces and inversion responses in domestic dogs (*Canis familiaris*). *Anim. Cogn.* 13:525–33. DOI:10.1007/s10071-009-0303-3.

Racca, A., K. Guo, K. Meints, D.S. Mills. 2012. Reading faces: differential lateral gaze bias in processing canine and human facial expressions in dogs and 4-year-old children. *PLOS ONE* 2012;7:e36076. DOI:10.1371/journal.pone.0036076.

Radjou, N., and P. Jaideep. 2012. *Jugaad Innovation: Think Frugal, Be Flexible, Generate Breakthrough Growth.* Jossey Bass, Wiley, Chichester.

Rand, A. S., G. C. Gorman, and W. M. Rand. 1975. Natural history, behavior, and ecology of *Anolis agassizii. Smithsonian Contrib. Zoo.* 176:27–38.

Rashevsky, N. 1948. *Mathematical Biophysics: Physico-mathematical Foundations of Biology.* 2nd ed. University of Chicago Press, Chicago. Originally published 1938.

Ravetz, J. 2006. When communication fails. In *Interfaces between Science and Society*, edited by A. G. Pereira, S. Guedes Vaz, and S. Tognetti, 16–34. Greenleaf, Sheffield, UK.

Rice, B., and M. Westoby. 1983a . Species richness in vascular vegetation of the West Head, New South Wales. *Austral Ecology* 8:163–68.

——. 1983 b. Plant species richness at the 0.1 hectare scale in Australian vegetation compared to other continents. *Vegetatio* 52:129–40.

Riley, G. A. 1963. Untitled comment. In *Marine Biology I: Proceedings of the 1st International Interdisciplinary Conference of the AIBS*, edited by G. A. Riley, 70–71. AIBS, Washington.

Rindler, W. 1969. *Essential Relativity: Special, General, and Cosmological.* Van Nostrand Reinhold, New York.

Root, R. 1974. Some consequences of ecosystem texture. In *Proceedings of the SIAMSIMS Conference on Ecosystem Analysis and Prediction,* edited by S. Levin, 83–97. Society for Industrial and Applied Mathematics, Philadelphia.

Rosen, Robert. 1972. Tribute to Nicolas Rashevsky 1899–1972. *Progress in Theoretical Biology* 2:xi–xiv.

——. 1975. Biological systems as paradigms for adaptation. In *Adaptive Economic Models*, edited by R. H. Day and T. Groves, 39–72. Academic, New York.

——.1979. Anticipatory systems in retrospect and prospect. *General Systems* 24:11–23.

——. 1981. The challenges of systems theory. *General Systems Bulletin* 11:2–5.

——. 1985. *Anticipatory Systems: Philosophical, Mathematical, and Methodological Foundations.* Pergamon, Oxford.

——. 1991. *Life Itself.* Columbia University Press, New York

——. 2000. *Essays on Life Itself.* Columbia University Press, New York.

——. 2012. *Anticipatory Systems: Philosophical, Mathematical, and Methodological Foundations.* 2nd ed. Springer, New York.

Rosenblueth, A., N. Wiener, and J. Bigelow. 1943. Behavior, purpose and teleology. *Philosophy of Science* 10:18–24.

Roughgarden, J., and S. Pacala. 1989. Taxon cycling among *Anolis* lizards populations: Review of the evidence. In *Speciation and Its Consequences*, edited by D. Otte and J. Endler. Sinauer, Sunderland, MA.

Sacher, G. 1967. The complementarity of entropy terms for the temperative dependence of development and aging. *Ann. NY Acad. Sci.* 138:680–712.

Sagar, G. R., and Ferdinandez, D. E. F. 1976. *Agropyron repens*—allelopathy? *Annals of Applied Biology* 83:341. DOI:10.1111/j.1744-7348.1976.tb00621.x.

Salthe, S. N. 1985. *Evolving Hierarchical Systems: Their Structure and Representation.* Columbia University Press, New York.

Saltelli, A., A. Benessia, S. Funtowicz, M. Giampietro, A. Guimaraes Pereira, J. Ravetz, R. Strand, and J. P. van der Sluijs. 2016. *The Rightful Place of Science: Science on the Verge.* Consortium for Science, Policy and Outcomes. Tempe.

Sandom, C., Søren Faurby, Brody Sandel, and Jens-Christian Svenning. 2014. Global late Quaternary megafauna extinctions linked to humans, not climate change. *Proc. R. Soc. B* 281:20133254.

Schmandt-Besserat, D. 1992. *Before Writing.* 2 vols. University of Texas Press, Austin.

——. 1996. *How Writing Came About.* University of Texas Press, Austin.

Schmidt-Nielsen, K. 1972. *How Animals Work.* Cambridge University Press, Cambridge.

Schneider, E. D., and J. J. Kay. 1994. Life as a manifestation of the Second Law of thermodynamics. *Math. Comput. Model.* 19:25–48.

Schrödinger, E. 1967. *What Is Life?* Cambridge University Press, Cambridge.

Schumacher, E. F. 1975. *Small Is Beautiful: Economics as if People Mattered.* Perennial, New York.

Seaton, A. P. C., and J. Antonovics. 1967. Population inter-relationships. I. Evolution in mixtures of *Drosophila* mutants. *Heredity* 22:19–33.

Seneff, S., G. Wainwright, and L. Mascitelli. 2011a. Is the metabolic syndrome caused by a high fructose, and relatively low fat, low cholesterol diet? *Arch. Med. Sci.* 7:8–20. DOI:10.5114/aoms.2011.20598.

——. 2011b. Nutrition and Alzheimer's disease: the detrimental role of a high carbohydrate diet. *Eur. J. Intern. Med.* 22:134–40.

Shepard, M. 2013. *Restoration Agriculture: Real World Permaculture for Farmers.* Acres, Austin, TX.

Shugart, H. H., Jr., and D. C. West. 1977. Development of an Appalachian deciduous forest succession model and its application to assessment of the impact of the Chestnut blight. *J. Env. Mgmt.* 5:161–79.

——. 1979. Size and pattern of simulated stands. *Forest Science* 25:120–22.

Simberloff, D. 1983. Competition theory, hypothesis-testing and other ecological buzz-words. *Amer. Nat.* 122: 626–35.

Simon, C. A. 1975. The influence of food abundance on territory size in the iguanid lizard *Sceloporus jarrovi. Ecology* 56:993–98.

Simon, H. A. 1962. The architecture of complexity. *Proc. Amer. Phil. Soc.* 106: 467–82.

——. 1973. The organization of complex systems. In *Hierarchy Theory: The Challenge of Complex Systems,* edited by H. H. Pattee, 1–28. Braziller, New York.

Smith, T. 1835. Conclusions on the results on the vegetation of Nova Scotia, and on vegetation in general, of certain natural and artificial causes deemed to actuate and affect them. *Mag. Nat. Hist.* 8:641–62.

Sokal, A. 1996. Transgressing the boundaries: Toward a transformative hermeneutics of quantum gravity. *Social Text* 46/47: 217–52.

——. 2008. *Beyond the Hoax: Science, Philosophy, and Culture.* Oxford University Press, Oxford.

Southwood, T. R. E. 1980. Ecology—A mixture of pattern and probabilism. *Synthese* 43:111–22.

Spencer, H. 1864. *Principles of Biology.* Vol 1. Williams and Norgate, London.

Stacey, R. D., D. Griffin, and P. Shaw. 2000. *Complexity and Management: Fad or Radical Challenge to Systems Thinking?* Routledge, London.

Stager, K. 1964. The role of olfaction in food location of the turkey vulture. *Los Angeles Co. Museum Contrib. Sci.* 81:1–63.

Stamps, J. A. 1973. Displays and social organization in female *Anolis aeneus. Copeia* 2:264–72.

Tainter, J. A. 1988. *The Collapse of Complex Societies.* Cambridge University Press, Cambridge.

Tainter, J. A., and T. F. H. Allen. 2015. Energy gain in historical anthropology: ants, empires, the evolution of organization. In *The Oxford Handbook of Historical Ecology and Applied Archeology,* edited by C. Isendahl and D. Stump. Oxford University Press, Oxford.

Tainter J. A., and G. Lucas. 1983. The epistemology of the significance concept. *American Antiquity* 48:707–19.

Thom, R. 1975. *An Outline of a General Theory of Models.* Benjamin, Reading, MA.

Thompson, D'Arcy W. 1917. *On Growth and Form.* Cambridge University Press, Cambridge.

Thomson, J. D. 1975. Some community-level aspects of a bog pollination system. Master's thesis, University of Wisconsin, Madison.

Ulanowicz, R. E. 1997. *Ecology, the Ascendant Approach.* Columbia University Press, New York.

———. 2003. Some steps toward a central theory of ecosystem dynamics. *Comp. Biol. Chem.* 27:523–530.

Vargas, A. 2009. Did Paul Kammerer discover epigenetic inheritance? A modern look at the controversial midwife toad experiments. *J. Exp. Zool. B Mol. Dev. Evol.* 312:667–78. DOI:10.1002/jez.b.21319.

Vavilov, N. I. 1926. *Studies on the Origin of Cultivated Plants.* Institute of Applied Botany and Plant Breeding, Leningrad.

———. 1950. *The Origin, Variation, Immunity, and Breeding of Cultivated Plants.* Translated by K. S. Chester. Chronica Botanica, Waltham, MA.

Voeller, B. 1971. Developmental physiology of fern gametophytes: Relevance for biology. *BioScience* 21:266–70.

Vonnegut, K. 1963. *Cat's Cradle.* Random House, New York.

Waddington, C. H. 1957. *The Strategy of the Genes.* Allen & Unwin, London.

Wade, M. J. 1977. An experimental study of group selection. *Evolution* 31:134–53.

———. 2016. *Adaptation in Metapopulations.* University of Chicago Press, Chicago.

Waltner-Toews, D., J. J. Kay, and N. Lister. 2008. *The Ecosystem Approach.* Columbia University Press, New York.

Weaver, I. C. G. 2009. Life at the interface between a dynamic environment and a fixed genome. In *Mammalian Brain Development*, edited by D. Janigro, 17–40. Springer, New York.

Weaver, I. C. G., N. Cervoni, F. A. Champagne, A. C. D'Alessio, S. Sharma, J. R. Seckl, S. Dymov, M. Szyf, and M. J. Meaney. 2004. Epigenetic programming by maternal behavior. *Nat. Neurosci.* 7:847–54.

Webb, T., III, and J. H. McAndrews. 1976. Corresponding patterns of contemporary pollen and vegetation in central North America. *Geol. Soc. Amer. Mem.* 145:267–99.

Weinberg, G. M. 1975. *An Introduction to General Systems Thinking.* Wiley, New York.

———. 2001. *An Introduction to General Systems Thinking.* Dorset House, New York.

Weiner, J. 1990. Asymmetric competition in plant populations. *TREE* 5: 360–364.

Whittaker, R. H. 1956. Vegetation of the Great Smoky Mountains. *Ecol. Monogr.* 26:1–80.

Whyte, L. L. 1973. *Accent on Form.* Greenwood, Westport, CT.

Whyte, L. L., A. G. Wilson, and D. Wilson, eds. 1969. *Hierarchical Structures.* American Elsevier, New York.

Widrow, B., J. R. Glover, Jr., J. M. McCool, J. Kaunitz, C. S. Williams, R. H.

Hearn, J. R. Zeidler, E. Dong, Jr., and R. C. Goodlin. 1975. Adaptive noise cancelling: principles and applications. *Proceedings of the IEEE* 63:1692–716.

Wiener, N. 1954. *Human Use of Human Beings: Cybernetics and Society*. Doubleday Anchor, Garden City, NY.

Wiens, J. A. 1977. On competition and variable environments. *Am. Sci.* 65:590–97.

Willerslev, E., J. Davison, M. Moora, M. Zobel, E. Coissac, M. E. Edwards, E. D. Lorenzen, M. Vestergård, G. Gussarova, J. Haile, J. Craine, L. Gielly, S. Boessenkool, L. S. Epp, P. B. Pearman, R. Cheddadi, D. Murray, K. A. Bråthen, N. Yoccoz, H. Binney, C. Cruaud, P. Wincker, T. Goslar, I. G. Alsos, E. Bellemain, A. K. Brysting, R. Elven, J. H. Sønstebø, J. Murton, A. Sher, M. Rasmussen, R. Rønn, T. Mourier, A. Cooper, J. Austin, P. Möller, D. Froese, G. Zazula, F. Pompanon, D. Rioux, V. Niderkorn, A. Tikhonov, G. Savvinov, R. G. Roberts, D. E. Ross MacPhee, M. T. P. Gilbert, K. H. Kjær, L. Orlando, Christian Brochmann, P. Taberlet. 2014. Fifty thousand years of Arctic vegetation and megafaunal diet. *Nature* 506:47. DOI:0.1038/nature12921.

Wilson, E. o. 1975. *Sociobiology: The New Synthesis*. Belknap, Cambridge, MA.

Wimsatt, W. C. 1980. Randomness and perceived-randomness in evolutionary biology. *Synthese* 43:287–329.

Yoon, C. K. 2005. Ernst Mayr, pioneer in tracing geography's role in the origin of species, dies at 100. *New York Times*, February 5, 2005.

Zadeh, L. A. 1965. Fuzzy sets. *Information and Control* 8:338.

Zellmer, A. J., T. F. H. Allen, and K. Kesseboehmer. 2006. The nature of ecological complexity: A protocol for building the narrative. *Ecol. Complex.* 3:171–82.

Zhu, Y., H. Chen, J. Fan, Y. Wang, Y. Li, J. Chen, J. Fan, S. Yang, L. Hu, H. Leung, T. W. Mew, S. P. Teng, Z. Wang, and C. C. Mundt. 2000. Genetic diversity and disease control in rice. *Nature* 406:718–22.

Zimmerer, K. S. 1998. The ecogeography of Andean potatoes. *BioScience* 48: 445–54.

Author Index

Subject Index